Lecture Notes in Mathematics 2202

More information about this series at http://www.springer.com/series/304

Tatsuo Nishitani

Cauchy Problem for Differential Operators with Double Characteristics

Non-Effectively Hyperbolic Characteristics

 Springer

Tatsuo Nishitani
Department of Mathematics
Osaka University
Toyonaka, Osaka
Japan

ISSN 0075-8434   ISSN 1617-9692   (electronic)
Lecture Notes in Mathematics
ISBN 978-3-319-67611-1   ISBN 978-3-319-67612-8   (eBook)
DOI 10.1007/978-3-319-67612-8

Library of Congress Control Number: 2017954399

Mathematics Subject Classification (2010): 35L15, 35L30, 35B30, 35S05, 34M40

Printed on acid-free paper

This Springer imprint is published by Springer Nature
The registered company is Springer International Publishing AG
The registered company address is: Gewerbestrasse 11, 6330 Cham, Switzerland

Preface

In the early 1970s, V.Ja. Ivrii and V.M. Petkov introduced the fundamental matrix F_p, which is now called the Hamilton map, at double characteristic points of the principal symbol p of a differential operator P and proved that if the Cauchy problem for P is C^∞ well-posed for any lower order term then at every double characteristic point F_p has non-zero real eigenvalues; such characteristic is now called *effectively hyperbolic*. If no real eigenvalue exist, that is *non-effectively hyperbolic*, they proved, under some restrictions, the subprincipal symbol must lie in some interval on the real line for the Cauchy problem to be C^∞ well-posed. This necessary condition for the C^∞ well-posedness at non-effectively hyperbolic characteristic point was completed soon afterwards and is now called the Ivrii–Petkov–Hörmander condition (IPH condition for short). In this monograph we provide a general picture of the Cauchy problem for differential operators with double characteristics exposing well/ill-posed results of the Cauchy problem with non-effectively hyperbolic characteristics obtained since 1980s, with detailed proofs.

This monograph is organized as follows. In Chap. 1, after giving a brief overview on the C^∞ well-posedness of the Cauchy problem and a quick introduction to pseudodifferential operators we review basic results and notion about hyperbolic double characteristics. In Chap. 2, we present detailed discussions on the behavior of principal symbols p near non-effectively hyperbolic characteristics. We prove that p admits a *nice* microlocal factorization for deriving energy estimates if the cube of some specified vector field annihilates p. In Chap. 3 we prove that p admits this factorization if and only if there is no bicharacteristic tangent to the double characteristic manifold. In Chap. 4 we propose energy estimates such that if at every point ρ in the phase space there is P_ρ coinciding with P in a conic neighborhood of ρ for which these proposed energy estimates hold then the Cauchy problem for P is locally solvable. In Chap. 5 we prove main results on the well-posedness of the Cauchy problem which could be stated: *if there is no tangent bicharacteristics and no transition of the spectral type of F_p then the Cauchy problem is C^∞ well-posed under the strict IHP condition*. In Chap. 6 we exhibit an example of second order differential operator with polynomial coefficients, verifying the Levi condition, with tangent bicharacteristic and no transition of the spectral type of F_p for which the Cauchy problem is ill-posed in the Gevrey class of order $s > 5$, and of course C^∞ ill-posed. In Chap. 7 we confirm the optimality of this Gevrey class proving that the Cauchy problem for P is well-posed in the Gevrey class of order 5 under the Levi condition, assuming no transition of the spectral type. Finally in Chap. 8, for the same operator studied in Chap. 6, we prove that the Cauchy problem is C^∞ ill-posed for *any choice of lower order term*, more strongly, ill-posed in the Gevrey class of order $s > 6$ for any lower order term.

Otsu in December 2016   Tatsuo Nishitani

Contents

Chapter 1
Introduction

Abstract In this chapter, after quickly reviewing the background which motivates to prepare this monograph we state basic facts on pseudodifferential operators without proofs, except for a few results. We then recall basic results on the Cauchy problem for differential operators with double characteristics, including basic notion and results about double characteristics of hyperbolic polynomials and hyperbolic quadratic forms which will be used throughout the monograph.

1.1 Cauchy Problem, an Overview

Let P be a differential operator of order m defined in a neighborhood of $\bar{x} \in \mathbb{R}^{n+1}$ and let $t = t(x)$ be a real valued smooth function given in a neighborhood of $\bar{x}$ with $t(\bar{x}) = 0$. We assume that P is non-characteristic with respect to $H = \{t(x) = 0\}$ at $\bar{x}$, that is $(Pt^m)(\bar{x}) \neq 0$. Let $u_0(x), \dots, u_{m-1}(x)$ be m-tuples of smooth functions on H defined near $\bar{x}$ then the Cauchy problem is that of finding u, in a neighborhood of $\bar{x}$, satisfying $Pu = 0$ near $\bar{x}$ and $(\partial/\partial v)^j u(x) = u_j(x), j = 0, \dots, m-1$ on H where v is the unit normal to H. Here $(u_0, \dots, u_{m-1})$ is called the initial data or the Cauchy data. Roughly speaking, the Cauchy problem is said to be E well-posed in the direction t if for any initial data in E, which is a function space given beforehand, there exists a unique solution to the Cauchy problem, and the differential operator for which the Cauchy problem is well-posed in the direction t is called hyperbolic in this direction. Choosing a system of local coordinates $x = (x_0, x') = (x_0, x_1, \dots, x_n)$ so that $t(x) = x_0, \bar{x} = 0$ and dividing P by a non vanishing function we have

$$P = D_0^m + \sum_{|\alpha| \leq m, \alpha_0 < m} a_\alpha(x) D^\alpha = \sum_{j=0}^{m} P_j$$

in these coordinates where $P_j = \sum_{|\alpha|=j} a_\alpha(x) D^\alpha$ denotes the homogeneous part of P of degree j and $D = (D_0, D') = (D_0, D_1, \dots, D_n)$, $D_j = -i\partial/\partial x_j$, $D^\alpha = D_0^{\alpha_0} \cdots D_n^{\alpha_n}$, $\alpha = (\alpha_0, \dots, \alpha_n) \in \mathbb{N}^{n+1}$. The homogeneous polynomial in

© Springer International Publishing AG 2017

T. Nishitani, *Cauchy Problem for Differential Operators with Double Characteristics*, Lecture Notes in Mathematics 2202,
DOI 10.1007/978-3-319-67612-8_1

$\xi = (\xi_0, \xi_1, \ldots, \xi_n)$ of degree m

$$p(x, \xi) = \xi_0^m + \sum_{|\alpha|=m, \alpha_0 < m} a_\alpha(x)\xi^\alpha$$

is called the principal symbol of P. We start with giving a concise definition of the well-posedness of the Cauchy problem which is equivalent to the classical definition of the well-posedness requiring the unique existence of solution with initial data on every $x_0 = \tau$, $|\tau| < \delta$ with a small $\delta > 0$.

Definition 1.1 We say that the Cauchy problem for P is C^∞ well-posed near the origin in x_0 direction if there exist a positive constant ϵ and a neighborhood ω of the origin such that for any $|\tau| \leq \epsilon$ and $f(x) \in C_0^\infty(\omega)$ vanishing in $x_0 < \tau$ one can find a unique $u(x) \in C^\infty(\omega)$ vanishing in $x_0 < \tau$ which satisfies $Pu = f$ in ω.

It follows easily from the definition that if $u \in C^\infty(\omega)$ vanishing in $x_0 < \tau$ satisfies $Pu \in C_0^\infty(\omega)$ and $Pu = 0$ in $x_0 < t$ ($\tau < t < \epsilon$) then it results $u = 0$ in $x_0 < t$. Indeed the equation $Pw = Pu$ has a solution $w(x) \in C^\infty(\omega)$ vanishing in $x_0 < t$. Since $w - u = 0$ in $x_0 < \tau$ and $P(w - u) = 0$ then by the uniqueness we get $w = u$ and hence $u = 0$ in $x_0 < t$. In this definition we require the causality that the future does not influence to the past, which is much weaker, at a glance, than the requirement of the finite propagation speed, but this requirement of the causality consists in the essential part of the hyperbolicity.

Remark 1.1 If we equip a topology which makes E a complete metric space then the continuous dependence of solutions to the Cauchy data, emphasized in the classical Hadamard's book [30], would be a consequence of Banach's closed graph theorem (see [66], for example).

Lemma 1.1 *Assume that the Cauchy problem for P is C^∞ well-posed near the origin in x_0 direction. Then for any $f(x) \in C_0^\infty(\omega)$ and any $u_j(x') \in C_0^\infty(\omega \cap \{x_0 = \tau\})$ there exists a unique solution $u \in C^\infty(\omega)$ to the following Cauchy problem*

$$\begin{cases} Pu = f & in \quad \omega \cap \{x_0 > \tau\}, \\ D_0^j u(\tau, x') = u_j(x'), \quad j = 0, 1, \ldots, m-1. \end{cases} \tag{1.1}$$

Proof Assume that u verifies (1.1) with $f \in C_0^\infty(\omega)$. Differentiating $Pu = f$ with respect to x_0 one can determine $u_j(x') = D_0^j u(\tau, x') \in C_0^\infty(\omega \cap \{x_0 = \tau\})$ for all $j \in \mathbb{N}$. From a classical theorem of Borel (see [33, Chapter 1] for example) we can find $\tilde{u} \in C_0^\infty(\omega)$ such that $D_0^j \tilde{u}(\tau, x') = u_j(x')$ for all $j \in \mathbb{N}$. It is clear that $D_0^j(P\tilde{u} - f) = 0$ on $\{x_0 = \tau\}$ for any $j \in \mathbb{N}$. Define g so that $g = P\tilde{u} - f$ in $x_0 > \tau$ and $g = 0$ in $x_0 < \tau$ then we have $g \in C_0^\infty(\omega)$. From the assumption there exists $v \in C^\infty(\omega)$ vanishing in $x_0 < \tau$ which satisfies $Pv = g$ in ω. Therefore we conclude that

$$\begin{cases} P(\tilde{u} - v) = f & \text{in} \quad \omega \cap \{x_0 > \tau\}, \\ D_0^j(\tilde{u} - v)(\tau, x') = u_j(x') & \text{on} \quad \omega \cap \{x_0 = \tau\} \end{cases}$$

and hence $\tilde{u} - v \in C^\infty(\omega)$ is a desired solution to (1.1).   □
Here we recall strictly hyperbolic operators.

Definition 1.2 We say that P is strictly hyperbolic near the origin in x_0 direction if the characteristic roots, that is the roots of $p(x, \xi_0, \xi') = 0$ with respect to ξ_0, are real distinct for any (x, ξ'), $\xi' = (\xi_1, \ldots, \xi_n) \neq 0$ and x in some neighborhood of the origin.
From now on we often omit "x_0 direction" and "C^∞" so that "well-posed" means C^∞ well-posed in x_0 direction. The Cauchy problem for higher order strictly hyperbolic systems was first studied by Petrovsky [91], in a quite general setting, and he derived energy estimates and proved the C^∞ well-posedness for any lower order term. The work was too hard to penetrate and the first simplification was made by Leray [55], he derived energy estimates by constructing a symmetrizer and constructed the solution by approximation from analytic case. Soon afterwards Gårding [24, 25] proved the existence of solutions by functional analysis alone without approximation process. Shortly afterwards Fourier analysis approach of Petrovsky reappeared by use of singular integral operators [63, 64].

Theorem 1.1 ([55, 91]) *Assume that P is strictly hyperbolic near the origin. Then the Cauchy problem for $P + Q$ with any differential operator of order at most $m - 1$ is C^∞ well-posed near the origin.*

Definition 1.3 P is said to be strongly hyperbolic near the origin if the Cauchy problem for $P + Q$ is C^∞ well-posed near the origin for any differential operator Q of order at most $m - 1$.
It seems that the term "strongly hyperbolic" was first used in [51]. See also [94]. According to this definition.

Corollary 1.1 *A strictly hyperbolic operator is a strongly hyperbolic operator.*
Meanwhile it was proved that the characteristic roots must be real for the Cauchy problem to be well-posed, in [54] for the case of simple characteristics and in [65] in full generality.

Theorem 1.2 ([54, 65]) *Assume that the Cauchy problem for P is well-posed near the origin. Then all characteristic roots ξ_0 are real for any $\xi' \in \mathbb{R}^n$ and any $x \in \omega$ with some neighborhood ω of the origin.*

Definition 1.4 We say that $p(x, \xi)$ is a hyperbolic polynomial with respect to ξ_0 if $p(x, \xi_0, \xi') = 0$ has only real roots for any x and ξ'.
After standing about 10 years, it was proved that the Levi condition is necessary and sufficient for the well-posedness of the Cauchy problem for differential operators with characteristics of constant multiplicity of at most two in Mizohata and Ohya [67, 68]. The word Levi condition stems from [58]. The case of one spatial

dimension was also studied in [53]. Subsequently the necessity of the Levi condition for the well-posedness was proved in Flaschka and Strang [23] and the sufficiency was proved in Chazarain [13] for differential operators with characteristics of constant multiplicities of any order.

As mentioned in the Preface, around the same period the work of Ivrii and Petkov [43] appeared, which introduced the Hamilton map F_p, the linearization of the Hamilton vector field H_p at a multiple characteristic and clarified some close relations between the well-posedness of the Cauchy problem and the spectral structure of F_p (note that p fails to be strictly hyperbolic polynomial at singular points of H_p). This was a breakthrough[1] in researches on hyperbolic operators with multiple characteristics. They conjectured that if every characteristic is at most double and effectively hyperbolic then the Cauchy problem is C^∞ well-posed for any lower order term. The word effective is chosen in [32] and stems from this conjecture. This conjecture has been proved affirmatively in the early 1980s. Since effectively hyperbolic characteristic has been well understood it remains to study non-effectively hyperbolic characteristics. If F_p has no Jordan block of size 4 corresponding to the eigenvalue 0 and there is no transition of the spectral type of F_p, then the Cauchy problem is C^∞ well-posed under the strict IPH condition which was proved in the late 1970s. The main remaining question was, of course concerning with the case that the Jordan canonical form of F_p contains a subspace of dimension 4 corresponding to the eigenvalue 0, whether the well-posedness assertion still holds or we need more necessary conditions on the subprincipal symbol other than IHP condition for the C^∞ well-posedness. The answer is unexpected one.

It has been recognized that what is crucial to the C^∞ well-posedness is not only F_p but also the behavior of bicharacteristics near the double characteristic manifold and F_p itself is not enough to determine completely the behavior of bicharacteristics. Strikingly enough, if there is a bicharacteristic tangent to the double characteristic manifold then the Cauchy problem is C^∞ ill-posed *for any lower order term*, that is seeking conditions other than IHP condition for the C^∞ well-posedness was off point. On the other hand if there is no bicharacteristic tangent to the double characteristic manifold and no transition of the spectral type of F_p then the above mentioned result still holds; the Cauchy problem is C^∞ well-posed under the strict IPH condition. Moreover, assuming no transition of spectral type of F_p the Cauchy problem is well-posed in the Gevrey class of order 5 under the Levi condition even if there is a tangent bicharacteristic.

For results on the Cauchy problem for differential operators before the 1980s, including an overview of hyperbolic differential operators with constant coefficients and those with characteristics of constant multiplicities, we refer to Gårding [26, 27], Melrose [62], Ivrii [42] and Mizohata-Ohya-Ikawa [69].

---

[1] See L. Gårding; Some recent results for hyperbolic differential equations, Proceeding of the 19th Nordic congress of mathematicians, Reykavik, 1985, pp. 50–59.

1.2 Sobolev Spaces

Here we recall some basic definitions about Sobolev spaces. For details we refer to [33]. In this section, to simplify notations we use ξ, x for ξ' and x'.

Definition 1.5 By $\mathscr{S}$ or $\mathscr{S}(\mathbb{R}^n)$ we denote the set of all $u \in C^\infty(\mathbb{R}^n)$ such that for all multi-indices α and β

$$\sup_x |x^\beta \partial_x^\alpha u(x)| < +\infty.$$

We denote by $\mathscr{S}'$ the set of all temperate distributions.
The Fourier transform $\hat{f}$ of $f \in \mathscr{S}$ is defined by

$$\hat{f}(\xi) = \int e^{-i\langle x, \xi \rangle} f(x) dx, \quad \xi \in \mathbb{R}^n.$$

We often write $x\xi$ for $\langle x, \xi \rangle$ and denote $\langle \xi \rangle = (1 + |\xi|^2)^{1/2}$.

Definition 1.6 For $s \in \mathbb{R}$ we denote by H^s or $H^s(\mathbb{R}^n)$ the space of all $u \in \mathscr{S}'$ such that $\langle \xi \rangle^s \hat{u}(\xi) \in L^2$. In H^s we define the inner product

$$(u, v)_s = (2\pi)^{-n} \int \langle \xi \rangle^{2s} \hat{u}(\xi) \overline{\hat{v}(\xi)} d\xi = (2\pi)^{-n} (\langle \xi \rangle^s \hat{u}, \langle \xi \rangle^s \hat{v})$$

hence $\|u\|_s = (2\pi)^{-n} \|\langle \xi \rangle^s \hat{u}\|_{L^2}$ and H^s is a Hilbert space equipped with the inner product $(\cdot, \cdot)_s$. When $s = 0$ we write simply $\| \cdot \|$ and $(\cdot, \cdot)$.
We define $\langle D \rangle^s u$ by

$$\langle D \rangle^s u = (2\pi)^{-n} \int e^{i\langle x, \xi \rangle} \langle \xi \rangle^s \hat{u}(\xi) d\xi$$

then from the Parseval's formula we have $(u, v)_s = (\langle D \rangle^s u, \langle D \rangle^s v)$. We denote $H^\infty = \cap_k H^k$, $H^{-\infty} = \cup_k H^k$.

Proposition 1.1 (A Special Case of Sobolev Embedding Theorem) *Let $p \in \mathbb{N}$. Then there exists $C_p > 0$ such that for $|\alpha| \leq p$ and $u \in H^{[n/2]+p+1}$ one has*

$$\sup_x |\partial_x^\alpha u(x)| \leq C_p \|u\|_{[n/2]+p+1}.$$

In particular $u \in C^p(\mathbb{R}^n)$ if $u \in H^{[n/2]+p+1}$.

1.3   Pseudodifferential Operators

Here we collect some basic facts about pseudodifferential operators of class $S_{1,0}$. For details we refer to [34]. In this section we still use ξ, x for ξ' and x'.

Definition 1.7 Let m be a real number. Then $S^m = S^m(\mathbb{R}^n \times \mathbb{R}^n)$ is the set of all $a \in C^\infty(\mathbb{R}^n \times \mathbb{R}^n)$ such that for any $\alpha, \beta \in \mathbb{N}^n$ we have

$$|a|_{\alpha,\beta} = \sup_{x,\xi} |(1 + |\xi|)^{-m+|\alpha|} a^{(\alpha)}_{(\beta)}(x, \xi)| < +\infty$$

where $a^{(\alpha)}_{(\beta)}(x, \xi) = \partial^\alpha_\xi \partial^\beta_x a(x, \xi)$. For $\ell \in \mathbb{N}$ we set $|a|_\ell = \sum_{|\alpha+\beta|\leq\ell} |a|_{\alpha,\beta}$. We write $S^{-\infty} = \cap S^m$, $S^\infty = \cup S^m$.

It is easy to check that $a^{(\alpha)}_{(\beta)} \in S^{m-|\alpha|}$ and $ab \in S^{m+m'}$ if $a \in S^m$ and $b \in S^{m'}$.

Definition 1.8 Let m be a real number and $a \in S^m$. We write $a \in S^m_{phg}$ if there exist $a_j \in S^{m-j}, j = 0, 1, \ldots$ which are homogeneous of degree $m - j$ in ξ when $|\xi| > 1$ such that

$$a - \sum_{j<k} a_j \in S^{m-k}.$$

From now on we simply say that a is homogeneous in ξ if a is homogeneous in ξ when $|\xi| > 1$ if otherwise stated.

Definition 1.9 Let $0 \leq t \leq 1$. For $a \in S^m$ we define the pseudodifferential operator $\text{Op}^t(a)$ by

$$(\text{Op}^t(a))u = (2\pi)^{-n} \int e^{i\langle x-y,\xi\rangle} a((1 - t)x + ty, \xi)u(y)dyd\xi, \quad u \in \mathscr{S}$$

and $\text{Op}^t(a)$ is called t-quantization of a. In particular $\text{Op}^{1/2}(a)$ is called Weyl quantization of a and usually we use $a(x, D)$ or $\text{Op}(a)$ denoting $\text{Op}^{1/2}(a)$. If $A = \text{Op}(a)$ we call a the Weyl symbol, or just symbol of A and denote $a = \sigma(A)$. We denote by $\text{Op}^t S^m$ the set of all $\text{Op}^t(a)$ with $a \in S^m$ and $\text{Op} S^m = \text{Op}^{1/2} S^m$.

In this monograph we employ the Weyl quantization unless otherwise stated.

Proposition 1.2 Let $a \in S^m$. Then

$$(a(x, D)u, v) = (u, \bar{a}(x, D)v), \quad u, v \in \mathscr{S}$$

that is, $a(x, D)^* = \bar{a}(x, D)$.

Definition 1.10 For $a(x, \xi)$, $b(x, \xi) \in C^\infty(\mathbb{R}^n \times \mathbb{R}^n)$ the Poisson bracket $\{a, b\}$ is defined by

$$\{a, b\} = \sum_{j=1}^{n} \left(\frac{\partial a}{\partial \xi_j} \frac{\partial b}{\partial x_j} - \frac{\partial a}{\partial x_j} \frac{\partial b}{\partial \xi_j} \right).$$

Theorem 1.3 Let $a_j \in S^{m_j}$, $j = 1, 2$. Then $a_1(x, D)a_2(x, D) \in \mathrm{Op}S^{m_1+m_2}$, that is there is $b \in S^{m_1+m_2}$ such that $a_1(x, D)a_2(x, D) = b(x, D)$ on $\mathscr{S}$ and $\mathscr{S}'$ where for any $N \in \mathbb{N}$ we have

$$b(x, \xi) - \sum_{|\alpha+\beta|<N} \frac{(-1)^{|\beta|}}{(2i)^{|\alpha+\beta|}\alpha!\beta!} a_{1(\beta)}^{(\alpha)} a_{2(\alpha)}^{(\beta)} \in S^{m_1+m_2-N}.$$

We write $b = a_1 \# a_2$. In particular $b(x, \xi) - \left(a_1 a_2 + \{a_1, a_2\}/2i \right) \in S^{m_1+m_2-2}$.

Corollary 1.2 Let $a_j \in S^{m_j}$, $j = 1, 2$. Then

$$\sigma([a_1(x, D), a_2(x, D)]) - \{a_1, a_2\}/i \in S^{m_1+m_2-2}$$

where $[A, B] = AB - BA$.

Corollary 1.3 Let $a \in S^m$ and $\phi \in S^k$ then

$$\sigma(a(x, D)^2) - a(x, \xi)^2 \in S^{2m-2},$$

$$\sigma(\phi(x, D)a(x, D)\phi(x, D)) - \phi(x, \xi)^2 a(x, \xi) \in S^{m+2k-2}.$$

Since we are mainly concerned with differential operators of order m, that is polynomials in D of order m with coefficients which are smooth functions, it is often convenient to use 0-quantization so that

$$\sum_{|\alpha|\leq m} a_\alpha(x)D^\alpha = \mathrm{Op}^0 \left(\sum_{|\alpha|\leq m} a_\alpha(x)\xi^\alpha \right).$$

Lemma 1.2 For $a(x, \xi) \in S^m$ there exists $b(x, \xi) \in S^m$ such that $b(x, D) = \mathrm{Op}^0(a)$ where $b(x, \xi)$ is given by

$$b(x, \xi) = (2\pi)^{-n} \int e^{iy\eta} a(x + y/\sqrt{2}, \xi + \eta/\sqrt{2})dyd\eta$$

and for any $N \in \mathbb{N}$ we have

$$b(x, \xi) = a(x, \xi) + \frac{i}{2} \sum_{j=1}^{n} (\partial^2 a/\partial x_j \partial \xi_j)(x, \xi) + r(x, \xi), \quad r \in S^{m-2}.$$

If $a(x, \xi)$ is a polynomial in ξ of order m then so is $b(x, \xi)$ and

$$b(x, \xi) = \sum_{|\alpha| \le m} \frac{i^{|\alpha|}}{2^{|\alpha|}\alpha!} \partial_x^\alpha \partial_\xi^\alpha a(x, \xi).$$

Theorem 1.4 *Let $a_j \in S^{m_j}$, $j = 1, 2$. Then $\mathrm{Op}^0(a_1)\mathrm{Op}^0(a_2) \in \mathrm{Op}^0 S^{m_1+m_2}$, that is there exists $b \in S^{m_1+m_2}$ such that $\mathrm{Op}^0(a_1)\mathrm{Op}^0(a_2) = \mathrm{Op}^0(b)$ on $\mathscr{S}$ and $\mathscr{S}'$ where for any $N \in \mathbb{N}$ we have*

$$b(x, \xi) - \sum_{|\alpha| < N} \frac{1}{i^{|\alpha|}\alpha!} a_1^{(\alpha)} a_{2(\alpha)} \in S^{m_1+m_2-N}.$$

With $P(x, \xi) = \sum_{|\alpha| \le m} a_\alpha(x)\xi^\alpha$ it follows from Lemma 1.2 that

$$\sum_{|\alpha| \le m} a_\alpha(x)D^\alpha = \mathrm{Op}\left(\sum_{|\alpha| \le m} \frac{i^{|\alpha|}}{2^{|\alpha|}\alpha!} \partial_x^\alpha \partial_\xi^\alpha P(x, \xi) \right). \tag{1.2}$$

We recall several results on estimates of pseudodifferential operators.

Theorem 1.5 *If $a(x, \xi) \in S^0$ then $a(x, D)$ is bounded on $L^2(\mathbb{R}^n)$.*

Theorem 1.6 (Sharp Gårding Inequality) *Assume that $a(x, \xi) \in S^m$ is non-negative $a(x, \xi) \ge 0$. Then there is $C > 0$ such that*

$$(a(x, D)u, u) \ge -C\|u\|_{(m-1)/2}^2, \quad u \in \mathscr{S}.$$

Corollary 1.4 *Let $a(x, \xi) \in S_{phg}^m$ and assume that $\mathrm{Re}\, a_0(x, \xi) \ge 0$. Then there is $C > 0$ such that*

$$\mathrm{Re}(a(x, D)u, u) \ge -C\|u\|_{(m-1)/2}^2, \quad u \in \mathscr{S}.$$

We finally state the Fefferman-Phong inequality.

Theorem 1.7 (Fefferman-Phong Inequality) *Assume $a(x, \xi) \in S^m$ is non-negative $a(x, \xi) \ge 0$. Then there is $C > 0$ such that*

$$(a(x, D)u, u) \ge -C\|u\|_{(m-2)/2}^2, \quad u \in \mathscr{S}.$$

Corollary 1.5 *Let $a \in S_{phg}^m$ and assume $a_0(x, \xi) + a_1(x, \xi) \ge 0$. Then there exists $C > 0$ such that*

$$\mathrm{Re}(a(x, D)u, u) \ge -C\|u\|_{(m-2)/2}^2, \quad u \in \mathscr{S}.$$

Before closing the section we make a brief look on pseudodifferential operators on $\mathbb{R}^n$ with nonnegative principal symbols. Let $a(x, \xi) \in S_{phg}^{2m}(\mathbb{R}^n \times \mathbb{R}^n)$ be such that $a_0(x, \xi) \geq 0$. Note that $\partial_x^\alpha \partial_\xi^\beta a_0(\bar{x}, \bar{\xi}) = 0$ for $|\alpha + \beta| \leq 1$ if $a_0(\bar{x}, \bar{\xi}) = 0$ and the Hamilton map F_{a_0} of a_0 at $(\bar{x}, \bar{\xi})$ is defined by Definition 1.13 below.

Lemma 1.3 *Let $a_0(x, \xi) = 0$ then all eigenvalues of the Hamilton map $F_{a_0}(x, \xi)$ are on the imaginary axis.*

Theorem 1.8 (Melin's Inequality [60])   *Let $a \in S_{phg}^{2m}$ and assume that $a_0 \geq 0$ and a_{-1} is real valued. If $a_{-1} + \mathrm{Tr}^+ F_{a_0} > 0$ when $a_0 = 0$ then for every compact set $K \subset \mathbb{R}^n$ one can find $c_K > 0$, $C_K > 0$ such that*

$$(\mathrm{Op}(a)u, u) \geq c_K \|u\|_{m-1/2}^2 - C_K \|u\|_{m-1}^2, \quad u \in C_0^\infty(K).$$

where the positive trace $\mathrm{Tr}^+ F_{a_0}$ of F_{a_0} is defined by Definition 1.15.

1.4   A Review on Hyperbolic Double Characteristics

By taking Theorem 1.2 into consideration we assume that the characteristic roots of $p(x, \xi)$ are all real. We start with recalling the definition of characteristics.

Definition 1.11 If $p(x, \xi)$ vanishes at $\rho = (\bar{x}, \bar{\xi}) \in \mathbb{R}^{2(n+1)}$, $\bar{\xi} \neq 0$ of order r, that is $\partial_x^\alpha \partial_\xi^\beta p(\rho) = 0$ for any $|\alpha + \beta| < r$ and $\partial_x^\alpha \partial_\xi^\beta p(\rho) \neq 0$ for some $|\alpha + \beta| = r$, we call ρ a characteristic of order r of p.

Lemma 1.4 ([43])   *If $p(x, \xi)$ is a hyperbolic polynomial verifying $(\partial/\partial \xi_0)^j p(\bar{x}, \bar{\xi}) = 0$ for $0 \leq j \leq r - 1$ and $(\partial/\partial \xi_0)^r p(\bar{x}, \bar{\xi}) \neq 0$ then $(\bar{x}, \bar{\xi})$ is a characteristic of order r.*

Strictly hyperbolic operators are those whose characteristics are real and simple by Lemma 1.4. If the Cauchy problem for differential operators with multiple characteristics is well-posed then the following necessary condition must be verified.

Theorem 1.9 ([43])   *Let $P = \mathrm{Op}^0\left(\sum_{j=0}^m P_j(x, \xi)\right)$ where $P_j(x, \xi)$ are homogeneous polynomials in ξ of degree j. Assume that the Cauchy problem for P is well-posed near the origin and let $(0, \bar{\xi})$ be a characteristic of order r. Then we have*

$$\partial_x^\alpha \partial_\xi^\beta P_{m-j}(0, \bar{\xi}) = 0, \quad |\alpha + \beta| < r - 2j, \quad j = 0, \ldots, [r/2]$$

where $[r/2]$ stands for the integer part of $r/2$.
Let $\sum_{j=0}^m \tilde{P}_j(x, \xi)$ be the Weyl symbol of P so that $P = \mathrm{Op}(\sum_{j=0}^m \tilde{P}_j(x, \xi))$. Then from (1.2) the same assertion of Theorem 1.9 holds for $\tilde{P}_j(x, \xi)$. In [43] we find some other necessary conditions for the well-posedness. Here we only cite a necessary condition which is independent of the choice of local coordinates system. For differential operators with simple characteristics we have Theorem 1.1 then, from

now on we are concerned with differential operators with double characteristics, which is the subject of this monograph.

Definition 1.12 One calls

$$H_p = \sum_{j=0}^{n} \left(\frac{\partial p}{\partial \xi_j} \frac{\partial}{\partial x_j} - \frac{\partial p}{\partial x_j} \frac{\partial}{\partial \xi_j} \right)$$

the Hamilton vector field of p. A bicharacteristic[2] of p is an integral curve of H_p, that is a solution to the Hamilton equation

$$\frac{dx_j}{ds} = \frac{\partial p}{\partial \xi_j}(x, \xi), \quad \frac{d\xi_j}{ds} = -\frac{\partial p}{\partial x_j}(x, \xi), \quad j = 0, 1, \ldots, n \qquad (1.3)$$

on which $p = 0$.

Multiple characteristics of p are singular (stationary) points of the Hamilton vector field H_p. Let $\rho = (\bar{x}, \bar{\xi})$ be a double characteristic of p. We linearize the Hamilton equation $\dot{X} = H_p(X)$ at ρ where $X = (x, \xi)$, that is inserting $X(s) = (\bar{x}, \bar{\xi}) + \epsilon Y(s)$ into the equation then the term linear in ϵ in the resulting equation yields $\dot{Y} = 2F_p(\rho)Y$ where $F_p(\rho)$ is given by

$$F_p(\rho) = \frac{1}{2} \begin{pmatrix} \dfrac{\partial^2 p}{\partial x \partial \xi}(\rho) & \dfrac{\partial^2 p}{\partial \xi \partial \xi}(\rho) \\ -\dfrac{\partial^2 p}{\partial x \partial x}(\rho) & -\dfrac{\partial^2 p}{\partial \xi \partial x}(\rho) \end{pmatrix}.$$

Definition 1.13 We call $F_p(\rho)$ the Hamilton map of p at ρ following [32]. In [43] F_p is called the fundamental matrix.

The following special structure of $F_p(\rho)$ results from the assumption that $p(x, \xi)$ is a hyperbolic polynomial.

Lemma 1.5 ([32, 43]) *All eigenvalues of $F_p(\rho)$ lie on the imaginary axis, possibly one exception of a pair of non-zero real eigenvalues $\pm\lambda$, $\lambda > 0$.*
We give a proof of the lemma in Sect. 1.5.

Definition 1.14 One says that ρ is an effectively hyperbolic characteristic if $F_p(\rho)$ has a non-zero real eigenvalue, we also say that $p(x, \xi)$ is effectively hyperbolic at ρ. Otherwise ρ is said to be non-effectively hyperbolic characteristic and $p(x, \xi)$ is called non-effectively hyperbolic at ρ.

Definition 1.15 The positive trace of $F_p(\rho)$ is defined by

$$\mathrm{Tr}^+ F_p(\rho) = \sum \mu_j$$

---

[2]Sometimes a bicharacteristic in this definition is called a null bicharacteristic.

where the sum is taken over all μ_j where $i\mu_j$ are the eigenvalues of $F_p(\rho)$ on the positive imaginary axis, counted with multiplicity.

Theorem 1.10 ([43]) *Assume that P is strongly hyperbolic near the origin. Then there is a neighborhood of the origin where every multiple characteristic of p is at most double and effectively hyperbolic.*

The converse was affirmatively answered in [37, 44, 61, 70] for special cases and in [45, 74] for general second order differential operators and in [47, 71] for general higher order differential operators.

Theorem 1.11 ([37, 44, 45, 47, 71, 74]) *Assume that every multiple characteristic of p is at most double and effectively hyperbolic. Then P is strongly hyperbolic near the origin.*

In [44, 45] the proofs are based on the transformation of the operator P to an operator with "nice" lower order terms by means of integro-pseudodifferential operators and on the energy estimates for the resulting operator, while in [71, 74] the proof is based on weighted energy estimates with pseudodifferential weights of which symbol is a power of (microlocal) time function, after some preliminary transformation by Fourier integral operators. For details we refer to [46, 50]. If we consider two or more differential operators with a same effectively hyperbolic characteristic which are not in involution we are forced to treat the problem without Fourier integral operators (see [49, 77]) and it is possible to avoid the use of Fourier integral operators in the latter method [79].

In what follows we are concerned with the case that p is non-effectively hyperbolic at double characteristics. The subprincipal symbol $P_{sub}(x, \xi)$ of P is defined as follows:

$$P - \mathrm{Op}\,(p + P_{sub}) \in \mathrm{Op}S^{m-2}$$

where $S^{m-2} = S^{m-2}(\mathbb{R}^{n+1} \times \mathbb{R}^{n+1})$ and P_{sub} is invariantly defined at double characteristics. Therefore from (1.2) if $P = \mathrm{Op}^0(P(x, \xi))$ then $P_{sub}(x, \xi)$ is given, reference to any local coordinates x, by

$$P_{sub}(x, \xi) = P_{m-1}(x, \xi) + \frac{i}{2} \sum_{j=0}^{n} \frac{\partial^2 p}{\partial x_j \partial \xi_j}(x, \xi).$$

Theorem 1.12 ([32, 43]) *Let $\rho = (0, \bar{\xi})$ be a non-effectively hyperbolic characteristic of p. If the Cauchy problem for P is C^∞ well-posed near the origin we have*

$$\mathsf{Im}\,P_{sub}(\rho) = 0, \quad -\mathrm{Tr}^+ F_p(\rho) \leq \mathsf{Re}\,P_{sub}(\rho) \leq \mathrm{Tr}^+ F_p(\rho). \tag{1.4}$$

The assertion (1.4) was proved in [43] for some cases corresponding to spectral property of F_p and the proof for the remaining cases was given in [32]. The condition (1.4) is called the Ivrii-Petkov-Hörmander condition (IPH condition, for short). If $\mathrm{Tr}^+ F_p(\rho) = 0$ the IPH condition is reduced to $P_{sub}(\rho) = 0$ and called the Levi condition. We call (1.4) with strict inequality the strict Ivrii-Petkov-Hörmander condition (strict IPH condition, for short).

Definition 1.16 Let ρ be a double characteristic of p. Then the localization p_ρ is the second order term in the Taylor expansion of p at ρ;

$$p(\rho + \epsilon X) = \epsilon^2 (p_\rho(X) + O(\epsilon)), \quad \epsilon \to 0, \quad X = (x, \xi) \in \mathbb{R}^{2(n+1)}.$$

Thus p_ρ is a homogeneous polynomial in (x, ξ) of degree 2 and would be considered as the first approximation of p near a double characteristic ρ.

Lemma 1.6 ([43]) $p_\rho(X)$ *is a quadratic hyperbolic form in* $X = (x, \xi) \in \mathbb{R}^{2(n+1)}$, *that is a quadratic form of signature* $(-1, 1, \ldots, 1, 0, \ldots, 0)$.

Proof Let $\rho = (\bar{x}, \bar{\xi})$ and denote $\rho' = (\bar{x}, \bar{\xi}')$. Without restrictions we can assume $\bar{\xi}_0 = 0$. One can find a conic neighborhood V of ρ' such that

$$p(x, \xi) = q(x, \xi) r(x, \xi), \quad (x, \xi') \in V$$

where q and r are hyperbolic polynomials of degree 2 and $m - 2$ with respect to ξ_0 and $r(\rho) \neq 0$, $(\partial/\partial\xi_0)^j q(\rho) = 0$ for $j = 0, 1$. Then from Lemma 1.4 it follows that $\partial_x^\alpha \partial_\xi^\beta q(\rho) = 0$ for any $|\alpha + \beta| \leq 1$ and it is clear that $p_\rho(X) = r(\rho) q_\rho(X)$ where

$$q_\rho(X) = \sum_{|\alpha+\beta|=2} \frac{1}{\alpha!\beta!} \partial_x^\alpha \partial_\xi^\beta q(\rho) x^\alpha \xi^\beta$$

is a quadratic form in (x, ξ). If $q_\rho(X) = 0$ would have a non-real root with respect to ξ_0 then $q(\rho + \mu X) = 0$ so does for small μ by Rouché's theorem contradicting that q is a hyperbolic polynomial in ξ_0. $\square$

Let $p_\rho(X, Y)$ be the polar form of $p_\rho(X)$, that is $p_\rho(X, Y) = p_\rho(Y, X)$ and $p_\rho(X, X) = p_\rho(X)$. Then it is clear that

$$p_\rho(X, Y) = \sigma(X, F_p(\rho) Y), \quad X, Y \in \mathbb{R}^{2(n+1)}$$

and in particular we have $p_\rho(X) = \sigma(X, F_p(\rho) X)$ where $\sigma = \sum_{j=0}^n d\xi_j \wedge dx_j$ is the standard symplectic 2 form on $\mathbb{R}^{n+1} \times \mathbb{R}^{n+1}$ and $\sigma((x, \xi), (y, \eta)) = \langle \xi, y \rangle - \langle x, \eta \rangle$ in local coordinates where $\langle x, y \rangle = \sum_{j=0}^n x_j y_j$. A linear subspace $V \subset \mathbb{R}^{n+1} \times \mathbb{R}^{n+1}$ has an annihilator V^σ with respect to the symplectic form:

$$V^\sigma = \{X \mid \sigma(X, Y) = 0, \forall Y \in V\}.$$

Lemma 1.7 ([32])   *Let $Q(X)$ be a quadratic hyperbolic form on $\mathbb{R}^{2(n+1)}$ and let $F \in M_{2(n+1)}(\mathbb{R})$ be the Hamilton map of Q, that is the map given by the formula*

$$\frac{1}{2}Q(X, Y) = \sigma(X, FY), \quad X, Y \in \mathbb{R}^{2(n+1)}.$$

Then Q takes one of the following forms with respect to a suitable symplectic basis on $\mathbb{R}^{2(n+1)}$.

(1) $Q = \lambda(x_0^2 - \xi_0^2) + \sum_{j=1}^{k} \mu_j(x_j^2 + \xi_j^2) + \sum_{j=k+1}^{\ell} \xi_j^2$,

(2) $Q = -\xi_0^2 + \sum_{j=1}^{k} \mu_j(x_j^2 + \xi_j^2) + \sum_{j=k+1}^{\ell} \xi_j^2$,

(3) $Q = -\xi_0^2 + 2\xi_0\xi_1 + x_1^2 + \sum_{j=2}^{k} \mu_j(x_j^2 + \xi_j^2) + \sum_{j=k+1}^{\ell} \xi_j^2$

where $\lambda > 0$, $\mu_j > 0$. In the case (1) F has non-zero real eigenvalues $\pm\lambda$ and in the cases (2) and (3) all eigenvalues of F are on the imaginary axis. In the cases (1) and (2) we have $\operatorname{Ker} F^2 \cap \operatorname{Im} F^2 = \{0\}$ while $\operatorname{Ker} F^2 \cap \operatorname{Im} F^2 \neq \{0\}$ in the case (3). The case (3) the Jordan canonical form of F contains a subspace of dimension 4 corresponding to the eigenvalue 0.

Although we find a highly sophisticated proof of the lemma in [34, Section 21.5] we present the proof in the next Sect. 1.5 to facilitate the reading and to make the monograph more self-contained.

Thanks to Lemma 1.7, in a suitable symplectic basis, the localization $p_\rho(X)$ of p will be one of (1)–(3) of Lemma 1.7. But in studying the well-posedness of the Cauchy problem, not all canonical transformations are allowed since one can only use the canonical transformation such that the associated Fourier integral operator preserves the causality, and hence the canonical form of p_ρ in Lemma 1.7 would not be always applicable. This is a main reason why the study of the Cauchy problem is not so straightforward. In [35], partly motivated by this observation, quadratic hyperbolic operators are intensively studied.

Corollary 1.6   *Assume that Q is a hyperbolic quadratic form. Then the following conditions are equivalent.*

(i) *F has non-zero real eigenvalues,*

(ii) *there is $v \in V_0^\sigma$ such that $Q(v) < 0$,*

(iii) *there is $v \in (\operatorname{Ker} F)^\sigma$ such that $Q(v) < 0$,*

(iv) *for any $v \in \operatorname{Ker} F$ there is $w \in \mathbb{R}^{2(n+1)}$ such that $\sigma(v, w) = 0$, $Q(w) < 0$*

where V_0 denotes the space of generalized eigenvectors belonging to the zero eigenvalue.

Proof The implication (i)$\Longrightarrow$(ii) follows from Lemma 1.7. Indeed assume the case (1) occurs. Then we have $V_0 = \{x_0 = \cdots = x_k = 0, \xi_0 = \cdots = \xi_\ell = 0\}$ and hence $V_0^\sigma = \{\xi_0, \ldots, \xi_k, x_0, \ldots, x_\ell\}$ so that $v = H_{x_0}$ is a desired vector. The implications (ii)$\Longrightarrow$(iii)$\Longrightarrow$(iv) are trivial. We now prove (iv)$\Longrightarrow$(i). By Lemma 1.7, Q has one of the forms (1)–(3) in suitable symplectic coordinates. Suppose now (3) occurs. Working in $\{x_0, x_1, \xi_0, \xi_1\}$ space $\operatorname{Ker} F$ is given by $\{x_1 = \xi_0 = \xi_1 = 0\}$.

If $\sigma(v, w) = 0$, $\forall v \in \mathrm{Ker}\, F$ it follows that the ξ_0 coordinate of w is zero. Hence we get $Q(w) \geq 0$ and this shows that if (iv) holds then (3) never occurs.

Suppose that the case (2) occurs. Working in $\{x_0, \xi_0\}$ space we have $\mathrm{Ker}\, F = \{\xi_0 = 0\}$. If $\sigma(v, w) = 0$, $\forall v \in \mathrm{Ker}\, F$ then we see that the ξ_0 coordinate of w is zero and hence $Q(w) \geq 0$. This shows that the case (2) also never happens if (iv) holds.

Thus we proved that (iv) implies that only the case (1) happens. This proves the assertion.   □

In this monograph we always assume that the doubly characteristic set $\Sigma = \{(x, \xi) \mid \partial_x^\alpha \partial_\xi^\beta p(\rho) = 0, \forall |\alpha + \beta| \leq 1\}$ of $p(x, \xi)$ is a C^∞ manifold and the following conditions are satisfied:

$$\begin{cases} F_p \text{ has no real eigenvalues on } \Sigma, \\ p(x, \xi) \text{ vanishes exactly of order 2 on } \Sigma, \\ \mathrm{rank}\left(\sum_{j=0}^n d\xi_j \wedge dx_j \big|_\Sigma \right) = \text{constant on } \Sigma. \end{cases} \tag{1.5}$$

It seems to be quite reasonable to assume these conditions when one makes a general study of differential operators with non-effectively hyperbolic characteristics. Here we note that the second condition is equivalent to $\dim T_\rho \Sigma = \dim \mathrm{Ker}\, F_p(\rho)$ for any $\rho \in \Sigma$, that is, the codimension of Σ is equal to the rank of the Hessian of p at every point on Σ. By the third condition we mean that for every $\rho \in \Sigma$ one can find a conic neighborhood V of ρ such that the rank of $\sum_{j=0}^n d\xi_j \wedge dx_j$ on $T_\kappa \Sigma$ is independent of $\kappa \in V \cap \Sigma$. According to the spectral type of $F_p(\rho)$, two different possible cases may arise

$$\mathrm{Ker}\, F_p^2(\rho) \cap \mathrm{Im}\, F_p^2(\rho) = \{0\}, \tag{1.6}$$

$$\mathrm{Ker}\, F_p^2(\rho) \cap \mathrm{Im}\, F_p^2(\rho) \neq \{0\}. \tag{1.7}$$

Definition 1.17 We say that p is of spectral type 1 near $\rho \in \Sigma$ if there is a conic neighborhood V of ρ such that (1.6) holds in $V \cap \Sigma$ and of spectral type 2 near ρ if (1.7) holds in $V \cap \Sigma$. We say that p is of spectral type 1 (resp. 2) on Σ if (1.6) (resp. (1.7)) holds on Σ everywhere.

Definition 1.18 We say that there is no transition (of spectral type) if for any $\rho \in \Sigma$ we can find a conic neighborhood V of ρ such that either (1.6) or (1.7) holds throughout $V \cap \Sigma$.

In this monograph we prove that if there is no transition of spectral type and there is no bicharacteristic tangent to Σ the Cauchy problem is C^∞ well-posed under the strict IPH condition. If the positive trace is zero, the IPH condition is reduced to the Levi condition and we prove that if the spectral type 2 on Σ and there is no bicharacteristics tangent to Σ, the Levi condition is necessary and sufficient in order that the Cauchy problem to be C^∞ well-posed. We also prove that the Cauchy problem is well-posed in the Gevrey class of order 5 under the Levi condition, assuming that the spectral type is 2 on Σ. Furthermore we present an example of second order differential operator with polynomial coefficients of spectral type 2 on Σ, verifying the Levi condition, with a tangent bicharacteristic for which the Cauchy problem is ill-posed in the Gevrey class of order $s > 5$, proving the optimality of the above result.

Before closing the section we give examples for operators of spectral type 1 and 2 after Lemma 1.7. The differential operator of symbol (2) in Lemma 1.7 is

$$P_a = -D_0^2 + \sum_{j=1}^{k} \mu_j(x_j D_n^2 + D_j^2) + \sum_{j=k+1}^{\ell} D_j^2 \tag{1.8}$$

which is of spectral type 1 with $\Sigma = \{\xi_0 = \xi_1 = \cdots = \xi_\ell = 0, \ x_1 = \cdots = x_k = 0\}$. Denoting the symbol by $P_a(x, \xi)$ it is clear that

$$P_a(x, \xi) = -\xi_0^2 + q(x, \xi'), \quad q(x, \xi') \geq 0, \quad \{\xi_0, q\} = 0.$$

The differential operator of symbol (3) in Lemma 1.7 with $k = 1$ is

$$P_b = -D_0^2 + 2D_0 D_1 + x_1^2 D_n^2 + \sum_{j=2}^{\ell} D_j^2 \tag{1.9}$$

which is of spectral type 2 with $\Sigma = \{\xi_0 = \xi_1 = \cdots = \xi_\ell = 0, \ x_1 = 0\}$. It is also clear that one can write

$$P_b(x, \xi) = -\xi_0(\xi_0 - 2\xi_1) + q(x, \xi'), \quad q(x, \xi') \geq 0, \quad \{\xi_0, q\} = 0. \tag{1.10}$$

Adding a term $x_1^3 D_n^2$ to P_b we get

$$\tilde{P}_b = -D_0^2 + 2D_0 D_1 + x_1^2 D_n^2 + \sum_{j=2}^{\ell} D_j^2 + x_1^3 D_n^2 \tag{1.11}$$

which is of course considered for small x_1 with the same Σ. Since $x_1^3 \xi_n^2$ vanishes of order 3 on Σ hence does not affect the Hamilton map so that both $\tilde{P}_b$ and P_b have the same Hamilton map and are of spectral type 2. Nevertheless the factorization such as (1.10) is impossible for $\tilde{P}_b$. Taking geometrical view point the difference becomes much clearer. Indeed for P_a and P_b there is no bicharacteristic tangent to Σ while there is a bicharacteristic of $\tilde{P}_b$ tangent to Σ.

Here we give an example for which the transition from spectral type 1 to spectral type 2 occurs:

$$P_c = -D_0^2 + D_1^2 + (x_0 + x_1 + x_1^3)^2 D_n^2 \quad (n \geq 2)$$

where $\Sigma = \{\xi_0 = \xi_1 = 0, x_0 + x_1 + x_1^3 = 0\}$. It is easy to check that P_c is of spectral type 1 in $\Sigma \setminus S$ and spectral type 2 on $S = \{\xi_0 = \xi_1 = 0, \ x_0 = x_1 = 0\} \subset \Sigma$. We also give examples for different types of transition, including the transition from non-effective type to effective type in Sect. 6.4.

At the end of this section we make some comments on two special classes of differential operators with non-effectively hyperbolic characteristics. Consider differential operators with coefficients depending only on the time variable $t = x_0$:

$$P = -\partial_t^2 + \sum_{i,j=1}^n a_{ij}(t)\partial_{x_i}\partial_{x_j} + \sum_{j=1}^n b_j(t)\partial_{x_j} + b_0(t)\partial_t + c(t)$$

where $a(t, \xi) = \sum_{i,j=1}^n a_{ij}(t)\xi_i\xi_j \geq 0$. It is easy to see that P is non-effectively hyperbolic at (t, ξ), $\xi \neq 0$ if and only if $\partial_t^j a(t, \xi) = 0$ for $j = 0, 1, 2$ and doubly characteristics are *necessarily of spectral type* 1 *with* 0 *positive trace*. For solvability of the Cauchy problem, see [70, Proposition 5.2], [15, 16, 96] and references given there. Another special class consists of differential operators in $\mathbb{R}^2$

$$P = -\partial_t^2 + a(t, x)\partial_x^2 + b_1(t, x)\partial_x + b_0(t, x)\partial_t + c(t, x), \quad (t, x) \in \mathbb{R}^2$$

where $a(t, x) \geq 0$. It is easy to examine that P is non-effectively hyperbolic at (t, x) if and only if $\partial_t^j a(t, x) = 0$ for $j = 0, 1, 2$ and doubly characteristics are *necessarily of spectral type* 1 *with* 0 *positive trace* again. If $a(t, x)$, $b_j(t, x)$, $c(t, x)$ are real analytic then a necessary and sufficient condition for the Cauchy problem to be C^∞ well-posed is obtained in [73]. For the smooth coefficients case see [17, 18, 31] and references given there. See also [88] for differential operators with more than two independent variables.

1.5   Hyperbolic Quadratic Forms on a Symplectic Vector Space

In this section we give a proof of Lemma 1.7. We start with

Definition 1.19 Let S be a finite dimensional vector space over $\mathbb{R}$ ($\mathbb{C}$) and let σ be a nondegenerate anti-symmetric bilinear form on S. Then we call S a (finite dimensional) real (complex) symplectic vector space. Let S_i ($i = 1, 2$) be two symplectic vector spaces with symplectic forms σ_i. If a linear bijection

$$T : S_1 \rightarrow S_2$$

verifies $T^*\sigma_2 = \sigma_1$, that is $\sigma_1(v, w) = \sigma_2(Tv, Tw)$, $v, w \in S$ then T is called a symplectomorphism.

That σ is said to be nondegenerate if

$$\sigma(v, w) = 0, \ \forall w \in S \Longrightarrow v = 0.$$

A typical example is $(T^*\mathbb{R}^{n+1}, \sigma)$ where $T^*\mathbb{R}^{n+1} = \{(x, \xi) \mid x, \xi \in \mathbb{R}^{n+1}\}$ and

$$\sigma((x, \xi), (y, \eta)) = \langle \xi, y \rangle - \langle x, \eta \rangle$$

which is a basic ingredient in Sect. 1.4.

Proposition 1.3 *Let S be a finite dimensional real symplectic vector space. Then the dimension of S is even and there is a symplectomorphism*

$$T : S \to T^*\mathbb{R}^n$$

with some n.

Proof Let e_j, f_j be the unit vector along x_j, ξ_j axis in $T^*\mathbb{R}^n$ respectively. It is clear that

$$\sigma(e_j, e_k) = \sigma(f_j, f_k) = 0, \ \ \sigma(f_j, e_k) = \delta_{jk} \tag{1.12}$$

where δ_{jk} is the Kronecker's delta. To prove this proposition it is enough to show that there exists a basis of S verifying (1.12). Take $f_1 \in S$, $f_1 \neq 0$. Since σ is nondegenerate one can take $e_1 \in S$ so that $\sigma(f_1, e_1) = 1$. Note that f_1 and e_1 are linearly independent. Let $S_0 = \text{span}\{f_1, e_1\}$ and

$$S_1 = S_0^\sigma = \{v \in S \mid \sigma(v, S_0) = 0\}.$$

Then we have $S = S_1 \oplus S_0$ for if $v \in S_1 \cap S_0$ then writing $v = af_1 + be_1$ one obtains

$$\sigma(v, f_1) = -b = 0, \ \sigma(v, e_1) = a = 0$$

and hence $v = 0$. We now show that S_1 is a symplectic vector space with the symplectic form σ. It is enough to check that σ is nondegenerate on S_1. Suppose $\sigma(v, S_1) = 0$, $v \in S_1$. By definition we see $\sigma(v, S_0) = 0$ hence $\sigma(v, S) = 0$ which gives $v = 0$. The rest of the proof is carried out by induction. $\qquad\square$

Definition 1.20 Let S be a symplectic vector space of dimension $2n$ with the symplectic form σ. A basis $\{f_j, e_j\}_{j=1}^n$ verifying (1.12) is called a symplectic basis.

Proposition 1.4 *Let S be a symplectic vector space of dimension $2n$ with the symplectic form σ. Let A, B be subsets of $J = \{1, 2, \ldots, n\}$. Assume that $\{e_j\}_{j \in A}$, $\{f_k\}_{k \in B}$ are linearly independent and verify (1.12). Then one can choose $\{e_j\}_{j \in J \setminus A}$, $\{f_k\}_{k \in J \setminus B}$ so that $\{e_j\}_{j \in J}$ and $\{f_k\}_{k \in J}$ become a full symplectic basis.*

Proof Assume $B \setminus A \neq \emptyset$. Take $l \in B \setminus A$. Then there exists $g \in S$ such that $\sigma(g, f_l) = -1$. With $V = \text{span}\{e_j, f_k \mid j \in A, k \in B\}$ we have $g \notin V$ because $\sigma(V, f_l) = 0$ by assumption. Choosing $\alpha_i, \beta_i, i \in A \cap B$ suitably one can assume that

$$e_l = g - \sum_{i \in A \cap B} \alpha_i e_i - \sum_{i \in A \cap B} \beta_i f_i$$

verifies

$$\sigma(e_l, e_j) = 0, \ j \in A, \ \ \sigma(e_l, f_k) = -\delta_{lk}, \ k \in B.$$

Repeating this argument we may assume that $B \subset A$. Applying the same arguments to $A \setminus B$ we may assume $A = B$. If $A = B \neq J$ then with

$$S_0 = \text{span}\{e_j, f_k \mid j \in A, k \in B\}$$

we consider $S_1 = S_0^\sigma$. Since S_1 is a symplectic vector space, then by Proposition 1.3 there is a symplectic basis for S_1 and hence it is enough to add this basis to $\{e_j, f_j\}_{j \in A = B}$. $\qquad\square$

Let Q be a quadratic form on S. Let F be the Hamilton map of Q. Since Q is symmetric we have $\sigma(FX, Y) = -\sigma(X, FY)$ and hence F is anti-symmetric with respect to σ. Let $S_\mathbb{C} = \{X + iY \mid X, Y \in S\}$ be the complexification of S and V_λ be the generalized eigenspace associated with the eigenvalue $\lambda \in \mathbb{C}$ of F.

Lemma 1.8 *If $\lambda + \mu \neq 0$ then $Q(V_\lambda, V_\mu) = 0$ and $\sigma(V_\lambda, V_\mu) = 0$. In particular $Q(V_\lambda, V_\lambda) = 0, \sigma(V_\lambda, V_\lambda) = 0$ if $\lambda \neq 0$.*

Proof Since $\lambda + \mu \neq 0$ then $F + \mu$ is bijective on V_λ. From

$$\sigma((F + \mu)^N V_\lambda, V_\mu) = \sigma(V_\lambda, (-F + \mu)^N V_\mu) = 0$$

for large N and hence $\sigma(V_\lambda, V_\mu) = 0$. Noticing that

$$Q(V_\lambda, V_\mu) = \sigma(V_\lambda, FV_\mu) = \sigma(V_\lambda, (F - \mu)V_\mu) + \mu\sigma(V_\lambda, V_\mu)$$
$$= \sigma(V_\lambda, (F - \mu)V_\lambda) = \cdots = \sigma(V_\lambda, (F - \mu)^N V_\mu) = 0$$

we get $Q(V_\lambda, V_\mu) = 0$. $\qquad\square$

Since F is a real map, we see that $V_{\bar\lambda} = \overline{V_\lambda}$. If $\lambda + \mu \neq 0, \bar\lambda + \mu \neq 0$ then

$$Q(V_\lambda, V_\mu) = Q(\overline{V_\lambda}, V_\mu) = Q(\overline{V_\lambda}, \overline{V_\mu}) = Q(V_\lambda, \overline{V_\mu}) = 0.$$

This shows that any two of $\text{Re}\, V_\lambda$, $\text{Im}\, V_\lambda$, $\text{Re}\, V_\mu$ and $\text{Im}\, V_\mu$ are Q orthogonal. Similar arguments prove that any pair of these spaces are also σ orthogonal. We next examine that

$$\dim_\mathbb{C} V_\lambda \leq \dim_\mathbb{R} \text{Re}\, V_\lambda. \tag{1.13}$$

To see this let $V_\lambda = \mathrm{span}_\mathbb{C}\{e_1, \ldots, e_s\}$. Suppose $\mathrm{Re}\, V_\lambda = \mathrm{span}_\mathbb{R}\{f_1, \ldots, f_r\}$ with $r < s$. Since $\mathrm{Re}\, e_i$, $\mathrm{Im}\, e_i \in \mathrm{Re}\, V_\lambda$ and hence $e_i \in \mathrm{span}_\mathbb{C}\{f_1, \ldots, f_r\}$ which is a contradiction. Note that

$$\mathrm{Ker} F = \{X \mid Q(X, Y) = 0, \ \forall Y \in S\}.$$

In what follows we assume that Q is nonnegative definite or hyperbolic, that is Q has the signature $(q, 1)$.

Lemma 1.9 *Let* $V \subset S$ *be a linear subspace of* S. *Assume that* $V \cap \mathrm{Ker} F = \{0\}$ *and* $Q(V) \le 0$. *Then* $\dim V \le 1$.

Proof Write $S = \mathrm{Ker} F \oplus S_0$ and let V_0 be the projection of V into S_0 along $\mathrm{Ker} F$. Since $\mathrm{Ker} F \cap V = \{0\}$ we have $\dim V_0 = \dim V$. Note that $Q(V_0) \le 0$ and Q is nondegenerate on S_0 and hence positive definite or has the Lorenz signature on S_0. Then it is clear $\dim V_0 \le 1$. □

Proof of Lemma 1.5 It suffices to show that if λ is an eigenvalue of F with $\mathrm{Re}\, \lambda \ne 0$ then λ is real and $\dim_\mathbb{C} V_\lambda = 1$. From Lemma 1.8 we see

$$Q(V_\lambda + \overline{V_\lambda}, V_\lambda + \overline{V_\lambda}) = Q(V_{\bar\lambda}, V_\lambda) + Q(V_\lambda, V_{\bar\lambda}) = 0$$

because $\lambda + \bar\lambda \ne 0$. This shows $Q(\mathrm{Re}\, V_\lambda) = 0$. Hence by Lemma 1.9 we have $\dim \mathrm{Re}\, V_\lambda \le 1$. On the other hand (1.13) shows that $\dim_\mathbb{C} V_\lambda \le \dim_\mathbb{R} \mathrm{Re}\, V_\lambda \le 1$ and hence $\dim_\mathbb{C} V_\lambda = \dim_\mathbb{R} \mathrm{Re}\, V_\lambda = 1$. Let $\mathrm{Re}\, V_\lambda = \mathrm{span}_\mathbb{R}\{f\}$ and $V_\lambda = \mathrm{span}_\mathbb{C}\{e\}$. Then it is clear that $e = \alpha f$ with some $\alpha \in \mathbb{C}$. Since $Fe = \lambda e$ and hence $Ff = \lambda f$ this shows that λ is real. □

We show that if λ, λ' are non-zero real eigenvalues of F then $\lambda = \pm\lambda'$. Assume $\lambda + \lambda' \ne 0$. Since V_λ, $V_{\lambda'}$ is one dimensional then $V_\lambda = \mathrm{span}_\mathbb{C}\{e\}$ and $V_{\lambda'} = \mathrm{span}_\mathbb{C}\{f\}$ with some $e, f \in S$. If $\mathrm{Ker} F \cap (V_\lambda + V_{\lambda'}) \ne \{0\}$ then e, f are linearly dependent. Otherwise we have

$$Q(\alpha e + \beta f, \alpha e + \beta f) = 2\alpha\beta Q(e, f) = 0.$$

From Lemma 1.9 it follows that $\dim(V_\lambda + V_{\lambda'}) \le 1$ and hence e and f are linearly dependent and then $\lambda = \lambda'$. We have thus proved

$$S_\mathbb{C} = \sum_{\mu > 0} \oplus (V_{i\mu} + V_{-i\mu}) \oplus (V_\lambda + V_{-\lambda}) \oplus V_0, \quad \lambda \in \mathbb{R}, \ \lambda \ne 0. \qquad (1.14)$$

Recall that the sum is Q and σ orthogonal. We first study $V_{\pm\lambda}$. Let $V_\lambda = \mathrm{span}\{e\}$, $V_{-\lambda} = \mathrm{span}\{f\}$, $e, f \in S$. Then we have $\sigma(e, f) \ne 0$ otherwise we would have $\sigma(e, S_\mathbb{C}) = 0$ and $e = 0$ which is a contradiction. Thus we may assume that

$\sigma(f, e) = 1$ and hence with $u = (f + e)/2$, $v = (f - e)/2$ one has

$$V_\lambda + V_{-\lambda} = \text{span}\{f, e\}, \quad Q(xu + \xi v) = \lambda(x^2 - \xi^2). \qquad (1.15)$$

We turn to pure imaginary eigenvalues.

Lemma 1.10 *Let λ be a pure imaginary eigenvalue of F. Then V_λ consists of simple eigenvectors and*

$$Q(v, \bar{v}) > 0 \quad if \quad v \in V_\lambda, \ v \neq 0.$$

Proof Let us fix $v \in V_{i\mu}$ ($\mu \neq 0$). With $v = v_1 + iv_2$, $v_i \in S$ we have $\bar{v} = v_1 - iv_2 \in V_{-i\mu}$ and

$$Q(v + \bar{v}, v + \bar{v}) = 2Q(v, \bar{v}) = 4Q(v_1, v_1) = 4Q(v_2, v_2) \qquad (1.16)$$

because $Q(v, v) = Q(\bar{v}, \bar{v}) = 0$. Suppose $Q(v, \bar{v}) \leq 0$ and hence $Q(v_1, v_1) \leq 0$, $Q(v_2, v_2) \leq 0$ by (1.16). Thus denoting $V = \text{span}_{\mathbb{R}}\{v_1, v_2\}$ we have $Q(V) \leq 0$. This proves that v_1 and v_2 are colinear by Lemma 1.9. Then one can write $v = \alpha f$ with some $\alpha \in \mathbb{C}, f \in S$. Since $Fv = i\mu v$ this gives a contradiction. This proves $Q(v, \bar{v}) > 0$. We now suppose that there are $v, w \in V_{i\mu}$ such that

$$Fv = i\mu v, \quad Fw = i\mu w + v.$$

Then we have $Q(v, \bar{v}) = \sigma(v, F\bar{v}) = -i\mu\sigma(v, \bar{v})$. On the other hand from

$$\sigma(v, F\bar{w}) = -i\mu\sigma(v, \bar{w}) + \sigma(v, \bar{v}) = -\sigma(Fv, \bar{w}) = -i\mu\sigma(v, \bar{w})$$

it follows that $\sigma(v, \bar{v}) = 0$. This implies $Q(v, \bar{v}) = 0$ which is a contradiction. Therefore eigenvalues are simple. $\qquad\qquad\qquad\qquad\qquad\qquad\qquad\qquad\qquad\quad\square$

Thanks to Lemma 1.10 it results that $Q(v, \bar{v})$ induces a inner product in $V_{i\mu}$. Choose a basis $\{e_1, \ldots, e_s\}$ for $V_{i\mu}$ which is orthogonal with respect to this inner product:

$$V_{i\mu} = \text{span}_{\mathbb{C}}\{e_1, \ldots, e_s\}, \quad Q(e_i, \bar{e}_j) = 0, \ i \neq j.$$

We put

$$V_i = \text{span}_{\mathbb{R}}\{\text{Re}\, e_i, \text{Im}\, e_i\}.$$

It is clear that $\dim V_i = 2$. Since $Q(\pm\bar{e}_i, \pm e_j) = 0$ ($i \neq j$) we see that V_i are Q orthogonal each other: $Q(V_i, V_j) = 0, i \neq j$. This proves that

$$\text{Re}\,(V_{i\mu} + V_{-i\mu}) = \sum_{j=1}^{s} \oplus V_j.$$

We now compute

$$
\begin{aligned}
Q(x\mathrm{Re}\, e_i + \xi\mathrm{Im}\, e_i) &= \sigma(x\mathrm{Re}\, e_i + \xi\mathrm{Im}\, e_i, F(x\mathrm{Re}\, e_i + \xi\mathrm{Im}\, e_i)) \\
&= \mu\sigma(x\mathrm{Re}\, e_i + \xi, -x\mathrm{Im}\, e_i + \xi\mathrm{Re}\, e_i) \\
&= -\mu\sigma(\mathrm{Re}\, e_i, \mathrm{Im}\, e_i)(x^2 + \xi^2)
\end{aligned}
$$

where $-\mu\sigma(\mathrm{Re}\, e_i, \mathrm{Im}\, e_i) > 0$ by Lemma 1.10. We now normalize e_i so that $\sigma(\mathrm{Re}\, e_i, \mathrm{Im}\, e_i) = -1$ and thus we obtain

$$
Q(x\mathrm{Re}\, e_i + \xi\mathrm{Im}\, e_i) = \mu(x^2 + \xi^2). \tag{1.17}
$$

We are left with $\mathrm{Re}\, V_0 = V$. From (1.14) it is sufficient to study Q on V and we may assume that $V = S$ and F is nilpotent on S.

Lemma 1.11 *Let F be nilpotent on S. Then there are symplectic subspaces V_i, $\dim V_i = 2$ such that*

$$
S = \left(\sum \oplus V_i\right) \oplus W
$$

where the sum being σ, Q orthogonal and one can choose symplectic basis $\{f_i, e_i\}$ for V_i such that

$$
Q(xe_i + \xi f_i) = \xi^2 \ \ or \ \ 0 \tag{1.18}
$$

and moreover F, restricted on W, verifies the followings:

(a) $\sigma(v, w) = 0, \quad \forall v, w \in \mathrm{Ker} F$
(b) $F^2 v = 0 \implies Q(v) = 0.$

Proof Assume that there is a $v \in S$ such that $F^2 v = 0$, $Q(v) = \sigma(v, Fv) \neq 0$. Set $V = \mathrm{span}\{v, Fv\}$ then V is a symplectic subspace. Decomposing $S = V \oplus V^\sigma$ we see that the sum is Q orthogonal for

$$
Q(v, w) = \sigma(v, Fw) = -\sigma(Fv, w) = 0, \quad v \in V, \ w \in V^\sigma
$$

because V is F invariant. Note that

$$
Q(\xi v + xFv) = \sigma(\xi v + xFv, \xi Fv) = \xi^2 Q(v).
$$

Since $Q(v) \neq 0$, normalizing $Q(v) = 1$ we obtain a desired symplectic basis $\{v, Fv\}$. We next assume that there are $v, w \in \mathrm{Ker} F$ such that $\sigma(v, w) \neq 0$. Denote $V = \mathrm{span}\{v, w\}$ which is symplectic. Then the same arguments as above show that

$S = V \oplus V^\sigma$ where the sum is Q orthogonal. Note that

$$Q(xv + \xi w) = \sigma(xv + \xi w, F(xv + \xi w)) = 0.$$

Repeating this argument one can remove all such V_i.   □

Lemma 1.12 *Let F be nilpotent on S and verifies* (a) *and* (b) *in Lemma 1.11. Then* $\dim S = 4$ *and one can choose a symplectic basis* $\{u_1, u_2, v_1, v_2\}$ *so that*

$$Q(x_1 u_1 + x_2 u_2 + \xi_1 v_1 + \xi_2 v_2) = -\xi_1^2 + 2\xi_1 \xi_2 + x_2^2. \tag{1.19}$$

Proof Study the map

$$F : \mathrm{Ker} F^2 \to \mathrm{Ker} F.$$

Note that $\mathrm{Im} F = (\mathrm{Ker} F)^\sigma \supset \mathrm{Ker} F$ by (a). Then for any $w \in \mathrm{Ker} F$ there is v such that $Fv = w$ and hence $F^2 v = 0$, that is $v \in \mathrm{Ker} F^2$. This shows that F is surjective on $\mathrm{Ker} F$. Thus we get

$$\mathrm{Ker} F \simeq \mathrm{Ker} F^2 / \mathrm{Ker} F.$$

On the other hand from (b) we have $Q(v) = 0$ for any $v \in \mathrm{Ker} F^2$. Then Lemma 1.9 shows that $\dim(\mathrm{Ker} F^2 / \mathrm{Ker} F) \leq 1$. Thus we get

$$\dim \mathrm{Ker} F = 1.$$

Since F is nilpotent on S one can choose v so that

$$S = \mathrm{span}\{v, Fv, \ldots, F^{N-1}v\}, \quad F^j v \neq 0, \ 1 \leq j \leq N - 1, \quad F^N v = 0.$$

It is clear that N is even and $N > 2$. We show that $N = 4$. To see this we first note that

$$Q(F^j v, F^k v) = \sigma(F^j v, F^{k+1} v) = (-1)^j \sigma(v, F^{j+k+1} v) = 0 \tag{1.20}$$

if $j + k + 1 \geq N$. Let $V = \mathrm{span}\{F^{N-4} v, F^{N-2} v\}$ then $V \cap \mathrm{Ker} F = \{0\}$ by definition. If $(N - 4) + (N - 2) + 1 \geq N$ then $Q(V) = 0$ by (1.20) hence which contradicts to $\dim V \leq 1$ by Lemma 1.9 which is a contradiction. Thus $(N-4)+(N-2)+1 \leq N-1$ and hence $N = 4$. We set

$$W = \mathrm{span}\{v, F^2 v\}.$$

If $Q(v, F^2 v) = \sigma(v, F^3 v) \leq 0$ then $Q(W) \leq 0$ by (1.20) which is a contradiction by Lemma 1.9 since $W \cap \mathrm{Ker} F = \{0\}$ is obvious. Therefore $Q(v, F^2 v) = \sigma(v, F^3 v) > 0$.

We normalize v so that $\sigma(v, F^3 v) = -1$. Put $w = v + tF^2 v$ with $t \in \mathbb{R}$. Then one has

$$\sigma(w, Fw) = \sigma(v + tF^2 v, Fv + tF^3 v)$$
$$= \sigma(v, Fv) + 2t\sigma(v, F^3 v) = \sigma(v, Fv) - 2t.$$

Then we can choose t so that $\sigma(w, Fw) = 0$. Since $\sigma(w, F^3 w) = -1$ with

$$u_1 = -F^3 w, \quad u_2 = F^3 w + Fw, \quad v_1 = w - F^2 w, \quad v_2 = F^2 w$$

it is easy to examine (1.19). □

Proof of Lemma 1.7 The proof is clear from (1.15), (1.17)–(1.19). □

Chapter 2
Non-effectively Hyperbolic Characteristics

Abstract In this chapter introducing the notion of local and microlocal elementary factorization of p, arising from standard techniques of energy integrals we prove that if p is of spectral type 1 on Σ then p always admits a local elementary factorization. On the other hand if p is of spectral type 2 even micolocal elementary factorization is not always possible. When p is of spectral type 2 near ρ we prove that p admits a "nice" microlocal factorization near ρ, which is also a microlocal elementary factorization at ρ, if the cube of some vector field H_S annihilates p on Σ near ρ. This factorization is crucial to deriving energy estimates.

2.1 Case of Spectral Type 1

In this chapter we follow the presentation of [84, Chapter 3]. As we will see in Chap. 5 the Cauchy problem for differential operators with characteristics of order at most double can be reduced to those for second order operators, differential in x_0 and pseudodifferential in x'. Therefore in this chapter we consider P of the form

$$P = -D_0^2 + A_1(x, D')D_0 + A_2(x, D')$$

with $A_j(x, D') \in \mathrm{Op}S_{phg}^j$ so that $A_j(x, \xi') \sim A_{j0} + A_{j1} + \cdots$. Here A_{jk} is homogeneous of degree $j - k$ in ξ'. Then the principal symbol $p(x, \xi)$ is

$$p(x, \xi) = -\xi_0^2 + A_{10}(x, \xi')\xi_0 + A_{20}(x, \xi'). \tag{2.1}$$

We start with the following definition.

Definition 2.1 ([40]) We say that $p(x, \xi)$ admits a local elementary factorization if there exist a neighborhood Ω of $x = 0$ and real valued $\lambda(x, \xi')$, $\mu(x, \xi')$ and $Q(x, \xi')$ which are C^∞ in $\Omega \times \mathbb{R}^n$, homogeneous in ξ' of degree 1, 1, 2 respectively and $Q(x, \xi') \geq 0$ such that

$$p(x, \xi) = -\Lambda(x, \xi)M(x, \xi) + Q(x, \xi')$$

© Springer International Publishing AG 2017

T. Nishitani, *Cauchy Problem for Differential Operators with Double Characteristics*, Lecture Notes in Mathematics 2202, DOI 10.1007/978-3-319-67612-8_2

with $\Lambda(x, \xi) = \xi_0 - \lambda(x, \xi')$ and $M(x, \xi) = \xi_0 - \mu(x, \xi')$ verifying with some $C > 0$

$$|\{\Lambda(x, \xi), Q(x, \xi')\}| \leq CQ(x, \xi'), \tag{2.2}$$

$$|\{\Lambda(x, \xi), M(x, \xi)\}| \leq C(\sqrt{Q(x, \xi')} + |\Lambda(x, \xi') - M(x, \xi')|). \tag{2.3}$$

If we can find such symbols defined in a conic neighborhood of ρ then we say that $p(x, \xi)$ admits a microlocal elementary factorization at ρ.

By conjugation with a Fourier integral operator with x_0 as a parameter, one can assume

$$p(x, \xi) = -\xi_0^2 + q(x, \xi'), \quad q(x, \xi') \geq 0.$$

For this p, the second condition of (1.5) is equivalent to that, at each $\bar{\rho} \in \Sigma$, one can find a conic neighborhood V of $\bar{\rho}$, $r \in \mathbb{N}$ and $\xi_0 = \phi_0$, $\phi_j(x, \xi')$, $j = 1, \ldots, r$ defined in V with linearly independent differentials such that

$$p = -\phi_0^2 + \sum_{j=1}^{r} \phi_j(x, \xi')^2 \tag{2.4}$$

where $V \cap \Sigma$ is given by $\phi_j = 0, j = 0, \ldots, r$. Let $Q(u, v)$ be the polar form of the localization $p_{\bar{\rho}}$ (see Definition 1.16). Since it is clear

$$Q(u, v)/2 = -d\phi_0(u)d\phi_0(v) + \sum_{j=1}^{r} d\phi_j(u)d\phi_j(v)$$

where $d\phi_j(u) = d\phi_j(\bar{\rho}; u)$, noting $d\phi_j(u) = \sigma(u, H_{\phi_j})$ it follows that

$$Q(u, v)/2 = -\sigma(u, H_{\phi_0})\sigma(v, H_{\phi_0}) + \sum_{j=1}^{r} \sigma(u, H_{\phi_j})\sigma(v, H_{\phi_j})$$

$$= \sigma\Big(u, -\sigma(v, H_{\phi_0})H_{\phi_0} + \sum_{j=1}^{r} \sigma(v, H_{\phi_j})H_{\phi_j}\Big) = \sigma(u, F_p(\bar{\rho})v).$$

Thus we have

$$F_p(\bar{\rho})v = -\sigma(v, H_{\phi_0})H_{\phi_0} + \sum_{j=1}^{r} \sigma(v, H_{\phi_j}(\bar{\rho}))H_{\phi_j}(\bar{\rho}). \tag{2.5}$$

In particular we see

$$\mathrm{Im}\, F_p(\bar{\rho}) = \langle H_{\phi_0}(\bar{\rho}), H_{\phi_1}(\bar{\rho}), \ldots, H_{\phi_r}(\bar{\rho})\rangle \tag{2.6}$$

where $\langle v_1, \ldots, v_r \rangle$ denotes the linear space spanned by $v_1, \ldots, v_r$. It is clear that

$$
\begin{aligned}
\operatorname{Ker} F_p(\bar{\rho}) &= \{ v \in \mathbb{R}^{2(n+1)} \mid \sigma(v, H_{\phi_j}) = 0, j = 0, 1, \ldots, r \} \\
&= \langle H_{\phi_0}, H_{\phi_1}, \ldots, H_{\phi_r} \rangle^\sigma = (\operatorname{Im} F_p(\bar{\rho}))^\sigma = T_{\bar{\rho}} \Sigma.
\end{aligned}
\tag{2.7}
$$

Here we note that $\{f, g\} = dg(H_f) = \sigma(H_f, H_g)$ (see Definition 1.10).

Lemma 2.1 *The last condition of (1.5) is equivalent to*

$$
\operatorname{rank}(\{\phi_i, \phi_j\})(\rho) = \text{const}, \quad \rho \in V \cap \Sigma
$$

where $(\{\phi_i, \phi_j\})$ denotes the $(r+1) \times (r+1)$ matrix with (i, j) entry $\{\phi_i, \phi_j\}$.

Proof Since $(T_\rho \Sigma)^\sigma = \langle H_{\phi_0}(\rho), \ldots, H_{\phi_r}(\rho) \rangle$ by (2.7) it is enough to show that the last condition of (1.5) is equivalent to $\operatorname{rank} \sigma |_{(T_\rho \Sigma)^\sigma} = \text{const}$. Consider the map

$$
L : T_\rho \Sigma \ni v \mapsto \sum_{j=1}^{s} \sigma(v, f_j(\rho)) f_j(\rho) \in T_\rho \Sigma
$$

where $T_\rho \Sigma = \langle f_1(\rho), \ldots, f_s(\rho) \rangle$. The last assumption of (1.5) implies that the rank of the matrix $(\sigma(f_i(\rho), f_j(\rho)))$ is constant and hence $\dim \operatorname{Ker} L = \dim(T_\rho \Sigma \cap (T_\rho \Sigma)^\sigma) = \text{const}$. This proves the desired assertion because the kernel of the linear map

$$
\tilde{L} : (T_\rho \Sigma)^\sigma \ni v \mapsto \sum_{j=0}^{r} \sigma(v, H_{\phi_j}(\rho)) H_{\phi_j}(\rho) \in (T_\rho \Sigma)^\sigma
$$

is just $\operatorname{Ker} L$.   □

Assume (1.6) then from Lemma 1.7, in a suitable symplectic coordinates system, the quadratic form p_ρ takes the form

$$
p_\rho = -\xi_0^2 + \sum_{j=1}^{k} \mu_j (x_j^2 + \xi_j^2) + \sum_{j=k+1}^{k+\ell} \xi_j^2.
\tag{2.8}
$$

Lemma 2.2 *The number k in (2.8) is independent of $\rho \in V \cap \Sigma$.*

Proof With $\{\psi_j\} = \{\xi_0, x_j, 1 \le j \le k, \xi_j, 1 \le j \le k + \ell\}$ it follows from Lemma 2.1 that the rank of $(\{\psi_i, \psi_j\})$ is constant on $V \cap \Sigma$. This shows that k is independent of $\rho \in V \cap \Sigma$.   □

Lemma 2.3 *There exist a conic neighborhood V of $\bar{\rho}$ and a smooth vector $h(\rho)$ defined in $V \cap \Sigma$ where we have*

$$
h(\rho) \in \operatorname{Ker} F_p^2(\rho), \quad p_\rho(h(\rho)) < 0, \quad \sigma(H_{x_0}, F_p(\rho) h(\rho)) = -1.
\tag{2.9}
$$

Proof Let p_ρ be of the form (2.8). Then from (2.5)

$$F_p^2(\rho)v = \sum_{j=1}^{k} \mu_j^2\big(\sigma(v, H_{\xi_j})H_{x_j} - \sigma(v, H_{x_j})H_{\xi_j}\big)$$

so that $\operatorname{Ker} F_p^2(\rho) = \{v \mid \sigma(v, H_{\xi_j}) = 0, \sigma(v, H_{x_j}) = 0, j = 1, \ldots, k\}$ and hence we see $\dim \operatorname{Ker} F_p^2(\rho) = 2n + 2 - 2k$ which is independent of $\rho \in \Sigma$ by Lemma 2.2. Let $p_{\bar\rho}$ be of the form (2.8). Since we have $F_p^2(\bar\rho)H_{x_0} = 0$, $p_{\bar\rho}(H_{x_0}) = -1$ and $\sigma(H_{x_0}, F_p(\bar\rho)H_{x_0}) = -1$ one can find a conic neighborhood V of $\bar\rho$ such that there exists a smooth $h(\rho)$ defined in $V \cap \Sigma$ such that for $\rho \in V \cap \Sigma$

$$h(\rho) \in \operatorname{Ker} F_p^2(\rho), \quad p_\rho(h(\rho)) < 0, \quad \sigma(H_{x_0}, F_p(\rho)h(\rho)) = -1. \tag{2.10}$$

We can assume that $h(\rho)$ is homogeneous of degree 0 in ξ, for if not we can just restrict to the sphere $|\xi| = 1$ and extend the restriction so that it becomes homogeneous of degree 0. $\square$

Lemma 2.4 *Assume that $h(\rho)$ satisfies (2.10). Then we have*

$$\sigma(v, F_p(\rho)h(\rho)) = 0, \quad v \neq 0 \Longrightarrow p_\rho(v) > 0.$$

Proof Let us fix $\rho \in V \cap \Sigma$. We can assume that p_ρ has the form (2.8). Set $w = F_p(\rho)h(\rho)$ so that $w \in \operatorname{Ker} F_p(\rho)$. One can write $h(\rho) = (y_0, \ldots, y_n, -1, \eta_1, \ldots, \eta_n)$ from (2.10) where $y_1 = \cdots = y_k = 0$, $\eta_1 = \cdots = \eta_k = 0$ then we see

$$w = H_{\xi_0} - \sum_{j=k+1}^{k+\ell} \eta_j H_{\xi_j}, \qquad 1 > \sum_{j=k+1}^{k+\ell} \eta_j^2.$$

Let $v = (x_0, \ldots, x_n, \xi_0, \ldots, \xi_n)$ and $\sigma(v, w) = 0$ so that $\xi_0 - \sum_{j=k+1}^{k+\ell} \eta_j \xi_j = 0$. If $\xi_0 = 0$ then $p_\rho(v) > 0$ is clear. Thus, assuming $\xi_0 \neq 0$, it is enough to show $\xi_0^2 < \sum_{j=k+1}^{k+\ell} \xi_j^2 + \sum_{j=1}^{k} \mu_j(x_j^2 + \xi_j^2)$. Indeed if otherwise we would have $\xi_0^2 \geq \sum_{j=k+1}^{k+\ell} \xi_j^2 + \sum_{j=1}^{k} \mu_j(x_j^2 + \xi_j^2)$ then

$$\xi_0^2 = \Big(\sum_{j=k+1}^{k+\ell} \eta_j \xi_j\Big)^2 \leq \Big(\sum_{j=k+1}^{k+\ell} \eta_j^2\Big)\Big(\sum_{j=k+1}^{k+\ell} \xi_j^2\Big) < \xi_0^2$$

which is a contradiction. $\square$

Proposition 2.1 *Assume that p is of spectral type 1 near $\rho \in \Sigma$. Then one can write p in a conic neighborhood V of ρ with some $\delta < 1$*

$$p = -\xi_0^2 + q = -\xi_0^2 + \sum_{i=1}^{r} \phi_i^2 = -(\xi_0 + \lambda)(\xi_0 - \lambda) + Q,$$

$$\lambda = \sum_{j=1}^{r} \gamma_j \phi_j, \quad |\lambda| \leq \delta \sqrt{Q} \quad \text{in } V,$$

$$\{\xi_0 - \lambda, \phi_i\} = 0 \quad \text{on } V \cap \Sigma, \quad i = 1, \dots, r.$$

In particular p admits a microlocal elementary factorization at ρ.

Proof We work in a conic neighborhood V of the reference point. Let $h(\rho)$ be in Lemma 2.3 and put $w(\rho) = F_p(\rho)h(\rho)$. Since $\operatorname{Im} F_p(\rho) = \langle H_{\xi_0}, H_{\phi_1}, \dots, H_{\phi_r}\rangle$ then one can write $w(\rho) = \gamma_0 H_{\xi_0} - \sum_{j=1}^{r} \gamma_j H_{\phi_j}$ where $\gamma_j(\rho)$ are smooth in $V \cap \Sigma$. From $\sigma(H_{x_0}, w(\rho)) = -1$ we have $\gamma_0 = 1$. As remarked above we can assume that γ_j are homogeneous of degree 0 in ξ. We denote

$$\lambda = \sum_{j=1}^{r} \gamma_j(x, \xi') \phi_j(x, \xi') \tag{2.11}$$

so that $w(\rho) = H_{\xi_0 - \lambda}$ on $V \cap \Sigma$. Write

$$p = -(\xi_0 + \lambda)(\xi_0 - \lambda) + \hat{q}, \quad \hat{q} = \sum_{j=1}^{r} \phi_j^2 - \left(\sum_{j=1}^{r} \gamma_j \phi_j\right)^2 = q - \lambda^2$$

and check that $\sum_{j=1}^{r} \gamma_j^2 < 1$. From Lemma 2.4 it follows that

$$\sigma(v, H_{\xi_0 - \lambda}) = 0, \ v \neq 0 \Longrightarrow \sum_{j=1}^{r} \sigma(v, H_{\phi_j})^2 - \left(\sum_{j=1}^{r} \gamma_j \sigma(v, H_{\phi_j})\right)^2 > 0$$

since $\sigma(v, F_p(\rho)v) = \sigma(v, F_{\hat{q}}(\rho)v)$. Note that the map

$$\langle H_{\xi_0 - \lambda}\rangle^\sigma / T_\rho \Sigma \ni v \mapsto (\sigma(v, H_{\phi_j}))_{j=1,\dots,r} \in \mathbb{R}^r$$

is surjective. Indeed $\dim \langle H_{\xi_0 - \lambda}\rangle^\sigma / T_\rho \Sigma = r$ and if $\sigma(v, H_{\xi_0 - \lambda}) = 0$, $\sigma(v, H_{\phi_j}) = 0$ for $j = 1, \dots, r$ then it follows that

$$v \in \langle H_{\xi_0 - \lambda}, H_{\phi_1}, \dots, H_{\phi_r}\rangle^\sigma = \langle H_{\xi_0}, H_{\phi_1}, \dots, H_{\phi_r}\rangle^\sigma = \operatorname{Ker} F_p(\rho) = T_\rho \Sigma.$$

From this it follows that $\langle \gamma, t \rangle^2 < |t|^2$ for any $0 \neq t \in \mathbb{R}^r$ and hence we conclude $|\gamma(\rho)| = (\sum_{j=1}^{r} \gamma_j(\rho)^2)^{1/2} < 1$. We extend $\gamma_j(\rho)$ ($\rho \in V \cap \Sigma$) to V such a way that $|\gamma| < 1$ in V. Therefore there exists $c > 0$ such that

$$\hat{q} \geq c \sum_{j=1}^{r} \phi_j(x, \xi')^2 \tag{2.12}$$

and hence we have $|\lambda|^2 \leq \delta q$ with some $\delta < 1$. Recall that $H_{\xi_0 - \lambda} \in \operatorname{Ker} F_p$ in $V \cap \Sigma$ which shows $\{\xi_0 - \lambda, \phi_j\} = 0$ in $V \cap \Sigma$ and hence

$$\{\xi_0 - \lambda, \lambda\} = 0 \quad \text{on} \quad V \cap \Sigma. \tag{2.13}$$

Then the proof follows immediately from (2.11)–(2.13). $\square$

Proposition 2.2 *Assume that p is of spectral type 1 on Σ. Then p admits a local elementary factorization $p = -M\Lambda + Q$ such that $|M - \Lambda| \leq C\sqrt{Q}$ with some $C > 0$.*

Proof We first note that $(0, \xi_0, \xi')$ is a double characteristic of p if and only if $q(0, \xi') = 0$ and $\xi' \neq 0$. If $q(0, \xi') \neq 0$ one can write $p = -\xi_0^2 + \phi^2$ nearby with $\phi = \sqrt{q}$. Together with this remark thanks to Proposition 2.1 we can find a finite number of conic open sets $\{V_i\}$ in $\mathbb{R}^{n+1} \times (\mathbb{R}^n \setminus \{0\})$ whose union covers $\omega \times (\mathbb{R}^n \setminus \{0\})$ where ω is a neighborhood of the origin of $\mathbb{R}^{n+1}$ and smooth $\{\lambda_i\}$ where λ_i is defined on V_i, homogeneous of degree 1 in ξ' such that one can write $p = -\xi_0^2 + q = -\xi_0^2 + \sum_{\alpha=1}^{r(i)} \phi_{i\alpha}^2 = -(\xi_0 + \lambda_i)(\xi_0 - \lambda_i) + q_i$ in V_i where $q_i = q - \lambda_i^2$ and $\lambda_i, \phi_{i\alpha}$ verify $|\lambda_i| \leq \delta\sqrt{q}$ with some $\delta < 1$ there and $\{\xi_0 - \lambda_i, \phi_{i\alpha}\}$ is a linear combination of $\phi_{i\alpha}, \alpha = 1, \ldots, r(i)$ on V_i. Take a partition of unity $\{\chi_i\}$, $\chi_i \in C_0^\infty(V_i)$ subordinate to $\{V_i\}$ such that $0 \leq \chi_i \leq 1$, homogeneous of degree 0 in ξ'. With $\lambda = \sum \chi_i \lambda_i$ we define $p = -(\xi_0 + \lambda)(\xi_0 - \lambda) + Q$ where $Q = q - \lambda^2$. Here we note that

$$|\lambda| \leq \sum \chi_i |\lambda_i| \leq \delta\sqrt{q} \sum \chi_i = \delta\sqrt{q},$$

$$Q = q - \lambda^2 \geq q - \delta^2 q = (1 - \delta^2)q \geq 0.$$

We now show that this gives a local elementary factorization. Note that

$$\{\xi_0 - \lambda, Q\} = \sum \chi_i \{\xi_0 - \lambda_i, Q\} + \sum (\xi_0 - \lambda_i)\{\chi_i, Q\}$$

$$= \sum \chi_i \{\xi_0 - \lambda_i, Q\} - \sum \lambda_i \{\chi_i, Q\}$$

because $\sum \{\chi_i, Q\} = 0$. Recall that $\{\xi_0 - \lambda_i, \phi_{i\alpha}\}$ is a linear combination of $\{\phi_{i\alpha}\}$ there. Since $c_1 \sum \phi_{i\alpha}^2 \geq Q \geq c_2 \sum \phi_{i\alpha}^2$ in V_i with some $c_i > 0$ hence

$Q = \sum Q_{\alpha\beta}\phi_{i\alpha}\phi_{i\beta}$ so that on the support of χ_i we have

$$|\{\xi_0 - \lambda_i, Q\}| \leq C \sum_\alpha \phi_{i\alpha}^2 \leq C'q_i \leq C'q \leq C''Q.$$

On the other hand we have $|\{\chi_i, Q\}| \leq C\sqrt{Q}$ for $Q \geq 0$ and $|\lambda_i| \leq \delta\sqrt{q} \leq C\sqrt{Q}$ which proves

$$|\{\xi_0 - \lambda, Q\}| \leq CQ. \tag{2.14}$$

We now study $|\{\xi_0 - \lambda, \xi_0 + \lambda\}| = 2|\{\xi_0 - \lambda, \lambda\}|$. Note that

$$\{\xi_0 - \lambda, \lambda\} = \sum \chi_i\{\xi_0 - \lambda, \lambda_i\} + \sum \lambda_i\{\xi_0 - \lambda, \chi_i\}$$

and $\chi_i\{\xi_0 - \lambda, \lambda_i\} = \chi_i \sum \chi_k\{\xi_0 - \lambda_k, \lambda_i\} - \chi_i \sum \lambda_k\{\chi_k, \lambda_i\}$. Since $\{\xi_0 - \lambda_k, \lambda_i\}$ is a linear combination of $\phi_{k\alpha}, \alpha = 1, \ldots, r(k)$ on $V_k \cap V_i$ the same arguments as above prove

$$|\{\xi_0 - \lambda_k, \lambda_i\}| \leq C\sqrt{q_k} \leq C\sqrt{q} \leq C'\sqrt{Q}$$

on the support of χ_k. It is clear that

$$|\lambda_k\{\chi_k, \lambda_i\}| \leq C\sqrt{q_k} \leq C\sqrt{q} \leq C'\sqrt{Q} \quad \text{on} \quad V_k,$$

$$|\lambda_i\{\xi_0 - \lambda, \chi_i\}| \leq C\sqrt{q_i} \leq C\sqrt{q} \leq C'\sqrt{Q} \quad \text{on} \quad V_i.$$

Hence we have $|\{\xi_0 - \lambda, \lambda\}| \leq C\sqrt{Q}$ which shows $|\{\xi_0 - \lambda, \xi_0 + \lambda\}| \leq C\sqrt{Q}$. This together with (2.14) proves the assertion. $\qquad\qquad\Box$

2.2 Case of Spectral Type 2

We next discuss the same problem studied in the preceding section when p is of spectral type 2 near $\bar{\rho} \in \Sigma$, that is $p(x, \xi)$ verifies (1.7) in a conic neighborhood V of $\bar{\rho}$. By Lemma 1.7, in a suitable symplectic coordinates system, the quadratic form $Q = p_\rho$ takes the form

$$Q = -\xi_0^2 + 2\xi_0\xi_1 + x_1^2 + \sum_{j=2}^{k} \mu_j(x_j^2 + \xi_j^2) + \sum_{j=k+1}^{k+\ell} \xi_j^2. \tag{2.15}$$

Recall that the Jordan canonical form of $F_p(\rho)$ contains a subspace of dimension 4 corresponding to the eigenvalue 0 at every $\rho \in V \cap \Sigma$.

Lemma 2.5 *The number k in (2.15) is independent of $\rho \in V \cap \Sigma$.*

Proof With $\{\psi_j\} = \{\xi_0, \xi_1, x_1, x_j, 2 \leq j \leq k, \xi_j, 2 \leq j \leq k + \ell\}$ it follows from Lemma 2.1 that the rank of $(\{\psi_i, \psi_j\})$ is constant on $V \cap \Sigma$. This shows that k is independent of $\rho \in V \cap \Sigma$. $\square$

Examining the canonical model (2.15) it is easy to see that

$$\dim \mathrm{Im}\, F_p^2(\rho) = 2 + 2(k-1), \quad \dim \mathrm{Im}\, F_p^3(\rho) = 1 + 2(k-1)$$

which are independent of ρ as we observed above. Since

$$\dim \left(\mathrm{Ker}\, F_p(\rho) \cap \mathrm{Im}\, F_p^3(\rho)\right) = 1, \quad \dim \left(\mathrm{Ker}\, F_p^2(\rho) \cap \mathrm{Im}\, F_p^2(\rho)\right) = 2$$

which is easily verified examining the canonical model (2.15) one can choose a smooth $z_1(\rho)$ defined in $V \cap \Sigma$ such that

$$\mathrm{Ker}\, F_p(\rho) \cap \mathrm{Im}\, F_p^3(\rho) = \langle z_1(\rho)\rangle, \quad \rho \in V \cap \Sigma.$$

Similarly there exist linearly independent smooth $h_j(\rho), j = 1, 2$ defined in $V \cap \Sigma$ such that $\mathrm{Ker}\, F_p^2(\rho) \cap \mathrm{Im}\, F_p^2(\rho) = \langle h_1(\rho), h_2(\rho)\rangle$. Since

$$F_p(\rho) : \mathrm{Ker}\, F_p^2(\rho) \cap \mathrm{Im}\, F_p^2(\rho) \to \mathrm{Ker}\, F_p(\rho) \cap \mathrm{Im}\, F_p^3(\rho)$$

is surjective we have $|F_p(\rho)h_1(\rho)| + |F_p(\rho)h_2(\rho)| \neq 0$. Note that there exist smooth $\alpha(\rho), \beta(\rho)$ with $\alpha^2(\rho) + \beta^2(\rho) \neq 0$ such that $z_1(\rho) = \alpha(\rho)h_1(\rho) + \beta(\rho)h_2(\rho)$. Denoting $z_2(\rho) = \beta(\rho)h_1(\rho) - \alpha(\rho)h_2(\rho)$ it is clear

$$F_p(\rho)z_2(\rho) \neq 0, \quad \rho \in V \cap \Sigma.$$

Note that $\mathrm{Ker}\, F_p^2(\rho) \cap \mathrm{Im}\, F_p^2(\rho) = \langle H_{\xi_0}, H_{x_1}\rangle$ and $z_2(\rho) = aH_{\xi_0} + bH_{x_1}$ with $b \neq 0$ in the canonical model (2.15).

Lemma 2.6 *We have for $\rho \in \Sigma$ near $\bar{\rho}$*

$$w \in \langle z_1(\rho)\rangle^\sigma \implies \sigma(w, F_p(\rho)w) \geq 0.$$

Proof Choose a system of symplectic coordinates so that p_ρ takes the form (2.15). It is easy to see that $\langle z_1(\rho)\rangle = \langle H_{\xi_0}\rangle$ and hence if $w \in \langle z_1(\rho)\rangle^\sigma$ then

$$\sigma(w, F_p(\rho)w) = Q(w) = x_1^2 + \sum_{j=2}^{k} \mu_j(x_j^2 + \xi_j^2) + \sum_{j=k+1}^{k+\ell} \xi_j^2 \geq 0$$

which is the assertion. $\square$

Thus we have proved

Proposition 2.3 *Assume that p satisfies* (1.5) *and is of spectral type 2 near $\bar{p}$ then there exist smooth non-zero $z_1(\rho)$, $z_2(\rho)$ defined on Σ near $\bar{p}$ on which one has*

$$z_1(\rho) \in \operatorname{Ker} F_p(\rho) \cap \operatorname{Im} F_p^3(\rho), \tag{2.16}$$

$$z_2(\rho) \in \operatorname{Ker} F_p^2(\rho) \cap \operatorname{Im} F_p^2(\rho), \quad F_p(\rho)z_2(\rho) \neq 0, \tag{2.17}$$

$$w \in \langle z_1(\rho) \rangle^\sigma \implies \sigma(w, F_p(\rho)w) \geq 0. \tag{2.18}$$

Since $F_p(\rho)z_2(\rho)$ is proportional to $z_1(\rho)$ on Σ near $\bar{p}$ we may assume, without restrictions

$$F_p(\rho)z_2(\rho) = -z_1(\rho). \tag{2.19}$$

We show that $z_j(\rho)$ are given as Hamilton vector fields of smooth functions near $\bar{p}$ vanishing on Σ.

Lemma 2.7 *There exists a smooth $S(x, \xi)$ defined near $\bar{p}$ vanishing on Σ such that $H_S(\rho) = z_2(\rho)$ on Σ near $\bar{p}$.*

Proof Note that from (2.5) it follows $F_p(\rho)v = \sum_{j=0}^{r} \epsilon_j \sigma(v, H_{\phi_j}(\rho)) H_{\phi_j}(\rho)$ where $\epsilon_0 = -1, \epsilon_j = 1, j \geq 1$ and hence

$$F_p^2(\rho)v = \sum_{k=0}^{r} \epsilon_k \sigma(v, H_{\phi_k}(\rho)) \sum_{j=0}^{r} \epsilon_j \sigma(H_{\phi_k}(\rho), H_{\phi_j}(\rho)) H_{\phi_j}(\rho).$$

This shows that $\operatorname{Im} F_p^2(\rho) = \langle \sum_{j=0}^{r} \epsilon_j \sigma(H_{\phi_k}(\rho), H_{\phi_j}(\rho)) H_{\phi_j}(\rho); k = 0, \ldots, r \rangle$ and with $A(\rho) = (a_{kj}(\rho)) = (\{\phi_k, \phi_j\}(\rho))$ we have $\operatorname{Im} F_p^2(\rho) = \langle f_1(\rho), \ldots, f_r(\rho) \rangle$ where $f(\rho) = A(\rho)H_\phi(\rho)$, $H_\phi = {}^t(-H_{\phi_0}, \ldots, H_{\phi_r})$. Since the rank of $A(\rho)$ is constant there exists $\beta_{ik}(\rho)$ such that with $g_i(\rho) = \sum_{k=0}^{r} \beta_{ik}(\rho) f_k(\rho)$ for $i = 1, \ldots, s$ we have $\operatorname{Im} F_p^2(\rho) = \langle g_1(\rho), \ldots, g_s(\rho) \rangle$. Since $z_2(\rho) \in \operatorname{Im} F_p^2(\rho)$ one can write $z_2(\rho) = \sum_{k=1}^{s} \alpha_k(\rho) g_k(\rho)$ with smooth $\alpha_k(\rho)$. Then

$$z_2(\rho) = \sum_{k=1}^{s} \alpha_k(\rho) \sum_{j=0}^{r} \beta_{kj}(\rho) f_j(\rho) = \sum_{k=1}^{s} \sum_{j=0}^{r} \sum_{\ell=0}^{r} \alpha_k(\rho) \beta_{kj}(\rho) a_{j\ell}(\rho) H_{\phi_\ell}(\rho).$$

Define S by

$$S = \sum_{k=1}^{s} \sum_{j=0}^{r} \sum_{\ell=0}^{r} \tilde{\alpha}_k \tilde{\beta}_{kj} \tilde{a}_{j\ell} \phi_\ell$$

where $\tilde{\alpha}_k$, $\tilde{\beta}_{kj}$ and $\tilde{a}_{j\ell}$ are smooth extensions of α_k, β_{kj} and $a_{j\ell}$ outside Σ. This is a desired one. $\qquad\square$

Lemma 2.8 *There exists a smooth* $\Lambda(x, \xi)$ *defined near* $\bar{\rho}$ *vanishing on* Σ *such that* $H_\Lambda(\rho) = z_1(\rho)$ *on* Σ *near* $\bar{\rho}$.

Proof It is enough to repeat the same arguments proving Lemma 2.7.   □

Proposition 2.4 *One can write* p *near* $\bar{\rho}$ *as*

$$p = -(\xi_0 + \phi_1(x, \xi'))(\xi_0 - \phi_1(x, \xi')) + \sum_{j=2}^{r} \phi_j(x, \xi')^2$$

where Σ *is given by* $\{\xi_0 = 0, \phi_1 = \cdots = \phi_r = 0\}$ *near* $\bar{\rho}$ *and we have*

$$\{\xi_0 - \phi_1, \phi_j\} = 0, \quad j = 1, \ldots, r, \quad \{\phi_1, \phi_2\} \neq 0 \tag{2.20}$$

and $H_{\xi_0 - \phi_1}$ *is proportional to* $z_1(\rho)$ *on* Σ *near* $\bar{\rho}$.

Proof In the proof we work in a conic neighborhood of $\bar{\rho}$. Let $\Lambda(x, \xi)$ be a smooth function vanishing on Σ such that $H_\Lambda(\rho)$ is proportional to $z_1(\rho)$ of which existence is assured by Lemma 2.8. Since $\sigma(H_{x_0}, F_p H_{x_0}) < 0$ and hence $\sigma(z_1, H_{x_0}) \neq 0$ by (2.18), without restrictions, we may assume that $\Lambda = \xi_0 - \lambda$ and $\lambda = \sum_{j=1}^{r} \gamma_j(x, \xi')\phi_j$ where ϕ_j are those in (2.4). Writing $p = -(\xi_0 - \lambda)(\xi_0 + \lambda) + \sum_{j=1}^{r} \phi_j^2 - (\sum_{j=1}^{r} \gamma_j \phi_j)^2$ one obtains

$$\sigma(v, F_p v) = -\sigma(v, H_\Lambda)\sigma(v, H_{\xi_0 + \lambda}) + \sum_{j=1}^{r} \sigma(v, H_{\phi_j})^2 - (\sum_{j=1}^{r} \gamma_j(\rho)\sigma(v, H_{\phi_j}))^2.$$

As observed in the proof of Proposition 2.1, the mapping

$$\langle H_\Lambda(\rho)\rangle^\sigma / T_\rho \Sigma \ni v \mapsto (\sigma(v, H_{\phi_j}))_{j=1,\ldots,r} \in \mathbb{R}^r$$

is surjective. Thanks to (2.18) we have $\sum_{j=1}^{r} \sigma(v, H_{\phi_j})^2 - (\sum_{j=1}^{r} \gamma_j(\rho)\sigma(v, H_{\phi_j}))^2 \geq 0$ if $v \in \langle H_\Lambda(\rho)\rangle^\sigma$ and hence one can conclude that $\sum_{j=1}^{r} \gamma_j(\rho)^2 = |\gamma(\rho)|^2 \leq 1$. We now show that

$$|\gamma(\rho)| = 1, \quad \rho \in \Sigma. \tag{2.21}$$

We first note that $\sigma(z_2, F_p z_2) = \sigma(z_1, z_2) = \sigma(F_p^3 w, z_2) = -\sigma(w, F_p^3 z_2) = 0$ because $z_1 = F_p^3 w$ with some w and $z_2 \in \mathrm{Ker}\, F_p^2$. Since $\sigma(z_2, z_1) = \sigma(z_2, H_\Lambda) = 0$ we have

$$0 = \sigma(z_2, F_p z_2) = \sum_{j=1}^{r} \sigma(z_2, H_{\phi_j})^2 - (\sum_{j=1}^{r} \gamma_j(\rho)\sigma(z_2, H_{\phi_j}))^2.$$

If $\sigma(z_2, H_{\phi_j}) = 0$ for $j = 1, \ldots, r$ then $z_2 \in \langle H_\Lambda, H_{\phi_1}, \ldots, H_{\phi_r} \rangle^\sigma = \operatorname{Ker} F_p$ which contradicts to $F_p z_2 = -z_1$. This proves that $\sigma(z_2(\rho), H_{\phi_j}(\rho))_{1 \leq j \leq r}$ is different from zero and hence one gets (2.21) because

$$\sum_{j=1}^r \sigma(z_2, H_{\phi_j})^2 = (\sum_{j=1}^r \gamma_j \sigma(z_2, H_{\phi_j}))^2 \leq |\gamma|^2 \sum_{j=1}^r \sigma(z_2, H_{\phi_j})^2.$$

We still denote by $\gamma(x, \xi')$ an extension of $\gamma(\rho)$ outside Σ such that $|\gamma(x, \xi')| = 1$. Thus we can write

$$p(x, \xi) = -(\xi_0 + \langle \gamma, \phi \rangle)(\xi_0 - \langle \gamma, \phi \rangle) + |\phi|^2 - \langle \gamma, \phi \rangle^2$$

where $\{\xi_0 - \langle \gamma, \phi \rangle, \phi_j\} = 0, j = 1, \ldots, r$ on Σ since $H_{\xi_0 - \langle \gamma, \phi \rangle} \in \operatorname{Ker} F_p$. Denote $\psi_1(x, \xi') = \sum_{j=1}^r \gamma_j(x, \xi') \phi_j(x, \xi')$ and taking a smooth orthonormal basis in $\mathbb{R}^r$; $\gamma(x, \xi'), e_2(x, \xi'), \ldots, e_r(x, \xi')$ where $e_j(x, \xi') = (e_{j1}, \ldots, e_{jr})$ and define

$$\psi_j(x, \xi') = \sum_{h=1}^r e_{jh}(x, \xi') \phi_h(x, \xi')$$

so that $\sum_{j=1}^r \psi_j(x, \xi')^2 = \sum_{j=1}^r \phi_j(x, \xi')^2$. Switching the notation from $\{\psi_j\}$ to $\{\phi_j\}$ we can thus write

$$p(x, \xi) = -(\xi_0 + \phi_1(x, \xi'))(\xi_0 - \phi_1(x, \xi')) + \sum_{j=2}^r \phi_j(x, \xi')^2$$

where $\{\xi_0 - \phi_1, \phi_j\} = 0$ on Σ for $j = 1, \ldots, r$. We finally check that $\{\phi_1, \phi_k\} \neq 0$ for some k. Indeed if otherwise we would have $\{\xi_0, \phi_j\} = 0, j = 1, \ldots, r$ and this would contradict (1.7). In fact if this would happen then we have

$$p_\rho = -\xi_0^2 + \sum_{j=1}^r \ell_j^2, \quad \{\xi_0, \ell_j\} = 0, \quad j = 1, \ldots, r.$$

Since $\sum_{j=1}^r \ell_j^2$ is a nonnegative definite quadratic form, in a suitable symplectic basis, p_ρ takes the form (2) of Lemma 1.7, a contradiction.   □

2.3 Vector Field H_S and Key Factorization

In Proposition 2.4 one can write $\{\xi_0 - \phi_1, \phi_j\} = \sum_{k=1}^r c_{jk} \phi_k$ but if $c_{j1} \neq 0$ then $\{\xi_0 - \phi_1, \phi_j\}$ could not be controlled by $\sum_{j=2}^r \phi_j^2$. This is a crucial point in the case that p is of spectral type 2. In this section we prove that if $H_S^3 p = 0$ on Σ near the reference point then p admits a "nice" microlocal factorization, where H_S is the Hamilton vector field of S specified as follows:

Let $S(x, \xi)$ be a smooth real valued function defined in a conic neighborhood of $\bar{\rho} \in \Sigma$, homogeneous of degree 0 in ξ, such that

$$S(x, \xi) = 0, \quad (x, \xi) \in \Sigma \tag{2.22}$$

near $\bar{\rho}$ and satisfying on Σ

$$H_S(\rho) \in \operatorname{Ker} F_p^2(\rho) \cap \operatorname{Im} F_p^2(\rho), \quad F_p(\rho)H_S(\rho) \neq 0. \tag{2.23}$$

We first remark that such a S exists in view of Lemma 2.7 and it is possible to choose S independent of ξ_0. In fact from Lemma 2.8 one can take $\Lambda(x, \xi)$ so that $\Lambda(\rho) = 0$, $H_\Lambda(\rho) = z_1(\rho)$ for $\rho \in \Sigma$. Since $\sigma(H_{x_0}, F_p(\rho)H_{x_0}) = -1$ it follows that $\sigma(H_{x_0}, H_\Lambda(\rho)) \neq 0$, $\rho \in \Sigma$ due to (2.18). This proves that one can write $\Lambda(x, \xi) = \xi_0 - \lambda(x, \xi')$ without restrictions. Write $S(x, \xi) = \alpha\xi_0 + f(x, \xi')$ and put $\tilde{S}(x, \xi') = S(x, \xi) - \alpha\Lambda(x, \xi)$. Then $\tilde{S}(x, \xi')$ verifies (2.22) and (2.23) since $H_\Lambda(\rho) \in \operatorname{Ker} F_p(\rho) \cap \operatorname{Im} F_p^3(\rho)$ on Σ.

The aim of this section is to prove the following factorization result.

Proposition 2.5 *Assume that p is of spectral type 2 near $\bar{\rho}$. Let S be a smooth function verifying (2.22) and (2.23) and assume that $H_S^3 p = 0$ near $\bar{\rho}$ on Σ. Then near $\bar{\rho}$ we can rewrite p as $p = -(\xi_0 + \lambda)(\xi_0 - \lambda) + Q$ with*

$$\lambda = \phi_1 + L(\phi')\phi_1 + \gamma\phi_1^3|\xi'|^{-2}, \quad \gamma \text{ is a real constant,}$$

$$Q = \sum_{j=2}^{r} \phi_j^2 + a(\phi)\phi_1^4|\xi'|^{-2} + b(\phi')L(\phi')\phi_1^2 \geq c(|\phi'|^2 + \phi_1^4|\xi'|^{-2}), \quad c > 0$$

where $L(\phi') = O(|\phi'||\xi'|^{-1})$ and $\phi(x, \xi') = (\phi_1(x, \xi'), \phi_2(x, \xi'), \ldots, \phi_r(x, \xi'))$, $\phi' = (\phi_2, \ldots, \phi_r)$. Here $\xi_0 - \lambda$ and ϕ_j satisfy near $\bar{\rho}$ the following conditions

$$|\{\xi_0 - \lambda, Q\}| \leq C(|\phi'|^2 + \phi_1^4|\xi'|^{-2}), \tag{2.24}$$

$$\{\xi_0 - \lambda, \phi_j\} = O(|\phi|), \quad j = 1, \ldots, r, \tag{2.25}$$

$$\{\phi_1, \phi_j\} = O(|\phi|), \quad j \geq 3, \tag{2.26}$$

$$\{\phi_1, \phi_2\} > 0. \tag{2.27}$$

We note that the condition $H_S^3 p = 0$ is independent of the choice of S which will be examined in the end of this section. To prove this proposition we first show

Lemma 2.9 *Assume that p admits a factorization $p = -M\Lambda + Q$ near $\bar{\rho}$ with $\Lambda = \xi_0 - \lambda$, $M = \xi_0 + \lambda$, $Q \geq 0$ such that H_Λ is proportional to $z_1(\rho)$ for $\rho \in \Sigma$. Let S be a smooth function verifying (2.22) and (2.23). Then we have $H_S^3 Q = 0$ and $\{S, M\} \neq 0$ on Σ near $\bar{\rho}$.*

Proof In the proof we work near $\bar{\rho}$, that is in a conic neighborhood of $\bar{\rho}$. We write $p = -\xi_0^2 + \sum_{j=1}^r \phi_j(x, \xi')^2$ then it is clear that $\Sigma = \{\xi_0 = 0, \lambda = 0, Q = 0\}$ and hence one can write $\Lambda = \xi_0 - \sum_{j=1}^r \gamma_j \phi_j$ and $Q = |\phi|^2 - \langle \gamma, \phi \rangle^2$. It is also clear that $|\gamma(x, \xi')| \leq 1$ near $\bar{\rho}$ because $Q \geq 0$ by assumption. Repeating the same arguments proving Proposition 2.4 we conclude that $|\gamma(\rho)| = 1$, $\rho \in \Sigma$ and $\gamma(\rho)$ is proportional to $\sigma(H_S(\rho), H_\phi(\rho))$ near $\bar{\rho}$;

$$H_S \phi(\rho) = \sigma(H_S(\rho), H_\phi(\rho)) = \alpha(\rho)\gamma(\rho), \quad \rho \in \Sigma$$

where $\sigma(H_S, H_\phi)$ stands for $(\sigma(H_S, \phi_1), \ldots, \sigma(H_S, H_{\phi_r}))$. Study $H_S^3(|\phi|^2 - \langle \gamma, \phi \rangle^2)$. It is clear that $H_S^3 \langle \phi, \phi \rangle = 6\langle H_S^2 \phi, H_S \phi \rangle$ on Σ and hence $H_S^3 \langle \phi, \phi \rangle = 6\alpha \langle H_S^2 \phi, \gamma \rangle$ on Σ. On the other hand one obtains

$$H_S^3 \langle \gamma, \phi \rangle^2 = 4(\langle H_S \gamma, \phi \rangle + \langle \gamma, H_S \phi \rangle)(2\langle H_S \gamma, H_S \phi \rangle + \langle \gamma, H_S^2 \phi \rangle)$$
$$+ 2\langle \gamma, H_S \phi \rangle(\langle H_S^2 \gamma, \phi \rangle + 2\langle H_S \gamma, H_S \phi \rangle + \langle \gamma, H_S^2 \phi \rangle).$$

On Σ this becomes

$$6\alpha \langle \gamma, H_S^2 \phi \rangle + 12\alpha^2 \langle H_S \gamma, \gamma \rangle. \tag{2.28}$$

Since $1 - |\gamma|^2 \geq 0$ near $\bar{\rho}$ and $1 - |\gamma|^2 = 0$ on Σ it follows that

$$H_S(1 - |\gamma|^2) = -H_S \langle \gamma, \gamma \rangle = -2\langle H_S \gamma, \gamma \rangle = 0 \text{ on } \Sigma.$$

Thus (2.28) is equal to $6\alpha \langle \gamma, H_S^2 \phi \rangle$ and hence the first assertion.

We turn to the second assertion. Since $H_S \in \text{Im} F_p$ then $\sigma(H_\Lambda, H_S) = \{\Lambda, S\} = 0$ on Σ hence $F_p H_S = -(1/2)\sigma(H_S, H_M)H_\Lambda + F_Q H_S$ which gives $\sigma(H_S, F_p H_S) = \sigma(H_S, F_Q H_S) = 0$ because $F_p H_S \in \text{Ker} F_p$ and $H_S \in \text{Im} F_p$. This proves $F_Q H_S = 0$ on Σ because $\sigma(H_S, F_Q H_S) = Q_\rho(H_S)$ and Q_ρ is nonnegative definite. Thus we have

$$F_p H_S = -(1/2)\sigma(H_S, H_M)H_\Lambda.$$

Since $F_p H_S \neq 0$ it follows that $\sigma(H_S, H_M) = \{S, M\} \neq 0$. □

Proof of Proposition 2.5 From Proposition 2.4 one can write

$$p(x, \xi) = -(\xi_0 + \phi_1(x, \xi'))(\xi_0 - \phi_1(x, \xi')) + |\phi'(x, \xi')|^2 \tag{2.29}$$

where

$$\{\xi_0 - \phi_1, \phi_j\}\big|_\Sigma = 0, \quad j = 1, \ldots, r, \quad \{\phi_1, \phi_2\}(\bar\rho) \neq 0. \tag{2.30}$$

Recall that $H_{\xi_0 - \phi_1}$ is proportional to $z_1(\rho)$ on Σ near $\bar\rho$. Consider

$$\tilde\phi_j = \sum_{k=2}^{r} O_{jk}\phi_k, \quad j = 2, \ldots, r$$

where $O = (O_{jk})$ is an orthogonal matrix which is smooth near $\bar\rho$. Choosing O suitably and switching the notation $\{\tilde\phi_j\}$ to $\{\phi_j\}$ again we can assume that $\{\phi_1, \phi_2\}(\bar\rho) \neq 0$ and $\{\phi_1, \phi_j\} = 0$ near $\bar\rho$ on Σ, $j = 3, \ldots, r$. Since we may assume $\{\phi_1, \phi_2\} > 0$ without restrictions then the assertions (2.26) and (2.27) are immediate.

Denote $L(\phi') = \langle \beta', \phi' \rangle$ where $\beta' = (\beta_2, \ldots, \beta_r)$ and β_j are smooth functions of (x, ξ'), homogeneous of degree -1 in ξ' and rewrite (2.29) as

$$p(x, \xi) = -(\xi_0 + \phi_1 + L(\phi')\phi_1 + \gamma\hat\phi_1^3|\xi'|^{-2})$$

$$\times(\xi_0 - \phi_1 - L(\phi')\phi_1 - \gamma\phi_1^3|\xi'|^{-2}) + |\phi'|^2 - L(\phi')^2\phi_1^2$$

$$-\gamma^2\phi_1^6|\xi'|^{-4} - 2\phi_1^2 L(\phi') - 2\gamma\phi_1^4|\xi'|^{-2} - 2\gamma L(\phi')\phi_1^4|\xi'|^{-2}$$

$$= -(\xi_0 + \phi_1 + L(\phi')\phi_1 + \gamma\phi_1^3|\xi'|^{-2}) \tag{2.31}$$

$$\times(\xi_0 - \phi_1 - L(\phi')\phi_1 - \gamma\phi_1^3|\xi'|^{-2})$$

$$+|\phi'|^2 - 2\gamma(1 + L(\phi') + \gamma\phi_1^2|\xi'|^{-2}/2)\phi_1^4|\xi'|^{-2}$$

$$-2L(\phi')(1 + L(\phi')/2)\phi_1^2 = -(\xi_0 + \lambda)(\xi_0 - \lambda) + Q$$

where $\lambda = \phi_1 + L(\phi')\phi_1 + \gamma\phi_1^3|\xi'|^{-2}$ and

$$Q = |\phi'|^2 - 2\gamma(1 + L(\phi') + \gamma\phi_1^2|\xi'|^{-2}/2)\phi_1^4|\xi'|^{-2} - 2L(\phi')(1 + L(\phi')/2)\phi_1^2.$$

The assertion (2.25) follows from (2.30) immediately. Taking γ negative and large enough it is clear that $Q \geq c(|\phi'|^2 + \phi_1^4|\xi'|^{-2})$ with some $c > 0$.

It remains to show (2.24). We prove that we can choose β' so that (2.24) holds, following the arguments in [9, 75]. Note that

$$\{\xi_0 - \lambda, Q\} = \{\xi_0 - \phi_1, |\phi'|^2 - 2L(\phi')(1 + L(\phi')/2)\phi_1^2\}$$

$$-\{L(\phi')\phi_1, |\phi'|^2\} + O(Q) \tag{2.32}$$

where one can write $\{\xi_0 - \phi_1, \phi_j\} = \sum_{k=1}^{r} \alpha_{jk}\phi_k$ for $j = 1, \ldots, r$ with smooth α_{jk}. Thus (2.32) reads as

$$\{\xi_0 - \lambda, Q\} = 2\sum_{\ell=2}^{r} \phi_\ell \sum_{k=1}^{r} \alpha_{\ell k}\phi_k - 2\phi_1^2 \sum_{\ell=2}^{r} \beta_\ell \sum_{k=1}^{r} \alpha_{\ell k}\phi_k(1 + L(\phi')/2)$$

$$-2\phi_1 \sum_{\ell=2}^{r} \phi_\ell \sum_{k=2}^{r} \beta_k\{\phi_k, \phi_\ell\} + O(Q). \tag{2.33}$$

Distinguishing the role of ϕ_1 from that of ϕ', one can write

$$\{\xi_0 - \lambda, Q\} = 2\sum_{\ell=2}^{r} \alpha_{\ell 1}\phi_\ell\phi_1 - 2\phi_1 \sum_{\ell=2}^{r} \phi_\ell \sum_{k=2}^{r} \beta_k\{\phi_k, \phi_\ell\}$$

$$-2\phi_1^3 \sum_{\ell=2}^{r} \beta_\ell\alpha_{\ell 1} + O(Q). \tag{2.34}$$

Put $\alpha_1' = (\alpha_{21}, \ldots, \alpha_{r1})$ then (2.34) becomes

$$\{\xi_0 - \lambda, Q\} = 2(\langle\alpha_1', \phi'\rangle + \langle\{\phi', \phi'\}\beta', \phi'\rangle)\phi_1 - 2\phi_1^3\langle\alpha_1', \beta'\rangle + O(Q). \tag{2.35}$$

We show that we can choose $\beta' = (\beta_2, \ldots, \beta_r)$ such that

$$\{\phi', \phi'\}\beta' + \alpha_1' = 0, \quad \langle\alpha_1', \beta'\rangle = 0 \tag{2.36}$$

on Σ so that the right-hand side of (2.35) is $O(Q)$.

Lemma 2.10 *For any $v \in \mathbb{C}^{r-1}$ satisfying $\{\phi', \phi'\}|_\Sigma v = 0$ we have $\langle\alpha_1'|_\Sigma, v\rangle = 0$ near $\bar{\rho}$.*

Proof We first make a closer look at our assumption $H_S^3 p = 0$ on Σ. Since S vanishes on Σ and one can assume that S is independent of ξ_0 then we can write

$$S(x, \xi') = \sum_{j=1}^{r} c_j(x, \xi')\phi_j(x, \xi'). \tag{2.37}$$

Since $H_{\xi_0 - \phi_1}$ is proportional to $z_1(\rho)$ on Σ then $F_p H_S$ is also proportional to $H_{\xi_0 - \phi_1}$ on Σ. Thanks to Proposition 2.6, multiplying S by a non-vanishing function if necessary, we may assume that

$$F_p H_S = -H_{\xi_0 - \phi_1} \quad \text{on} \quad \Sigma. \tag{2.38}$$

We study the identity (2.38). Plugging (2.37) into (2.38) to get

$$
F_p H_S(\rho) = -\frac{1}{2}\{S, \xi_0 + \phi_1\} H_{\xi_0 - \phi_1} + \sum_{j=2}^{r} \{S, \phi_j\} H_{\phi_j}
$$

$$
= -\frac{1}{2} \sum_{h=1}^{r} c_h \{\phi_h, \xi_0 + \phi_1\} H_{\xi_0 - \phi_1} + \sum_{j=2}^{r} \sum_{h=1}^{r} c_h \{\phi_h, \phi_j\} H_{\phi_j}
$$

$$
= -H_{\xi_0 - \phi_1}
$$

on Σ because $\{S, \xi_0 - \phi_1\} = 0$. Hence we have on Σ

$$
(1/2) \sum_{h=1}^{r} c_h \{\phi_h, \xi_0 + \phi_1\} = 1, \tag{2.39}
$$

$$
c_1 \{\phi_1, \phi_j\} + \sum_{h=2}^{r} c_h \{\phi_h, \phi_j\} = 0, \quad j = 2, \ldots, r. \tag{2.40}
$$

Taking $\{\phi_h, \xi_0 + \phi_1\} = \{\phi_h, \xi_0 - \phi_1\} + 2\{\phi_h, \phi_1\}$ into account, we have from (2.39)

$$
c_2 \{\phi_2, \phi_1\} = 1 \tag{2.41}
$$

because $\{\phi_j, \phi_1\} = 0$ for $j \geq 3$. We multiply (2.40) by c_j and sum up over $j = 2, \ldots, r$ which yields $-c_1 + \sum_{h=2}^{r} \sum_{j=2}^{r} c_j c_h \{\phi_h, \phi_j\} = 0$. The second term on the left-hand side vanishes because $(\{\phi_k, \phi_h\})$ is anti-symmetric and thus we get $c_1 = 0$ and (2.40) gives

$$
\{S, \phi_j\} = 0, \quad j = 2, \ldots, r, \quad S = \sum_{h=2}^{r} c_h \phi_h \tag{2.42}
$$

near $\bar{\rho}$ on Σ where $c_2 = \{\phi_2, \phi_1\}^{-1} \neq 0$. By Lemma 2.9 one obtains

$$
H_S^3 p = -3\{S, \xi_0 + \phi_1\}\{S, \{S, \xi_0 - \phi_1\}\} = c\{S, \{S, \xi_0 - \phi_1\}\}
$$

with some $c \neq 0$ which follows from Lemma 2.9 again. Take (2.41) and (2.42) into account we see that $H_S^3 p = 0$ on Σ implies that

$$
\{S, \xi_0 - \phi_1\} = O(|\phi'| + \phi_1^2). \tag{2.43}
$$

Since $\{S, \phi_1\} = 1$ then from (2.42) it follows that $\alpha_{j1} = \{S, \{\xi_0 - \phi_1, \phi_j\}\}$. Thanks to the Jacobi identity we get for $j \geq 2$

$$
\alpha_{j1} = -\{\xi_0 - \phi_1, \{\phi_j, S\}\} - \{\phi_j, \{S, \xi_0 - \phi_1\}\} = -\{\phi_j, \{S, \xi_0 - \phi_1\}\}
$$

on Σ because of (2.42). Thus from (2.43) we can write $\alpha_{j1} = \sum_{k=2}^{r} w_k \{\phi_j, \phi_k\}$ on Σ with some w_k. Then one has

$$\sum_{j=2}^{r} v_j \alpha_{j1} = \sum_{k=2}^{r} w_k \sum_{j=2}^{r} \{\phi_j, \phi_k\} v_j = 0$$

on Σ which is the desired assertion.   □

Thanks to Lemma 2.10 the first equation $\{\phi', \phi'\}\beta' = -\alpha_1'$ in (2.36) has a smooth solution β' defined on Σ near $\bar{\rho}$ and the second equation of (2.36) always holds since $\{\phi', \phi'\}$ is anti-symmetric. Thus we have proved the assertion (2.24) and the proof of Proposition 2.5 is completed.   □

Corollary 2.1 *Assume that p is of spectral type 2 near $\bar{\rho}$ and $H_S^3 p(\rho) = 0$ on Σ with a smooth function S verifying (2.22) and (2.23) near $\bar{\rho}$. Then p admits a microlocal elementary factorization at $\bar{\rho}$.*

Proof It suffices to note that the factorization given in Proposition 2.5 is also a microlocal elementary factorization at $\bar{\rho}$. In fact in a small conic neighborhood of $\bar{\rho}$ one has $|\lambda| \geq c |\phi_1|$ with some $c > 0$. Then taking (2.25) into account one has

$$|\{\xi_0 - \lambda, \xi_0 + \lambda\} = 2|\{\xi_0 - \lambda, \lambda\}| \leq C|\phi| \leq C'(|\lambda| + \sqrt{Q})$$

which proves the desired assertion since $|\{\xi_0 - \lambda, Q\}| \leq CQ$ is clear from Proposition 2.5.   □

Before closing the section we examine that $H_S^3 p = 0$ is independent of the choice of S.

Proposition 2.6 ([9]) *Let S_1, S_2 be two smooth functions verifying (2.22) and (2.23). Then there exists a non-zero smooth function C such that $H_{S_1}^3 p|_{\Sigma} = C H_{S_2}^3 p|_{\Sigma}$ near $\bar{\rho}$.*

Proof Let S_1, S_2 be two functions verifying the assumptions. From Proposition 2.4 we can write $p = -M\Lambda + Q$ with $Q \geq 0$ where H_Λ is proportional to $z_1(\rho)$ and $\{\Lambda, Q\}$ vanishes of second order on Σ. By (2.23) one can write $F_p H_{S_j} = c_j H_\Lambda$ with $c_j \neq 0, j = 1, 2$. Now

$$H_{S_j}^3 p = \{S_j, \{S_j, \{S_j, -\Lambda M + Q\}\}\} = -3\{S_j, M\}\{S_j, \{S_j, \Lambda\}\}$$

on Σ because $\{S_j, \Lambda\} = 0$ and $H_{S_j}^3 Q = 0$ on Σ by Lemma 2.9. Since one can write

$$H_{S_j} = \theta_j z_2(\rho) + H_{f_j}(\rho), \quad \rho \in \Sigma, \ j = 1, 2$$

with $H_{f_j} \in \text{Ker} F_p \cap \text{Im} F_p^3$ where f_j vanishes on Σ then we obtain

$$H_{S_1}(\rho) = (\theta_1/\theta_2) H_{S_2}(\rho) + H_f(\rho)$$

where $H_f(\rho) \in \operatorname{Ker} F_p$ and f vanishes on Σ. Write $\alpha_j = -3\{S_j, M\}, j = 1, 2$ which is different from zero by Lemma 2.9. Then we have

$$
\begin{aligned}
H^3_{S_1} p &= \alpha_1 \{S_1, \{S_1, \Lambda\}\} = \alpha_1 \{(\theta_1/\theta_2)S_2 + f, \{(\theta_1/\theta_2)S_2 + f, \Lambda\}\} \\
&= \alpha_1 \big[(\theta_1/\theta_2)^2 \{S_2, \{S_2, \Lambda\}\} + (\theta_1/\theta_2)\{S_2, \{f, \Lambda\}\} \\
&\quad + (\theta_1/\theta_2)\{f, \{S_2, \Lambda\}\} + \{f, \{f, \Lambda\}\}\big].
\end{aligned}
$$

Since $\{S_j, \Lambda\} = 0$, $\{f, \Lambda\} = 0$ on Σ then $\{f, \{S_2, \Lambda\}\} = 0$ and $\{f, \{f, \Lambda\}\} = 0$ on Σ. This shows that

$$
H^3_{S_1} p = \alpha_1 \big[(\theta_1/\theta_2)^2 \{S_2, \{S_2, \Lambda\}\} + (\theta_1/\theta_2)\{S_2, \{f, \Lambda\}\}\big].
$$

Since $H_f \in \operatorname{Im} F_p \cap \operatorname{Ker} F_p$ and hence $\{S_2, f\} = 0$ on Σ. Thus we see $\{S_2, \{f, \Lambda\}\} = 0$ on Σ thanks to the Jacobi identity

$$
\{S_2, \{f, \Lambda\}\} = -\{f, \{\Lambda, S_2\}\} - \{\Lambda, \{S_2, f\}\}.
$$

Therefore one has $H^3_{S_1} p\big|_{\Sigma} = (\alpha_1/\alpha_2)(\theta_1/\theta_2)^2 H^3_{S_2} p\big|_{\Sigma}$ which is the assertion. □

Chapter 3
Geometry of Bicharacteristics

Abstract If p is of spectral type 1 on Σ there is no bicharacteristic with a limit point in Σ. When p is of spectral type 2 the spectral property of F_p itself is not enough to determine completely the behavior of bicharacteristics and we need to look at the third order term of the Taylor expansion of p around the reference characteristic to obtain a complete picture of the behavior of bicharacteristics. We prove that there is no bicharacteristic with a limit point in Σ near ρ if and only if the cube of the vector field H_S, introduced in Chap. 2, annihilates p on Σ near ρ. This suggests that the behavior of bicharacteristics near Σ closely relates to the possibility of microlocal factorization discussed in Chap. 2. In the last section, we prove that at every effectively hyperbolic characteristic there exists exactly two bicharacteristics that are transversal to Σ having which as a limit point. This makes the difference of the geometry of bicharacteristics clearer.

3.1 Behaviors of Bicharacteristics

Although our main concern is the geometry of bicharacteristics near non-effectively hyperbolic characteristics we start with stating the behaviors of bicharacteristics near effectively hyperbolic characteristics, of which proof [52] is given in Sect. 3.5 to make the monograph more self-contained. Assume that p is effectively hyperbolic at ρ and we consider bicharacteristics of p tending to ρ as $s \uparrow +\infty$ or $s \downarrow -\infty$, that is bicharacteristics with the limit point ρ.

Proposition 3.1 ([48, 52]) *There are exactly four such bicharacteristics. Two of them are incoming toward ρ with respect to the parameter s, and the other two are outgoing. Each one of the incoming (resp. outgoing) bicharacteristics is naturally continued to the other one, and the resulting two curves are regular, C^∞ or analytic corresponding to the assumption on the principal symbol. The tangent lines of the resulting two smooth curves at ρ are parallel to the eigenvectors corresponding to the non-zero real eigenvalues $\pm\lambda$ of $F_p(\rho)$ respectively.*

Proposition 3.2 *Assume that the double characteristic set Σ is a smooth manifold on which p is effectively hyperbolic. Then such continued bicharacteristics are transversal to Σ.*

© Springer International Publishing AG 2017

T. Nishitani, *Cauchy Problem for Differential Operators with Double Characteristics*, Lecture Notes in Mathematics 2202, DOI 10.1007/978-3-319-67612-8_3

Proof Let $X \neq 0$ be a tangent to such a bicharacteristic at ρ and hence $F_p(\rho)X = \lambda X$ with a real $\lambda \neq 0$ due to Proposition 3.1. If $X \in T_\rho \Sigma$ and hence $X \in \text{Ker} F_p(\rho)$ by (2.7) this would give $X = 0$ which is a contradiction. □

We turn to consider the case that p is non-effectively hyperbolic on Σ. If p is of spectral type 1 on Σ there is no bicharacteristic landing to Σ as $s \uparrow +\infty$ or $s \downarrow -\infty$. Indeed we have

Proposition 3.3 *Assume* (1.5) *and* p *is of spectral type 1 on* Σ. *Then there is no bicharacteristic emanating from a simple characteristic which have a limit point in* Σ.

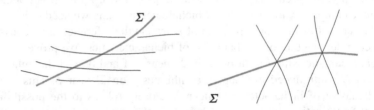

We will see in Chap. 5 that the geometry of bicharacteristics of p near Σ is reduced to that of p of second order. Thus from now on we assume that $p(x, \xi)$ has the form (2.1). In view of Proposition 2.1 to prove Proposition 3.3 it suffices to show the following

Lemma 3.1 ([40]) *Assume that* p *admits a microlocal elementary factorization at* $\bar{\rho} \in \Sigma$. *Then there is a conic neighborhood* V *of* $\bar{\rho}$ *such that there is no bicharacteristic with a limit point in* $\Sigma \cap V$.

Proof Let $p = -M\Lambda + Q$ be a microlocal elementary factorization at $\bar{\rho}$. Note that $\Sigma = \{(x, \xi) \mid \Lambda(x, \xi) = M(x, \xi) = Q(x, \xi') = 0\}$ because $\partial_{\xi_0} p = -(\Lambda(x, \xi) + M(x, \xi)) = 0$ and $p(x, \xi) = 0$ implies $\Lambda^2(x, \xi) + Q(x, \xi') = 0$. Let $\gamma(s)$ be a bicharacteristic of p which lies outside Σ for $-\infty < s < +\infty$. Since $M\Lambda = Q \geq 0$ on $\gamma(s)$ we may assume that $M \geq 0$, $\Lambda \geq 0$ and $M + \Lambda > 0$ on $\gamma(s)$. Thus we have $dx_0(s)/ds = -\Lambda(\gamma(s)) - M(\gamma(s)) < 0$ so that we can take x_0 as a new parameter;

$$\frac{d}{dx_0} \Lambda(\gamma(x_0)) = \frac{d}{ds} \Lambda(\gamma(s)) \frac{ds}{dx_0} = \{p, \Lambda\}(\gamma(s)) \frac{ds}{dx_0}.$$

On $\gamma(s)$ we have

$$|\{p, \Lambda\}| \leq C(Q + \Lambda\sqrt{Q} + \Lambda|\Lambda - M|) = C\Lambda(M + \sqrt{\Lambda M} + |\Lambda - M|).$$

Since $(M + \sqrt{\Lambda M} + |\Lambda - M|)/(\Lambda + M) \leq 3$ one has

$$|d\Lambda(\gamma(x_0))/dx_0| \leq C\Lambda(\gamma(x_0)). \tag{3.1}$$

Suppose that $\gamma(x_0) \notin \Sigma$ for $x_0 \neq 0$ and $\lim_{x_0 \to 0} \gamma(x_0) \in \Sigma$ so that $\Lambda(\gamma(0)) = 0$. From (3.1) it follows that $\Lambda(\gamma(x_0)) = 0$ and hence $Q(\gamma(x_0)) = 0$. Since Q is nonnegative it follows that $\{Q, M\}(\gamma(x_0)) = 0$. This shows $|\{p, M\}| \leq CM|\Lambda - M|$ on $\gamma(x_0)$ and then we have $|dM(\gamma(x_0))/dx_0| \leq CM(\gamma(x_0))$ hence $M(\gamma(x_0)) = 0$ so that $\gamma(x_0) \in \Sigma$ for all x_0 near 0 which is a contradiction. $\qquad \square$

In the case that p is of spectral type 2 the situation is completely different. Indeed there could exist bicharacteristics of p with a limit point in Σ. We give examples. Let $1 \leq k \leq n-1$ and $q_i, r_i, i = 1, 2, \ldots, k$ be positive constants. Consider

$$p(x, \xi) = -\xi_0^2 + \sum_{i=1}^{k} q_i(x_{i-1} - x_i)^2 \xi_n^2 + \sum_{i=1}^{k} r_i \xi_i^2 + \xi_n^{-1} \sum_{i=1}^{k} \epsilon_i \xi_i \xi_k^2 \qquad (3.2)$$

near $x = 0$ so that p is a hyperbolic polynomial if $(|\xi_1| + \cdots + |\xi_k|)/|\xi_n|$ is small and $|\epsilon_i|$ are bounded. The double characteristic manifold of p is given by $\Sigma = \{\xi_i = 0, 0 \leq i \leq k, x_i = x_{i+1}, 0 \leq i \leq k-1\}$. It is easily checked that p is of spectral type 2 on Σ if and only if

$$\sum_{i=1}^{k} r_i^{-1} = 1. \qquad (3.3)$$

Note that the term $f(\xi') = \xi_n^{-1} \sum_{i=1}^{k} \epsilon_i \xi_i \xi_k^2$ does not affect the Hamilton map H_p since $f(\xi')$ vanishes of order 3 on Σ.

Lemma 3.2 ([72]) *Assume (3.3) then we can choose $\{\epsilon_i\}$ so that there exists a bicharacteristic of p with a limit point in Σ.*

Proof Thanks to (3.3) one can write p as follows;

$$p = -\xi_0^2 + \left(\sum_{i=1}^{k} \xi_i\right)^2 + \sum_{i=1}^{k} q_i(x_{i-1} - x_i)^2 \xi_n^2 + \sum_{i=1}^{k-1} \gamma_i \phi_i^2 + f(\xi')$$

where $\phi_i(\xi') = \xi_i - \delta_i(\xi_{i+1} + \cdots + \xi_k)$ and $\gamma_i > 0, \delta_i > 0$. On bicharacteristics we have $d\phi_i/ds = \{p, \phi_i\} = 2\sum_{j=1}^{k} q_j \xi_n^2 \{x_{j-1} - x_j, \phi_i\}(x_{j-1} - x_j)$. Here we note that the condition $\sum_{j=1}^{k} q_j\{x_{j-1} - x_j, \phi_i\}(x_{j-1} - x_j) = 0, 1 \leq i \leq k-1$ can be written $(\partial \phi / \partial \tilde{\xi})(\partial q / \partial \tilde{x}) = 0$ with $\phi = (\phi_1, \ldots, \phi_{k-1}), q = \sum_{i=1}^{k} q_i(x_{i-1} - x_i)^2$, $\tilde{\xi} = (\xi_1, \ldots, \xi_k)$ and $\tilde{x} = (x_1, \ldots, x_k)$ which is equivalent to

$$x_{j-1} - x_j = a_j(x_{k-1} - x_k), \quad 1 \leq j \leq k-1 \qquad (3.4)$$

where a_j are positive constants determined by $\{q_j\}$ and $\{\delta_j\}$. We look for bicharacteristics on which we have $\phi_i = 0, 1 \leq i \leq k-1$ and $\xi_n = 1$. To do so we consider an auxiliary equation

$$\begin{cases} \dot{z} = \alpha \zeta^2, \\ \dot{\zeta} = 2q_k z \end{cases} \qquad (3.5)$$

where α is a non-zero real constant which will be determined later. Note that (3.5) has a solution of the form $(z, \zeta) = (As^{-3}, Bs^{-2})$ with non-zero constants A, B. Taking the fact $d(\sum_{i=0}^{k} \xi_i)/ds = -\sum_{i=0}^{k} \partial p/\partial x_i = 0$ into account we set $\xi_k(s) = \zeta(s)$ and determine $\xi_i(s)$, $0 \le i \le k-1$ successively by

$$\phi_i(\tilde{\xi}) = 0, \ 1 \le i \le k-1, \quad \sum_{i=0}^{k} \xi_i = 0. \tag{3.6}$$

Then it is clear that $\xi_i(s) = c_i s^{-2}$ with some c_i. Assuming that (3.6) is verified the Hamilton equation for $x_i(s)$ are reduced to

$$\dot{x}_0 = -2\xi_0, \quad \dot{x}_\mu = 2\sum_{i=1}^{k} \xi_i + \partial f(\xi')/\partial \xi_\mu, \ 1 \le \mu \le k. \tag{3.7}$$

We take solutions $x_i(s)$ to (3.7) satisfying $x_i(s) = O(|s|^{-1})$ as $|s| \to \infty$. Then from (3.6) and (3.7), assuming $\xi_n(s) = 1$, it follows that

$$\dot{x}_0 - \dot{x}_1 = -\epsilon_1 \xi_k^2, \quad \dot{x}_{\mu-1} - \dot{x}_\mu = (\epsilon_{\mu-1} - \epsilon_\mu)\xi_k^2, \ 2 \le \mu \le k,$$
$$\dot{x}_{k-1} - \dot{x}_k = (\epsilon_{k-1} - 3\epsilon_k)\xi_k^2. \tag{3.8}$$

We choose $\{\epsilon_i\}$ so that

$$\epsilon_{i-1} - \epsilon_i = a_i(\epsilon_{k-1} - 3\epsilon_k), \quad \epsilon_0 = 0, \quad \epsilon_k \ne 0, \ 1 \le i \le k-1.$$

This gives $\epsilon_{k-1} - 3\epsilon_k = -3\epsilon_k(1 + a_1 + \cdots + a_{k-1})^{-1}$. We take $\alpha = -3\epsilon_k(1 + a_1 + \cdots + a_{k-1})^{-1}$ so that $\dot{x}_{k-1} - \dot{x}_k = \alpha \xi_k^2$. Since $\xi_k(s) = \zeta(s)$ and $(x_{k-1} - x_k) = 0$ as $s \to -\infty$ then we conclude that $x_{k-1}(s) - x_k(s) = z(s)$. Similarly we have $x_{i-1}(s) - x_i(s) = a_i(x_{k-1}(s) - x_k(s))$ for $1 \le i \le k-1$. In view of (3.6) we see $d\phi_i(\tilde{\xi}(s))/ds = 0$ for $1 \le i \le k-1$ which proves that $(\partial \phi/\partial \tilde{\xi})d\tilde{\xi}(s)/ds = 0$. Since the kernel of $\partial \phi/\partial \tilde{\xi}$ is one dimensional then we have

$$d\tilde{\xi}/ds = \kappa(s)\partial q/\partial \tilde{x}$$

with some $\kappa(s)$. The second equation in (3.5) shows that $\kappa(s) = -1$ and hence $(x(s), \xi(s))$ is a solution to the Hamilton equation, where of course $x_j = \xi_j = 0$ for $k+1 \le j \le n-1$ and $x_n = 0$, $\xi_n = 1$. It is clear that $\xi_i(s) \to 0$, $x_i(s) \to 0$ for $0 \le i \le k$ as $s \to -\infty$ and that $A \ne 0, B \ne 0$ implies $(x(s), \xi(s)) \notin \Sigma$ for any s. $\square$

If $k = 1$ then $p(x, \xi) = -\xi_0^2 + q(x_0 - x_1)^2\xi_n^2 + \xi_1^2 + \epsilon\xi_n^{-1}\xi_1^3$ and (3.3) is verified. Choosing a new system of local coordinates $y_1 = x_1 - x_0$ and $y_j = x_j$ for $j \ne 1$ we have $p(x, \xi) = -\xi_0^2 + 2\xi_1\xi_0 + qx_1^2\xi_n^2 + \epsilon\xi_n^{-1}\xi_1^3$. We choose a new system of homogeneous symplectic coordinates $y_1 = \xi_1\xi_n^{-1}$, $\eta_1 = -x_1\xi_n$ and $y_j = x_j$, $\eta_j = \xi_j$

for $j \neq 1$, which leaves the Hamilton field H_p invariant, then $p(x, \xi)$ becomes

$$p(x, \xi) = -\xi_0^2 + 2x_1 \xi_0 \xi_n + q\xi_1^2 + \epsilon x_1^3 \xi_n^2. \tag{3.9}$$

Theorem 3.1 ([78]) *Assume (1.5) and that p is of spectral type 2 near $\bar{\rho}$. Then the following assertions are equivalent.*

 (i) $H_S^3 p = 0$ *on Σ near $\bar{\rho}$ for some S satisfying (2.22) and (2.23) near $\bar{\rho}$,*
 (ii) *there is a conic neighborhood of $\bar{\rho}$ on which no bicharacteristic of p emanating from a simple characteristic has a limit point.*

Corollary 3.1 *Assume (1.5) and that p is of spectral type 2 near $\bar{\rho}$. Then the following assertions are equivalent.*

 (i) *p admits the factorization given in Proposition 2.5 near $\bar{\rho}$,*
 (ii) *there is a conic neighborhood of $\bar{\rho}$ on which no bicharacteristic of p emanating from a simple characteristic has a limit point.*

Proof Since the factorization given in Proposition 2.5 is also a microlocal elementary factorization at $\bar{\rho}$, as observed in the proof of Corollary 2.1, then Lemma 3.1 shows that (i) implies (ii). Conversely in view of Theorem 3.1 and Proposition 2.5 we see that (ii) implies (i).   □

Corollary 3.2 *Assume (1.5) and that p is of spectral type 2 on Σ. Then the following assertions are equivalent*

 (i) $H_S^3 p = 0$ *on Σ near ρ with some S verifying (2.22) and (2.23) at every $\rho \in \Sigma$,*
 (ii) *there is no bicharacteristic of p with a limit point in Σ.*

The proof of Theorem 3.1 goes as follows. If $H_S^3 p = 0$ on Σ near $\bar{\rho}$ for some such S then p admits a microlocal elementary factorization at $\bar{\rho}$ by Corollary 2.1. Then thanks to Lemma 3.1 we conclude that (i) implies (ii). Thus to prove Theorem 3.1 it suffices to show that there is a bicharacteristic of p with a limit point in Σ near $\bar{\rho}$ if the condition (i) fails at $\bar{\rho}$. We prove this assertion following [87] in the next three sections.

3.2   Expression of p as Almost Symplectically Independent Sums

From the assumption (1.5), for any $\bar{\rho} \in \Sigma$, one can find $\phi_j(x, \xi'), j = 1, \ldots, r$ such that we have

$$\begin{cases} p = -\xi_0^2 + \sum_{j=1}^r \phi_j^2(x, \xi'), & \Sigma = \{\phi_j = 0, j = 0, \ldots, r\}, \\ \text{rank}\left(\{\phi_i, \phi_j\}\right)_{0 \leq i, j \leq r} = \text{constant on } \Sigma \end{cases}$$

in a conic neighborhood of $\bar{\rho}$ with linearly independent differentials $d\phi_j(\bar{\rho})$, $j = 0, \ldots, r$ where we have set $\xi_0 = \phi_0$.

In this section we write $f = O(|\phi|)$ if f is a linear combination of $\phi_1, \ldots, \phi_r$ in some open set. It is also understood that every open set has non empty intersection with Σ. To simplify notations we often use the same $\{\phi_j\}_{1 \leq j \leq r}$ denoting $\{\tilde{\phi}_j\}_{1 \leq j \leq r}$ which is related to $\{\phi_j\}_{1 \leq j \leq r}$ by a smooth orthogonal transformation if it is clear from the context.

Definition 3.1 Let I_k, $k = 1, \ldots, t$ be subsets of a finite index set $\hat{I}$ which are mutually disjoint. We say that $\{\phi_j\}_{j \in I_k}$, ($k = 1, \ldots, t$) are symplectically independent in U if $\{\phi_i, \phi_j\} = O(|\phi|)$ in U for any $i \in I_p, j \in I_q, p \neq q$.

Let $A = (a_{ij})$ be an $m \times m$ anti-symmetric matrix of the form

$$\begin{cases} a_{ij} \neq 0 & \text{if} \quad |i-j| = 1, \\ a_{ij} = 0 & \text{if} \quad |i-j| \neq 1. \end{cases} \tag{3.10}$$

Then the next lemma is easily checked.

Lemma 3.3 *Let A be an $m \times m$ anti-symmetric matrix satisfying* (3.10). *Then* $\det A \neq 0$ *if m is even while* $\operatorname{rank} A = m - 1$ *if m is odd.*

Let us consider

$$Q = \sum_{j=1}^{r} \phi_j^2$$

where it is assumed that $\phi_j(x, \xi)$ are defined in U and the differentials $\{d\phi_j\}$ are linearly independent there. Then

Lemma 3.4 *Assume that there exist $i, j \in \hat{I} = \{1, \ldots, r\}$ and $\rho \in U \cap \Sigma$ such that* $\{\phi_i, \phi_j\}(\rho) \neq 0$. *Then there are a partition of $\hat{I}$, $\hat{I} = I \cup J$, an open set $V \subset U$ and* $\{\phi_i\}_{i \in I}$, $\{\phi_j\}_{j \in J}$ *which are symplectically independent such that one can write*

$$Q = \sum_{i \in I} \phi_i^2 + \sum_{j \in J} \phi_j^2, \tag{3.11}$$

$$\det\left(\{\phi_i, \phi_j\}\right)_{i,j \in I} \neq 0 \quad \text{in } V. \tag{3.12}$$

Proof We first prove that one can find an open set $V \subset U$ and $\{\phi_i\}_{i \in I}$, $\{\phi_j\}_{j \in J}$ which are symplectically independent and satisfies (3.11) and one has in V

$$\begin{cases} \{\phi_i, \phi_j\} \neq 0 & \text{if} \quad |i-j| = 1, \ i, j \in I, \\ \{\phi_i, \phi_j\} = O(|\phi|) & \text{if} \quad |i-j| \neq 1, \ i, j \in I. \end{cases} \tag{3.13}$$

Without restrictions, we may assume $\{\phi_1, \phi_j\}(\rho_1) \neq 0$ with some $\rho_1 \in U \cap \Sigma$ and j. Consider a smooth orthogonal transformation sending $\{\phi_2, \ldots, \phi_r\}$ to $\{\tilde{\phi}_2, \ldots, \tilde{\phi}_r\}$

such that $\tilde{\phi}_i = \sum_{k=2}^r O_{ik}\phi_k$ for $i = 2, \ldots, r$. Noting

$$\{\phi_1, \sum_{k=2}^r O_{ik}\phi_k\} = \sum_{k=2}^r O_{ik}\{\phi_1, \phi_k\} + O(|\phi|)$$

we choose O_{ik} so that $\sum_{k=2}^r O_{2k}\{\phi_1, \phi_k\} \neq 0$ and $\sum_{k=2}^r O_{ik}\{\phi_1, \phi_k\} = 0$ for $i = 3, \ldots, r$ in some open set $U_1 \subset U$. Switching the notation from $\{\tilde{\phi}_j\}_{j=2}^r$ to $\{\phi_j\}_{j=2}^r$ we may assume that $Q = \sum \phi_j^2$ and

$$\{\phi_1, \phi_2\} \neq 0, \quad \{\phi_1, \phi_j\} = O(|\phi|), \quad j = 3, \ldots, r$$

in U_1. Consider $\{\phi_2, \phi_j\}, j \geq 3$. If $\{\phi_2, \phi_j\} = 0$ in $U_1 \cap \Sigma$ for all $j \geq 3$ then it is enough to take $I = \{1, 2\}$ and $J = \{3, \ldots, r\}$. If not then there exist $\rho_2 \in U_1 \cap \Sigma$ and $j_2 \geq 3$ such that $\{\phi_2, \phi_{j_2}\}(\rho_1) \neq 0$. Continuing this procedure we can conclude that there exist an open set $V \subset U$ and $\{\phi_i\}_{i \in I}, \{\phi_j\}_{j \in J}$ which are symplectically independent and verifies (3.11) and (3.13) in V.

We turn to the next step. Take $\rho \in V \cap \Sigma$. If $|I|$ is even then from Lemma 3.3 and (3.13) it follows that $(\{\phi_i, \phi_j\}(\rho))_{i,j \in I}$ is non singular and hence so is near ρ. Thus (3.12) holds. If $|I|$ is odd then from Lemma 3.3 and (3.13) it follows that $\mathrm{rank}(\{\phi_i, \phi_j\}(\rho))_{i,j \in I} = |I| - 1$. Note that $\mathrm{rank}(\{\phi_i, \phi_j\})_{i,j \in I} \leq |I| - 1$ near ρ because $(\{\phi_i, \phi_j\})_{i,j \in I}$ is an anti-symmetric matrix of odd order then we have

$$\mathrm{rank}(\{\phi_i, \phi_j\})_{i,j \in I} = |I| - 1 \tag{3.14}$$

in some neighborhood V' of ρ. Let $I = \{i_1, i_2, \ldots, i_\ell\}$. From (3.14) it follows that $\dim \mathrm{Ker}(\{\phi_i, \phi_j\})_{i,j \in I} = 1$ and hence we can choose smooth $c_i(x, \xi), i \in I$ such that $\sum_{j \in I} c_j^2 = 1$ and $\sum_{j \in I}\{\phi_i, \phi_j\}c_j = 0, i \in I$ holds in V'. Choosing a smooth orthogonal matrix $(O_{ij})_{ij \in I}$ so that $O_{i_1 j} = c_j$ and considering

$$\tilde{\phi}_i = \sum_{j \in I} O_{ij}\phi_j, \quad i \in I$$

we may assume that $\{\tilde{\phi}_j, \tilde{\phi}_{i_1}\} = O(|\phi|)$ in V' for all $j \in I$. Therefore noting that $\mathrm{rank}(\{\phi_i, \phi_j\})_{i,j \in I} = \mathrm{rank}(\{\tilde{\phi}_i, \tilde{\phi}_j\})_{i,j \in I}$ it follows from (3.14) that

$$\det(\{\tilde{\phi}_i, \tilde{\phi}_j\})_{i,j \in I'} \neq 0$$

where $I' = I \setminus \{i_1\}$. Thus $\{\tilde{\phi}_i\}_{i \in I'}$ and $\{\tilde{\phi}_j\}_{j \in J'}, J' = J \cup \{i_1\}$ verify the desired assertion. $\square$

Lemma 3.5 *There exist a partition of* $\hat{I}$, $\hat{I} = I \cup K$, *an open set* $V \subset U$ *and* $\{\phi_i\}_{i \in I}$, $\{\phi_j\}_{j \in K}$ *which are symplectically independent such that we can write*

$$Q = \sum_{i \in I} \phi_i^2 + \sum_{j \in K} \phi_j^2$$

where $\det((\{\phi_i, \phi_j\})_{i,j \in I} \neq 0$ *while* $\{\phi_i, \phi_j\} = O(|\phi|)$ *in* V *for all* $i, j \in K$.

Proof From Lemma 3.4 there are an open set $V_1 \subset U$ and $\{\phi_i\}_{i \in I_1}$, $\{\phi_j\}_{j \in J_1}$, symplectically independent in V_1, which verify (3.11) and (3.12). If $\{\phi_i, \phi_j\} = 0$ in $V_1 \cap \Sigma$ for all $i, j \in J_1$ then it is enough to choose $I = I_1$ and $K = J_1$. Otherwise applying Lemma 3.4 to $Q_1 = \sum_{j \in J_1} \phi_j^2$ we find a partition $J_1 = I_2 \cup J_2$, an open set $V_2 \subset V_1$ and $\{\phi_i\}_{i \in I_2}$, $\{\phi_j\}_{j \in J_2}$ which are symplectically independent in V_2 verifying

$$Q_1 = \sum_{i \in I_2} \phi_i^2 + \sum_{j \in J_2} \phi_j^2, \quad \det((\{\phi_i, \phi_j\})_{i,j \in I_2} \neq 0.$$

Repeating this argument at most $[r/2]$ times we conclude that there are an open set $V \subset U$ and $\{\phi_j\}_{j \in I_k}$ $(k = 1, \ldots, t)$, $\{\phi_j\}_{j \in K}$, which are symplectically independent in V and satisfy in V that $Q = \sum_{i=1}^{t} \sum_{j \in I_i} \phi_j^2 + \sum_{j \in K} \phi_j^2$ and

$$\det((\{\phi_i, \phi_j\})_{i,j \in I_p} \neq 0, \quad p = 1, \ldots, t, \quad \{\phi_i, \phi_j\} = 0, \ \forall i, j \in K.$$

Let us set $I = \cup_{i=1}^{t} I_i$ then it is obvious that $\{\phi_i\}_{i \in I}$, $\{\phi_j\}_{j \in K}$ are symplectically independent in V. Note that $((\{\phi_i, \phi_j\}(\rho))_{i,j \in I}$ is the direct sum of $((\{\phi_i, \phi_j\}(\rho))_{i,j \in I_k}$ $(k = 1, \ldots, t)$ if $\rho \in V \cap \Sigma$ and hence $\det((\{\phi_i, \phi_j\})_{i,j \in I} \neq 0$ in some open set which proves the assertion. $\qquad \Box$

Proposition 3.4 *Assume* (1.5) *and that* p *is of spectral type 2 near* $\rho \in \Sigma$. *Let* U *be any neighborhood of* ρ. *Then there exist an open set* $V \subset U$ *and* $\{\phi_j\}_{j \in I_0}$, $\{\phi_j\}_{j \in I_1}$, $\{\phi_j\}_{j \in K}$ *which are symplectically independent in* V *where* $\{0, 1, \ldots, r\} = I_0 \cup I_1 \cup K$, $I_0 = \{0, 1, \ldots, l\}$ *with even* $l (\geq 2)$, *such that one can write*

$$p = -(\xi_0 + \phi_1)(\xi_0 - \phi_1) + \sum_{k=2}^{l} \phi_j^2 + \sum_{j \in I_1} \phi_j^2 + \sum_{j \in K} \phi_j^2$$

and we have in V

$$\begin{cases} \{\xi_0 - \phi_1, \phi_j\} = O(|\phi|), \quad j = 0, \ldots, r, \\ \{\phi_1, \phi_2\} \neq 0 \quad \text{if } l = 2, \\ \text{rank}((\{\phi_i, \phi_j\})_{2 \leq i,j \leq l} = l - 2 \quad \text{if } l \geq 4, \\ \det((\{\phi_i, \phi_j\})_{i,j \in I_1} \neq 0, \\ \{\phi_i, \phi_j\} = O(|\phi|), \quad \forall i, j \in K. \end{cases}$$

Proof As the first step we prove that one can write

$$p = -\xi_0^2 + \sum_{j=1}^{l} \phi_j^2 + \sum_{j \in I_1} \phi_j^2 + \sum_{j \in K} \phi_j^2 \tag{3.15}$$

where $\{\phi_j\}_{j \in I_0}$, $\{\phi_j\}_{j \in I_1}$, $\{\phi_j\}_{j \in K}$ are symplectically independent, $\{0, 1, \ldots, r\} = I_0 \cup I_1 \cup K$, $I_0 = \{0, \ldots, l\}$ with even l (≥ 2) and

$$\begin{cases} \dim \mathrm{Ker}(\{\phi_i, \phi_j\})_{0 \leq i,j \leq l} = 1, \\ \det(\{\phi_i, \phi_j\})_{i,j \in I_1} \neq 0, \\ \{\phi_i, \phi_j\} = O(|\phi|), \quad \forall i, j \in K. \end{cases} \tag{3.16}$$

Recall that one can write

$$p = -\xi_0^2 + \sum_{j=1}^{r} \phi_j^2$$

near ρ. We write $\phi_0 = \xi_0$ as before. Suppose $\{\phi_0, \phi_j\}(\rho) = 0$ for all j. Then with $q = \sum_{j=1}^{r} \phi_j^2$ we see easily that $\mathrm{Ker}\, F_p^2(\rho) \cap \mathrm{Im}\, F_p^2(\rho) = \mathrm{Ker}\, F_q^2(\rho) \cap \mathrm{Im}\, F_q^2(\rho)$ which is the trivial subspace because q is nonnegative. This contradicts (1.7). Thus we have $\{\phi_0, \phi_j\}(\rho) \neq 0$ with some $j \geq 1$. Now repeating the same arguments employed in the proof of Lemma 3.4 we conclude that there exist an open set $V \subset U$ and $\{\phi_j\}_{j \in \{0, \ldots, l\}}$, $\{\phi_j\}_{j \in \{l+1, \ldots, r\}}$, symplectically independent in V satisfying (3.11) and (3.13) with $I = \{0, \ldots, l\}$, $l \geq 1$.

We now show that l is even by contradiction. Suppose that l is odd and recall $p_\rho(X) = \sigma(X, F_p(\rho)X)$ with $X = (x, \xi) \in \mathbb{R}^{2(n+1)}$. Set $\psi = \sum_{1 \leq 2j+1 \leq l} c_{2j+1} \phi_{2j+1}$ with $c_{2j+1} \in \mathbb{R}$. We note that

$$p_\rho(H_\psi) = -\{\phi_0, \psi\}^2(\rho) + \sum_{j=1}^{r} \{\phi_j, \psi\}^2(\rho)$$

$$= -\{\phi_0, c_1 \phi_1\}^2(\rho) + \sum_{2 \leq 2i < l} \{\phi_{2i}, \psi\}^2(\rho).$$

Since l is odd, thanks to (3.13) we can choose c_{2j+1} so that $\{\phi_{2i}, \psi\}(\rho) = 0$ for $2 \leq 2i < l$ and $c_1 = 1$. This implies that $p_\rho(H_\psi) = -\{\phi_0, \phi_1\}^2(\rho) < 0$ and hence $F_p(\rho)$ has non-zero real eigenvalues by Corollary 1.6 which contradicts the assumption (1.5). Thus we have proved that l is even. Since l is even $\dim \mathrm{Ker}(\{\phi_i, \phi_j\})_{0 \leq i,j \leq l} = 1$ follows from (3.13) easily. If $l = r$ then the proof is complete. Otherwise it suffices to apply Lemma 3.5 to $\sum_{l+1}^{r} \phi_j^2$.

We turn to the second step. Write $\bar{p} = -\xi_0^2 + \sum_{j=1}^{l} \phi_j^2$ and $\bar{q} = \sum_{j=l+1}^{r} \phi_j^2$. We remark that

$$\operatorname{Ker} F_{\bar{p}}^2(\rho) \cap \operatorname{Im} F_{\bar{p}}^2(\rho) \neq \{0\}, \quad \rho \in \Sigma. \tag{3.17}$$

Indeed since $\{\phi_j\}_{0 \leq j \leq l}$ and $\{\phi_j\}_{l+1 \leq j \leq r}$ are symplectically independent then $F_p(\rho) = F_{\bar{p}}(\rho) \oplus F_{\bar{q}}(\rho)$ (direct sum) in a suitable symplectic basis in $\mathbb{R}^{2(n+1)}$. Noticing that $\operatorname{Ker} F_{\bar{q}}^2(\rho) \cap \operatorname{Im} F_{\bar{q}}^2(\rho) = \{0\}$ we obtain the assertion since p is of spectral type 2.

Since $\{\phi_j\}_{0 \leq j \leq l}$ satisfies (3.13) then we see that $(\{\phi_i, \phi_j\})_{1 \leq i,j \leq l}$ is non singular in some open set from Lemma 3.3 and then there are smooth $c_j, j = 1, \ldots, l$ such that

$$\sum_{j=1}^{l} \{\phi_k, \phi_j\} c_j = \{\phi_k, \phi_0\}, \quad k = 1, \ldots, l. \tag{3.18}$$

Write $c_j = C_j(\phi', \theta)$ where $\theta = (\theta_{r+1}, \ldots, \theta_{2n+2})$ is chosen so that (ϕ_0, ϕ', θ), $\phi' = (\phi_1, \ldots, \phi_r)$ is a system of local coordinates and define

$$\bar{c}_j = C_j(0, \theta)$$

so that $c_j = \bar{c}_j(\theta) + O(|\phi|)$. Thus $(1, -\bar{c}_1, \ldots, -\bar{c}_l)$ is in $\operatorname{Ker}(\{\phi_i, \phi_j\})_{0 \leq i,j \leq l}$ modulo $O(|\phi|)$ then noting (3.16) we see that $\operatorname{Ker}(\{\phi_i, \phi_j\}(\rho))_{0 \leq i,j \leq l}$ is spanned by $(1, -\bar{c}_1(\theta), \ldots, -\bar{c}_l(\theta))$ for $\rho = (0, \theta) \in \Sigma$. From (3.17) there exists $0 \neq X \in \operatorname{Ker} F_{\bar{p}}^2(\rho) \cap \operatorname{Im} F_{\bar{p}}^2(\rho)$ such that $F_{\bar{p}}(\rho)X \in \operatorname{Ker} F_{\bar{p}}(\rho)$. Since $X \in \operatorname{Im} F_{\bar{p}}(\rho)$ we can put

$$X = H_f(\rho), \quad f = \sum_{j=0}^{l} a_j \phi_j$$

with some $a = (a_0, \ldots, a_l) \in \mathbb{R}^{l+1}$ and note that $F_{\bar{p}}(\rho)X \in \operatorname{Ker} F_{\bar{p}}(\rho)$ implies that $(-\{\phi_0, f\}(\rho), \{\phi_1, f\}(\rho), \ldots, \{\phi_l, f\}(\rho)) = k(-1, \bar{c}_1(\theta), \ldots, \bar{c}_l(\theta))$ with some $k \in \mathbb{R}$. With $A = (\{\phi_i, \phi_j\}(\rho))_{0 \leq i,j \leq l}$ we have $A\,{}^t a = k\,{}^t(1, \bar{c}_1(\theta), \ldots, \bar{c}_l(\theta))$. Such $a \in \mathbb{R}^{l+1}$ exists if and only if

$$ {}^t A\,v = 0, \quad v = (v_0, \ldots, v_l) \in \mathbb{R}^{l+1} \implies v_0 + \sum_{j=1}^{l} v_j \bar{c}_j(\theta) = 0. \tag{3.19}$$

Since ${}^t A\,v = -A\,v = 0$ and hence v is proportional to $(-1, \bar{c}_1(\theta), \ldots, \bar{c}_l(\theta))$ if ${}^t A\,v = 0$. Thus (3.19) proves that $1 - \sum_{j=1}^{l} \bar{c}_j(\theta)^2 = 0$. Therefore denoting $\tilde{\phi}_1(x, \xi') = \sum_{j=1}^{l} \bar{c}_j(x, \xi') \phi_j(x, \xi')$ with $\bar{c}_j(x, \xi') = \bar{c}_j(\theta(x, \xi'))$ we can take a smooth orthogonal matrix $O = (O_{ij})_{1 \leq i,j \leq l}$ of which first row is $(\bar{c}_1, \ldots, \bar{c}_l)$ such that with

$\tilde{\phi}_k = \sum_{j=1}^{l} O_{kj}\phi_j$ we have

$$-\xi_0^2 + \sum_{j=1}^{l} \phi_j^2 = -(\xi_0 + \tilde{\phi}_1)(\xi_0 - \tilde{\phi}_1) + \sum_{j=2}^{l} \tilde{\phi}_j^2.$$

It is clear that

$$\{\xi_0 - \tilde{\phi}_1, \tilde{\phi}_j\} = O(|\phi|), \quad j = 0, 1, \ldots, r \tag{3.20}$$

because $\{\phi_k, \phi_0 - \sum_{j=1}^{l} \bar{c}_j\phi_j\} = O(|\phi|)$ for $k = 0, 1, \ldots, l$ which follows from (3.18), and hence the assertion for $j = 0, 1, \ldots, l$. The assertion for $j = l + 1, \ldots, r$ is obvious since $\{\phi_j\}_{0 \leq j \leq l}$ and $\{\phi_j\}_{l+1 \leq j \leq r}$ are symplectically independent. With $\tilde{\phi}_0 = \xi_0 - \tilde{\phi}_1$ it is clear that

$$\operatorname{rank}(\{\tilde{\phi}_i, \tilde{\phi}_j\})_{0 \leq i,j \leq l} = \operatorname{rank}(\{\phi_i, \phi_j\})_{0 \leq i,j \leq l} = l$$

and hence $\operatorname{rank}(\{\tilde{\phi}_i, \tilde{\phi}_j\})_{1 \leq i,j \leq l} = l$ by (3.20). When $l = 2$ this shows that $\{\tilde{\phi}_1, \tilde{\phi}_2\} \neq 0$. Let $l \geq 4$. Note that $\operatorname{rank}(\{\tilde{\phi}_i, \tilde{\phi}_j\})_{2 \leq i,j \leq l} \leq l - 2$ since $l - 1$ is odd. Suppose that $\operatorname{rank}(\{\tilde{\phi}_i, \tilde{\phi}_j\}(\rho))_{2 \leq i,j \leq l} \leq l - 3$ at some ρ. Then it is easy to check that $\operatorname{rank}(\{\tilde{\phi}_i, \tilde{\phi}_j\}(\rho))_{1 \leq i,j \leq l} \leq l - 1$ which is a contradiction. Thus switching the notation from $\tilde{\phi}_j$ to ϕ_j ($j = 1, \ldots, l$) we get the desired assertion.   □

3.3   Reduction of the Hamilton Equation to a Coupling System of ODE

In this section assuming that (i) fails we reduce the Hamilton equation to a coupling of two ODE systems with a singular point of the first and the second kind respectively. Assume that

$$H_S^3 p \neq 0$$

in some open set U with a smooth S satisfying (2.22) and (2.23). We choose an open set $V \subset U, V \cap \Sigma \neq \emptyset$ where Proposition 3.4 holds. We fix a $\bar{\rho} \in V \cap \Sigma$ and work near $\bar{\rho}$. Since the case $l = 2$ is easier than the case $l \geq 4$ so we assume $l \geq 4$. Choose a system of symplectic coordinates (X, Ξ) such that $X_0 = x_0$ and $\Xi_0 = \xi_0 - \phi_1$. Switching the notation from (X, Ξ) to (x, ξ) one can write

$$p = -\xi_0^2 - 2\xi_0\phi_1 + \sum_{j=2}^{l} \phi_j^2 + \sum_{j=l+1}^{\ell} \phi_j^2 + \sum_{j=\ell+1}^{r} \phi_j^2.$$

Here we recall

$$\text{rank}(\{\phi_i, \phi_j\})_{0 \le i,j \le l} = l, \quad \phi_0 = \xi_0 \tag{3.21}$$

near $\bar{\rho}$. Since $\dim \text{Ker}(\{\phi_i, \phi_j\})_{2 \le i,j \le l} = 1$ by Proposition 3.4 one can choose a smooth $c = (c_2, \ldots, c_l)$ with $\sum c_j^2 = 1$ so that c spans $\text{Ker}(\{\phi_i, \phi_j\})_{2 \le i,j \le l}$. We make a smooth orthogonal transformation sending $\{\phi_j\}_{2 \le j \le l}$ to $\{\tilde{\phi}_j\}_{2 \le j \le l}$ such that $\tilde{\phi}_2 = \sum c_j \phi_j$ and switching the notation from $\{\tilde{\phi}_j\}$ to $\{\phi_j\}$ again we obtain

Proposition 3.5 *Choosing a suitable system of symplectic coordinates, p can be written in the form*

$$p = -\xi_0^2 - 2\xi_0 \phi_1 + \phi_2^2 + \sum_{j=3}^{\ell} \phi_j^2 + \sum_{j=\ell+1}^{r} \phi_j^2$$

in some open set V such that we have in $V \cap \Sigma$

$$\begin{cases} \{\xi_0, \phi_j\} = 0, \ 0 \le j \le r, \\ \{\phi_2, \phi_j\} = 0, \ j \ne 1, \ \{\phi_2, \phi_1\} \ne 0, \\ \{\phi_i, \phi_j\} = 0, \ 0 \le i \le r, \ \ell+1 \le j \le r, \\ \det\left(\{\phi_i, \phi_j\}\right)_{3 \le i,j \le \ell} \ne 0. \end{cases}$$

Proof The first assertion follows from (3.20). It is clear $\{\phi_2, \phi_j\} = 0$ in $V \cap \Sigma$ for $j = 2, \ldots, l$ by the definition of ϕ_2. The assertion $\{\phi_2, \phi_j\} = 0$ for $j = l+1, \ldots, r$ is clear because the original $\{\phi_j\}_{2 \le j \le l}$ and $\{\phi_j\}_{l+1 \le j \le r}$ are symplectically independent and the new ϕ_2 is a linear combination of the original $\{\phi_j\}_{2 \le j \le l}$. If $\{\phi_2, \phi_1\}(\rho) = 0$ then it is obvious that $\text{rank}(\{\phi_i, \phi_j\}(\rho))_{0 \le i,j \le l} \le l - 1$ which is a contradiction and hence the second assertion. The third assertion is clear. Since $\{\phi_0, \phi_j\} = 0, 0 \le j \le r$ and $\{\phi_2, \phi_j\} = 0$ unless $j = 1$ we see easily that $\text{rank}(\{\phi_i, \phi_j\})_{0 \le i,j \le l} \le l - 1$ which contradicts (3.21) if $\text{rank}(\{\phi_i, \phi_j\})_{3 \le i,j \le l} \le l - 3$. This proves that $\det(\{\phi_i, \phi_j\})_{3 \le i,j \le l} \ne 0$. Since $(\{\phi_i, \phi_j\}(\rho))_{3 \le i,j \le \ell}$ is a direct sum $(\{\phi_i, \phi_j\}(\rho))_{3 \le i,j \le l} \oplus (\{\phi_i, \phi_j\}(\rho))_{l+1 \le i,j \le \ell}$ for $\rho \in V \cap \Sigma$ then we have the last assertion. $\qquad\square$

Take $\xi_0, x_0, \phi_1, \ldots, \phi_r, \psi_1, \ldots, \psi_k$ $(r + k = 2n)$ to be a system of local coordinates around $\bar{\rho}$. Note that we can assume that ψ_j are independent of x_0 taking $\psi_j(0, x', \xi')$ as new ψ_j. Moreover we can assume that $\{\phi_2, \psi_j\} = 0$ and $\{\phi_1, \psi_j\} = 0$ on $V \cap \Sigma$ taking $\psi_j - \{\psi_j, \phi_2\}\phi_1/\{\phi_1, \phi_2\} - \{\psi_j, \phi_1\}\phi_2/\{\phi_2, \phi_1\}$ as new ψ_j. Thus it can be assumed that

$$\{\xi_0, \psi_j\} = 0, \ \{\phi_2, \psi_j\} = 0, \ \{\phi_1, \psi_j\} = 0, \ 1 \le j \le k$$

holds in $V \cap \Sigma$. Thanks to Jacobi identity one can assume

$$\{\phi_2, \{\phi_j, \xi_0\}\} = 0, \quad j = \ell+1, \ldots, r \tag{3.22}$$

in $V \cap \Sigma$ since we have $\{\phi_j, \{\xi_0, \phi_2\}\} = O(|\phi|)$ and $\{\xi_0, \{\phi_2, \phi_j\}\} = O(|\phi|)$ for $\ell + 1 \le j \le r$ by Proposition 3.5.

Let $\gamma(s) = (x(s), \xi(s))$ be a solution to the Hamilton equation (1.3) and recall $df(\gamma(s))/ds = \{p, f\}(\gamma(s))$. We change the parameter from s to t:

$$t = s^{-1}$$

so that we have $d/ds = -tD$ and $D = t(d/dt)$ and hence

$$\frac{d}{ds}(t^p F) = -t^{p+1}(DF + pF).$$

We now introduce new unknowns

$$\begin{cases} \xi_0(s) = t^4 \Xi_0(t), \quad x_0(s) = tX_0(t), \\ \phi_1(\gamma(s)) = t^2 \Phi_1(t), \quad \phi_2(\gamma(s)) = t^3 \Phi_2(t), \\ \phi_j(\gamma(s)) = t^4 \Phi_j(t), \quad 3 \le j \le \ell, \\ \phi_j(\gamma(s)) = t^3 \Phi_j, \quad \ell + 1 \le j \le r, \\ \psi_j(\gamma(s)) = t^2 \Psi_j(t), \quad 1 \le j \le k \end{cases} \tag{3.23}$$

and write $w = (\Xi_0, X_0, \Phi_1, \ldots, \Phi_r, \Psi_1, \ldots, \Psi_k)$. Denote

$$\{\phi_j, \xi_0\} = \sum_{i=1}^{r} C_i^j \phi_i, \quad \kappa_j = C_1^j(\bar\rho), \quad \delta = \{\phi_1, \phi_2\}(\bar\rho)$$

then from (3.22) we get

$$\kappa_j = 0, \quad j = \ell + 1, \ldots, r. \tag{3.24}$$

Thanks to Proposition 3.5 and (3.24) the Hamilton equations (1.3) is reduced to

$$\begin{cases} D\Xi_0 = -4\Xi_0 - 2\kappa_2 \Phi_1 \Phi_2 + tG(t, w), \\ DX_0 = -X_0 + 2\Phi_1 + tG(t, w), \\ D\Phi_1 = -2\Phi_1 + 2\delta\Phi_2 + tG(t, w), \\ D\Phi_2 = -3\Phi_2 - 2\kappa_2 \Phi_1^2 + 2\delta\Xi_0 + tG(t, w), \\ tD\Phi_j = -4t\Phi_j - 2\kappa_j \Phi_1^2 \\ \qquad -2\sum_{k=3}^{\ell}\{\phi_k, \phi_j\}(\bar\rho)\Phi_k + tG(t, w), \quad 3 \le j \le \ell, \\ D\Phi_j = -3\Phi_j + tG(t, w), \quad \ell + 1 \le j \le r, \\ D\Psi_j = -2\Psi_j - 2\sum_{k=\ell+1}^{r}\{\phi_k, \psi_j\}(\bar\rho)\Phi_k + tG(t, w), \quad 1 \le j \le k \end{cases} \tag{3.25}$$

where $G(t, w)$, may change from line to line, denotes a smooth function in (t, w) defined near $(0, 0)$ such that $G(t, 0) = 0$.

Lemma 3.6 *We have*

$$H_{\phi_2}(\bar{\rho}) \in \operatorname{Ker} F_p^2(\bar{\rho}) \cap \operatorname{Im} F_p^2(\bar{\rho}), \quad F_p(\bar{\rho})H_{\phi_2}(\bar{\rho}) \neq 0.$$

Proof From Proposition 3.5 it is easy to check that $F_p(\bar{\rho})H_{\phi_2} = \delta H_{\xi_0}$ and $F_p(\bar{\rho})H_{\xi_0} = 0$ so that $F_p^2(\bar{\rho})H_{\phi_2}(\bar{\rho}) = 0$. Since $\det(\{\phi_i, \phi_j\}(\bar{\rho}))_{3 \leq i,j \leq l} \neq 0$ we can choose $f = \phi_1 + \sum_{j=3}^{l} c_j \phi_j$ so that $H_p(\bar{\rho})H_f(\bar{\rho}) = H_{\phi_2}(\bar{\rho})$ which proves $H_{\phi_2}(\bar{\rho}) \in \operatorname{Im} F_p^2(\bar{\rho})$ since $H_f(\bar{\rho}) \in \operatorname{Im} F_p(\bar{\rho})$. $\square$

From Lemma 3.6 we can take $S = \phi_2$ and we conclude

$$\kappa_2 = C_1^2(\bar{\rho}) = \frac{\{\phi_2, \{\phi_2, \xi_0\}\}(\bar{\rho})}{\{\phi_2, \phi_1\}(\bar{\rho})} = \frac{-H_{\phi_2}^3 p(\bar{\rho})}{2\delta} \neq 0. \tag{3.26}$$

We introduce a class of formal power series in $(t, \log t)$

$$\mathscr{E} = \{ \sum_{0 \leq j \leq i} t^i (\log t)^j w_{ij} \mid w_{ij} \in \mathbb{C}^N \}$$

in which we look for a formal solution to the reduced Hamilton equation (3.25).

Lemma 3.7 *Assume that $w = (\Xi_0, X_0, \Phi_1, \ldots, \Phi_r, \Psi_1, \ldots, \Psi_k) \in \mathscr{E}$ satisfies (3.25) formally and $\Phi_2(0) \neq 0$. Then necessarily $\Phi_2(0) = -1/(\kappa_2 \delta^2)$ and $w(0)$ is uniquely determined. In particular $X_0(0) \neq 0$.*

Proof Taking $\det(\{\phi_i, \phi_j\}(\bar{\rho}))_{3 \leq i,j \leq \ell} \neq 0$ into account, the assertion follows from the special form (3.25) and (3.26). $\square$

Let $\bar{w} = (\bar{\Xi}_0, \bar{X}_0, \bar{\Phi}_1, \ldots, \bar{\Phi}_r, \bar{\Psi}_1, \ldots, \bar{\Psi}_k)$ be the uniquely determined $w(0)$ given by Lemma 3.7 and look for a formal solution to (3.25) of the form $\bar{w} + w$ with $w \in \mathscr{E}^\#$ where $\mathscr{E}^\# = \{\sum_{1 \leq i, 0 \leq j \leq i} t^i (\log t)^j w_{ij} \mid w_{ij} \in \mathbb{C}^N\}$. To simplify notation we set

$$\begin{cases} w^I = (X_0, \Phi_2, \Xi_0, \Phi_1), & w^{II} = (\Phi_3, \ldots, \Phi_\ell), \\ w^{III} = (\Phi_{\ell+1}, \ldots, \Phi_r), & w^{IV} = (\Psi_1, \ldots, \Psi_k) \end{cases}$$

then $w = {}^t(w^I, w^{II}, w^{III}, w^{IV})$ satisfies

$$HDw = Aw + tF + G(t, w), \quad A = \begin{bmatrix} A_I & O & O & O \\ B_{II} & A_{II} & O & O \\ O & O & -3E & O \\ O & O & B_{IV} & -2E \end{bmatrix} \tag{3.27}$$

with $H = E \oplus O \oplus E \oplus E$ where E is the identity matrix and O is the zero matrix. Moreover F is a constant vector and

$$G(t, w) = \sum_{2 \leq i, 0 \leq j \leq i} G_{ij} t^i (\log t)^j, \quad G_{ij} = G_{ij}(w_{pq} \mid q \leq p \leq i - 1).$$

Noting that

$$A_I = \begin{bmatrix} -1 & 0 & 0 & 2 \\ 0 & -3 & 2\delta & -4\kappa_2 \bar{\Phi}_1 \\ 0 & -2\kappa_2 \bar{\Phi}_1 & -4 & -2\kappa_2 \bar{\Phi}_2 \\ 0 & 2\delta & 0 & -2 \end{bmatrix}$$

and taking into account $\kappa_2 \delta^2 \bar{\Phi}_2 = -1$, $\kappa_2 \delta \bar{\Phi}_1 = -1$, which follows from Lemma 3.7, we have

Lemma 3.8 *The eigenvalues of A_I are $\{-6, -4, -1, 1\}$ while A_{II} is a non-singular anti-symmetric matrix $(\{\phi_i, \phi_j\}(\bar{\rho}))_{3 \leq i,j \leq \ell}$ so that A_{II} is diagonalizable with non-zero pure imaginary eigenvalues.*

Proposition 3.6 *There exists $w = (\Xi_0, X_0, \Phi_1, \ldots, \Phi_r, \Psi_1, \ldots, \Psi_k) \in \mathcal{E}$ such that $\Phi_2(0) \neq 0$ and $X_0(0) \neq 0$ satisfying (3.25).*

Proof Note that (3.27) implies that $H(iw_{ij} - (j + 1)w_{ij+1}) = Aw_{ij} + \delta_{i1}\delta_{j0}F + G_{ij}$ where $G_{ij} = 0$ for $i = 0, 1$. Then we have

$$\begin{cases} (H - A)w_{11} = 0, \\ (H - A)w_{10} = w_{11} + F. \end{cases}$$

Choose $w_{11} \in \text{Ker}\,(H - A)$ so that $F + w_{11} \in \text{Im}\,(H - A)$. Then we can take $w_{10} \neq 0$ so that $(H - A)w_{10} = F + w_{11}$ since $\text{Ker}\,(H - A) \neq \{0\}$ by Lemma 3.8. We turn to the case $i \geq 2$

$$(iH - A)w_{ij} = (j + 1)w_{ij+1} + G_{ij}. \tag{3.28}$$

With $j = i$, (3.28) turns to $(iI - A)w_{ii} = G_{ii}(w_{pq} \mid q \leq p \leq i - 1)$. Since $iH - A$ is non singular for $i \geq 2$ by Lemma 3.8 one has $w_{ii} = (iH - A)^{-1}G_{ii}(w_{pq} \mid q \leq p \leq i - 1)$. Recurrently one can solve w_{ij} by

$$w_{ij} = (iH - A)^{-1}((j + 1)w_{ij+1} + G_{ij}(w_{pq} \mid q \leq p \leq i - 1))$$

for $j = i - 1, i - 2, \ldots, 0$. This proves the assertion. $\qquad \square$

Note that this formal solution is uniquely determined up to a term vt, $v \in$ Ker $(H - A)$. Since $A_{II} = (\{\phi_i, \phi_j\}(\bar{\rho}))_{3 \leq i, j \leq l}$ is diagonalizable, choosing a non singular constant matrix S one can assume

$$S^{-1} A_{II} S = i\Lambda$$

where Λ is a diagonal matrix with non-zero real entries. Denote $u = S^{-1} w^{II}$ and $v = (w^I, w^{III}, w^{IV})$ then (3.25) becomes

$$\begin{cases} tDu = tK_1 u + i\Lambda u + Q_1(v) + tG_1(t, u, v), \\ Dv = K_2 v + Q_2(v) + tG_2(t, u, v) \end{cases} \tag{3.29}$$

where K_j are constant matrices, $Q_j(v)$ are quadratic forms in v and $G_j(t, u, v)$ are smooth functions such that $G_j(t, 0, 0) = 0$. Let

$$u = \sum_{0 \leq j \leq i} u_{ij} t^i (\log t)^j, \quad v = \sum_{0 \leq j \leq i} v_{ij} t^i (\log t)^j \tag{3.30}$$

be a formal solution obtained in Proposition 3.6. Denote by u_N, v_N which are obtained from (3.30) getting rid of the terms $t^i (\log t)^j$ with $i \geq N + 1$ then for any given $m \in \mathbb{N}$ there is $N = N(m)$ such that u_N, v_N satisfies (3.29) modulo $O(t^{m+2})$. We look for a solution to (3.29) in the form $(u_N, v_N) + t^m(u, v)$. Then the equations satisfied by (u, v) is (after dividing by t^m)

$$\begin{cases} \left(t^2 \dfrac{d}{dt} - i\Lambda\right) u = -t(mI - K_1)u + L_1(t)v + tR_1(t, u, v) + t^2 F_1(t), \\ t \dfrac{d}{dt} v = -(mI - K_2)v + L_2(t)v + tR_2(t, u, v) + tF_2(t) \end{cases} \tag{3.31}$$

where $R_j(t, u, v)$ are C^1 functions defined near $(0, 0, 0) \in \mathbb{R} \times \mathbb{C}^{N_1} \times \mathbb{C}^{N_2}$ satisfying

$$|R_j(t, u, v)| \leq B_j(|u| + |v|) \tag{3.32}$$

for $(t, u, v) \in \{|t| \leq T\} \times \{|u| \leq CT\} \times \{|v| \leq CT\}$ where $L_j(t) \in C^1((0, T])$ and $F_j(t) \in C((0, T])$ satisfy

$$\|L_j(t)\|_{C([0,T])}, \quad \|F_j(t)\|_{C([0,T])}, \quad \|tL_j'(t)\|_{C([0,T])} \leq B.$$

Equation (3.31) is a coupling of two ODE systems which has $t = 0$ as a singular point of the first and the second kind respectively.

3.4   Existence of Tangent Bicharacteristics

In this section we prove that there exists a solutions to (3.31) with $u(0) = v(0) = 0$ which implies the existence of tangent bicharacteristics. Recall that Λ is a constant non-singular real diagonal matrix;

$$\Lambda = \text{diag}\,(\lambda_1, \dots, \lambda_{N_1}), \quad \lambda_j \in \mathbb{R} \setminus \{0\}. \tag{3.33}$$

Our aim in this section is to prove:

Proposition 3.7 *If $m \in \mathbb{R}$ is sufficiently large then (3.31) has a solution (u, v) such that $u(0) = 0$, $v(0) = 0$.*

We first introduce integral operators $\mathcal{H}$ and $\mathcal{G}$ defined by

$$\mathcal{H}[f] = \int_0^t e^{-\frac{i}{t}\Lambda + \frac{i}{s}\Lambda}(t/s)^{-m} s^{-2} f(s)ds, \quad \mathcal{G}[h] = \int_0^t (t/s)^{-m} s^{-1} h(s)ds$$

for $f \in C([0, T])$ with $f(t) = O(t)$ as $t \downarrow 0$ and for $h \in C([0, T])$ so that we have

$$(t^2 d/dt - i\Lambda)\mathcal{H}[f] = -mt\mathcal{H}[f] + f, \quad t(d/dt)\mathcal{G}[h] = -m\mathcal{G}[h] + h. \tag{3.34}$$

We start with

Lemma 3.9 *Let $m > 0$ be large enough. Then we have*

$$\mathcal{H}[f](t) = -(i\Lambda)^{-1} f(t) + m(i\Lambda)^{-1} \mathcal{H}[tf](t) + (i\Lambda)^{-1} \mathcal{H}[t^2 f'](t),$$

$$|\mathcal{H}[f](t)| \le (1/m)\|s^{-1}f\|_{C([0,t])}, \quad |\mathcal{G}[h](t)| \le (1/m)\|h\|_{C([0,t])}.$$

Proof Let $m > 0$. Note that

$$\mathcal{H}[f] = e^{-\frac{i}{t}\Lambda} \int_{1/t}^{\infty} e^{i\rho\Lambda}(t\rho)^{-m} f(1/\rho)d\rho.$$

Then integration by parts shows

$$\mathcal{H}[f] = -(i\Lambda)^{-1} f(t) + m(i\Lambda)^{-1} e^{-\frac{i}{t}\Lambda} \int_{1/t}^{\infty} e^{i\rho\Lambda}(t\rho)^{-m-1} t f(1/\rho)d\rho$$

$$+ (i\Lambda)^{-1} e^{-\frac{i}{t}\Lambda} \int_{1/t}^{\infty} e^{i\rho\Lambda}(t\rho)^{-m} \rho^{-2} f'(1/\rho)d\rho$$

$$= -(i\Lambda)^{-1} f(t) + m(i\Lambda)^{-1} \int_0^t e^{-\frac{i}{t}\Lambda + \frac{i}{s}\Lambda}(t/s)^{-m} s^{-1} f(s)ds$$

$$+ (i\Lambda)^{-1} \int_0^t e^{-\frac{i}{t}\Lambda + \frac{i}{s}\Lambda}(t/s)^{-m} f'(s)ds$$

which proves the first assertion. Since $|e^{-\frac{i}{t}\Lambda+\frac{i}{s}\Lambda}| \leq 1$ we have

$$|\mathcal{H}[f](t)| \leq \int_0^1 s^{m-1}|(ts)^{-1}f(ts)|ds \leq \frac{1}{m}\|t^{-1}f\|_{C([0,t])}$$

which is the second assertion. The third assertion is clear.   □

Using (3.34) we rewrite (3.31) as an integral equation;

$$\begin{cases} u = \mathcal{H}[tK_1 u + L_1(t)v + tR_1(t,u,v) + t^2 F_1], \\ v = \mathcal{G}[K_2 v + L_2(t)v + tR_2(t,u,v) + t^2 F_2]. \end{cases}$$

Let $u_0(t) = 0$, $v_0(t) = 0$ and define $u_n(t)$, $v_n(t)$ successively by

$$\begin{cases} u_{n+1}(t) = \mathcal{H}[tK_1 u_n + L_1(t)v_n + tR_1(t,u_n,v_n) + t^2 F_1], \\ v_{n+1}(t) = \mathcal{G}[K_2 v_n + L_2(t)v_n + tR_2(t,u_n,v_n) + t^2 F_2]. \end{cases}$$

From now on, to simplify notations we write $\|f\|_T$ for $\|f\|_{C([0,T])}$.

Lemma 3.10 *There exist positive constants C, C^* ($C^* < C$) and $T > 0$ such that we have for $n = 0, 1, 2, \ldots$*

$$|u_n(t)| \leq Ct, \quad |v_n(t)| \leq C^*t \quad for \ 0 \leq t \leq T. \tag{3.35}$$

Proof Assume (3.35) holds for n and $n - 1$. Write

$$u_{n+1} = \mathcal{H}[tK_1 u_n] + \mathcal{H}[L_1(t)v_n] + \mathcal{H}[tR_1(t,u_n,v_n)] + \mathcal{H}[tF_1].$$

From Lemma 3.9 and the inductive hypothesis we see

$$|\mathcal{H}[t^2 F_1]| \leq (B/m)t, \quad |\mathcal{H}[tK_1 u_n]| \leq (|K_2|C/m)t. \tag{3.36}$$

Noting that $|tR_1(t,u_n,v_n)| \leq 2B_1 Ct^2$ which follows from the inductive hypothesis, we have from Lemma 3.9 that

$$|\mathcal{H}[tR_1(t,u_n,v_n)]| \leq (2B_1 C/m)t. \tag{3.37}$$

We next estimate $\mathcal{H}[L_1(t)v_n]$. By Lemma 3.9 one can write

$$\mathcal{H}[L_1(t)v_n] = -(i\Lambda)^{-1}L_1(t)v_n + m(i\Lambda)^{-1}\mathcal{H}[tL_1(t)v_n]$$
$$+ (i\Lambda)^{-1}\mathcal{H}[t^2 L_1'(t)v_n] + (i\Lambda)^{-1}\mathcal{H}[t^2 L_1(t)v_n'].$$

Denote $|\Lambda^{-1}| = \lambda$ and $A_k = \|L_k\|_T + 2B_k CT$ for $k = 1, 2$. It is clear

$$|(i\Lambda)^{-1}L_1 v_n| \leq \lambda\|L_1\|_T C^*t \tag{3.38}$$

while Lemma 3.9 gives

$$|m(i\Lambda)^{-1}\mathscr{H}[tL_1(t)v_n]| \le \lambda\|L_1(t)v_n\|_{C([0,t])} \le \lambda\|L_1\|_T C^* t,$$
$$|(i\Lambda)^{-1}\mathscr{H}[t^2 L_1'(t)v_n]| \le (\lambda/m)\|tL_1'(t)\|_T Ct. \tag{3.39}$$

Recall that $tv_n' = -mv_n + K_2 v_{n-1} + L_2(t)v_{n-1} + tR_2(t, u_{n-1}, v_{n-1}) + tF_2$. This together with the induction hypothesis gives that

$$|tv_n'| \le m|v_n| + |K_2|Ct + \|L_2\|_T Ct + 2B_2 Ct^2 + Bt \le mC^* t + A_1 Ct + Bt.$$

Then thanks to Lemma 3.9 one gets

$$|(i\Lambda)^{-1}\mathscr{H}[t^2 L_1(t)v_n']| \le (\lambda/m)\|L_1\|_T\{mC^* + A_1 C + B\}t$$
$$\le \lambda\|L_1\|_T C^* t + (\lambda/m)(A_1 C + B)t. \tag{3.40}$$

From (3.38)–(3.40) it follows that

$$|\mathscr{H}[L_1(t)v_n]| \le 3\lambda\|L_1\|_T C^* t + (\lambda/m)(\|tL_1'\|_T C + A_1 C + B)t. \tag{3.41}$$

Combining the estimates (3.36), (3.37) and (3.41) one can conclude that $|u_{n+1}(t)| \le 3\lambda\|L_1\|_T C^* t + \mathscr{A}_2 t/m$ where $\mathscr{A}_2 = |K_2|C + B + 2B_1 C + \lambda(\|tL_1'\|_T C + A_1 C + B)$. Fix a $C^* > 0$ and choose $C > C^*$ so that $C/2 > 3\lambda\|L_1\|_T C^*$. Then if m is chosen such that $\mathscr{A}_2/m \le C/2$ then we have $|u_{n+1}(t)| \le Ct$. We turn to

$$v_{n+1} = \mathscr{G}[K_2 v_n] + \mathscr{G}[L_2(t)v_n] + \mathscr{G}[tR_2(t, u_n, v_n)] + \mathscr{G}[tF_2].$$

By Lemma 3.9 and the induction hypothesis one has

$$|\mathscr{G}[K_2 v_n]| \le (|K_2|C/m)t, \quad |\mathscr{G}[L_2(t)v_n]| \le (\|L_2\|_T C/m)t, \quad |\mathscr{G}[tF_2]| \le (B/m)t.$$

Since $|tR_2(t, u_n, v_n)| \le 2B_2 Ct^2$ we have by Lemma 3.9 that

$$|v_{n+1}| \le (\|L_2\|_T C + |K_2|C + B)t/m.$$

Hence to conclude the proof it suffices to take m so that $\mathscr{A}_2/m \le C/2$ and $(\|L_2\|_T C + |K_2|C + B)t/m \le C^*$ holds. $\qquad\square$

Let $\tilde{B}_j$ be such that $|\partial R_j/\partial u| + |\partial R_j/\partial v| \le \tilde{B}_{2j}$ for $(t, u, v) \in \{|t| \le T\} \times \{|u| \le CT\} \times \{|v| \le CT\}$. We now show

Lemma 3.11 *For large m we have*

$$|v_n - v_{n-1}| \le (1/m)A_2(\|u_{n-1} - u_{n-2}\|_{C([0,t])} + \|v_{n-1} - v_{n-2}\|_{C([0,t])}),$$
$$t|v_n' - v_{n-1}'| \le 2A_2(\|u_{n-1} - u_{n-2}\|_{C([0,t])} + \|v_{n-1} - v_{n-2}\|_{C([0,t])}).$$

Proof We first note that

$$|tR_j(t, u_{n-1}, v_{n-1}) - tR_j(t, u_{n-2}, v_{n-2})|$$
$$\leq \tilde{B}_j t(|u_{n-1} - u_{n-2}| + |v_{n-1} - v_{n-2}|) \tag{3.42}$$

from which one gets

$$|\mathscr{G}[tR_2(t, u_{n-1}, v_{n-1}) - tR_2(t, u_{n-2}, v_{n-2})]|$$
$$\leq (2\tilde{B}_2 T/m)(\|u_{n-1} - u_{n-2}\|_{C([0,t])} + \|v_{n-1} - v_{n-2}\|_{C([0,t])}). \tag{3.43}$$

It is also clear that

$$|\mathscr{G}[L_2(t)(v_{n-1} - v_{n-2})]| \leq (\|L_2\|_T/m)\|v_{n-1} - v_{n-2}\|_{C([0,t])},$$
$$|\mathscr{G}[K_2(v_{n-1} - v_{n-2})]| \leq (|K_2|/m)\|v_{n-1} - v_{n-2}\|_{C([0,t])}. \tag{3.44}$$

The first assertion follows from (3.43) and (3.44).
   We study $t(v_n' - v_{n-1}')$. Recall that

$$t(v_n' - v_{n-1}') = -m(v_n - v_{n-1}) + K_2(v_{n-1} - v_{n-2}) + L_2(t)(v_{n-1} - v_{n-2})$$
$$+tR_2(t, u_{n-1}, v_{n-1}) - tR_2(t, u_{n-2}, v_{n-2})$$

which shows that

$$|t(v_n' - v_{n-1}')| \leq m|v_n - v_{n-1}| + (|K_2| + \|L_2\|_T)|v_{n-1} - v_{n-2}|$$
$$+\tilde{B}_2 t(|u_{n-1} - u_{n-2}| + |v_{n-1} - v_{n-2}|)$$
$$\leq m|v_n - v_{n-1}| + A_2(|u_{n-1} - u_{n-2}| + |v_{n-1} - v_{n-2}|).$$

Here we apply the first assertion to estimate $|v_n - v_{n-1}|$ and we get

$$|t(v_n' - v_{n-1}')| \leq 2A_2(\|u_{n-1} - u_{n-2}\|_{C([0,t])} + \|v_{n-1} - v_{n-2}\|_{C([0,t])})$$

which is the desired assertion.   □

Proof of Theorem 3.7 We show that u_n, v_n converges to some u, v in $C([0, T])$.
Write

$$u_{n+1} - u_n = \mathscr{H}[K_1(u_n - u_{n-1})] + \mathscr{H}[L_1(t)(v_n - v_{n-1})]$$
$$+ \mathscr{H}[tR_1(t, u_n, v_n) - tR_1(t, u_{n-1}, v_{n-1})]$$

and set $W_n(t) = \|u_n - u_{n-1}\|_{C([0,t])} + \|v_n - v_{n-1}\|_{C([0,t])}$. From (3.42) and Lemma 3.9 it follows $|\mathscr{H}[tR_1(t, u_n, v_n) - tR_1(t, u_{n-1}, v_{n-1})]| \leq (\tilde{B}_2/m)W_n(t)$. By Lemma 3.9 one can write

$$\mathscr{H}[L_1(t)(v_n - v_{n-1})] = -(i\Lambda)^{-1}L_1(t)(v_n - v_{n-1}) + m(i\Lambda)^{-1}\mathscr{H}[tL_1(t)(v_n - v_{n-1})]$$
$$+ (i\Lambda)^{-1}\mathscr{H}[t^2 L_1'(t)(v_n - v_{n-1})]$$
$$+ (i\Lambda)^{-1}\mathscr{H}[t^2 L_1(t)(v_n' - v_{n-1}')].$$

From Lemma 3.11 one has $|(i\Lambda)^{-1}L_1(t)(v_n - v_{n-1})| \leq (\lambda/m)\|L_1\|_T A_2 W_{n-1}(t)$ while

$$|m(i\Lambda)^{-1}\mathscr{H}[tL_1(v_n - v_{n-1})]| + |(i\Lambda)^{-1}\mathscr{H}[t^2 L_1'(t)(v_n - v_{n-1})]|$$
$$\leq (\lambda\|L_1\|_T + (\lambda/m)\|tL_1'\|_T)\|v_n - v_{n-1}\|_{C([0,t])}$$
$$\leq (\lambda(\|L_1\|_T + \|tL_1'\|_T)A_2/m)W_{n-1}(t)$$

where the last inequality follows from Lemma 3.11. Finally we see that by Lemmas 3.7 and 3.11

$$|(i\Lambda)^{-1}\mathscr{H}[t^2 L_1(t)(v_n' - v_{n-1}')]| \leq (\lambda/m)\|tL_1(t)(v_n' - v_{n-1}')\|_{C([0,t])}$$
$$\leq (2\lambda A_2/m)\|L_1\|_T W_{n-1}(t).$$

Combining these estimates one concludes that $|u_{n+1} - u_n|$ is bounded by

$$(2/m)(B_{21}C + \tilde{B}_{21})W_n(t) + (\lambda A_2/m)(4\|L_1\|_T + \|tL_1'\|_T)W_{n-1}(t).$$

We turn to $v_{n+1} - v_n$: Recall that

$$v_{n+1} - v_n = \mathscr{G}[K_2(v_n - v_{n-1})] + \mathscr{G}[L_2(t)(v_n - v_{n-1})]$$
$$+ \mathscr{G}[tR_2(t, u_n, v_n) - tR_2(t, u_{n-1}, v_{n-1})].$$

From (3.42) it is easy to see that $|v_{n+1} - v_n| \leq A_2 W_n(t)/m$. We now take m large so that we have $W_{n+1}(t) \leq \delta\{W_n(t) + W_{n-1}(t)\}$ for $0 \leq t \leq T$ with $0 < \delta < 1/2$. It is easy to check that

$$W_n(t) \leq \sum_{k=1}^{n-2}(2\delta)^k(W_2 + W_1).$$

This proves that $\{u_n\}$, $\{v_n\}$ converges in $C([0, T])$ to some $u(t)$, $v(t) \in C([0, T])$ which completes the proof of Proposition 3.7.

We now complete the proof of Theorem 3.1. Thanks to Proposition 3.7 there exists w satisfying (3.25). Switching to the original coordinates this shows that the

Hamilton equations (1.3) has a solution $(x(s), \xi(s))$ such that $\lim_{s \to \infty} (x(s), \xi(s)) \in \Sigma$. From (3.23) we have

$$\frac{d\phi_j}{dx_0}\bigg|_{x_0=0} = \left(\frac{d\phi_j}{dt} \bigg/ \frac{dx_0}{dt}\right)_{x_0=0} = 0$$

and hence the curve $(x(s), \xi(s))$ is actually tangent to Σ.   □

Another approach to study the existence of tangent bicharacteristics is found in [72] and [3].

3.5   Transversal Bicharacteristics

In this section we give a proof of Proposition 3.1. As remarked just before Lemma 3.1 it suffices to study $p(x, \xi)$ of the form (2.1). Without restrictions we can assume that $A_{10}(x, \xi') = 0$:

$$p(x, \xi) = -\xi_0^2 + q(x, \xi')$$

where $q(x, \xi') \geq 0$. Let $\bar{\rho} \in \Sigma$ be a double characteristic of p and set

$$\Sigma_1 = \{(x, \xi) \mid p(x, \xi) = 0, dp(x, \xi) \neq 0\}$$

so that Σ_1 is the set of simple characteristics of p.

Definition 3.2 Let

$$\gamma : s \mapsto \gamma(s) = \big(x(s), \xi(s)\big) \in \Sigma_1$$

be a bicharacteristic of p defined in $[s_0, +\infty)$ (resp. $(-\infty, s_0]$) with some s_0. We say that γ is incoming (resp. outgoing) relative to $\bar{\rho}$ if

$$\gamma(s) \to \bar{\rho} \quad \text{as} \quad s \uparrow +\infty \quad (\text{resp. } s \downarrow -\infty).$$

Let Q be a hyperbolic quadratic form on $\mathbb{R}^{2(n+1)}$. We say that Q is effectively hyperbolic if F has a non-zero real eigenvalue (see Definition 1.14). In what follows we shall not use the homogeneity in ξ of $p(x, \xi)$; we thus introduce a system of local coordinates (y, η) by setting

$$(y, \eta) = (x, \xi) - \bar{\rho}$$

so that $\bar{\rho} = (0, 0)$. We write (x, ξ) instead of (y, η).

Lemma 3.12 *Assume that $p_{\bar{\rho}}(x, \xi) = Q(x, \xi)$ be an effectively hyperbolic quadratic form. Then after a linear symplectic change of coordinates around $\bar{\rho}$, the symbol $p(x, \xi)$ takes the form*

$$-p(x, \xi) = \lambda \{\xi_0^2 - E(x, \xi')\} + O(|(x, \xi)|^3) \quad as \quad (x, \xi) \to \bar{\rho}$$

where

$$E(x, \xi') = x_0^2 + \sum_{j=1}^{k} \mu_j(x_j^2 + \xi_j^2) + \sum_{j=k+1}^{\ell} \mu_0 x_j^2$$

with some positive constants μ_j and μ_0. Furthermore the hyperbolicity of $p(x, \cdot)$ with respect to ξ_0 near $x = 0$ is preserved after the change of coordinates.

Proof Note that $p(x, \xi) = p_{\bar{\rho}}(x, \xi) + O(|(x, \xi)|^3)$ when $(x, \xi) \to \bar{\rho}$. Since $p_{\bar{\rho}}(x, \xi)$ is an effectively hyperbolic quadratic form, by virtue of Lemma 1.7 there is a linear symplectic change of coordinates $S: (y, \eta) \mapsto (x, \xi)$ such that

$$-p_{\bar{\rho}}(S(y, \eta)) = \lambda\{\eta_0^2 - E(y, \eta')\}.$$

Writing (x, ξ) instead of (y, η) we have the first assertion. It remains to show that $p(S(y, \eta)) = \tilde{p}(y, \eta)$ is hyperbolic polynomial with respect to η_0 near $y = 0$. Setting

$$p_{\pm}(x, \xi) = \xi_0 \mp q(x, \xi')^{1/2}, \qquad \tilde{p}_{\pm}(y, \eta) = p_{\pm}(S(y, \eta))$$

we show that each one of $\tilde{p}_{\pm}(y, \eta)$ has a real zero $\eta_0^{\pm} = \eta_0^{\pm}(y, \eta')$ with $|\eta_0^{\pm}| \leq \epsilon$ in a closed region $|(y, \eta')| \leq \epsilon^2$, as far as the constant $\epsilon > 0$ is small enough. If we write

$$\xi_0 = \alpha\eta_0 + O(\epsilon^2), \qquad q(x, \xi') = \beta\eta_0^2 + O(\epsilon^3)$$

in a closed region $\{|\eta_0| \leq \epsilon, \ |(y, \eta')| \leq \epsilon^2\}$ with constants α and β, then $\beta \geq 0$ and

$$-\tilde{p}(y, \eta) = (\alpha^2 - \beta)\eta_0^2 + O(\epsilon^3)$$

so that $\alpha^2 - \beta = c > 0$ since the quadratic part of $p(S(y, \eta))$ starts with $c\,\eta_0^2$. On the other hand, if in addition $|\eta_0| = \epsilon$ then

$$\tilde{p}_{\pm}(y, \eta) = \alpha\eta_0 \mp \beta^{1/2}|\eta_0| + O(\epsilon^2)$$

so that each one of $\tilde{p}_{\pm}(y, \eta)$ has different sign at $\eta_0 = \pm\epsilon$. Hence the desired conclusion follows from the continuity of $\tilde{p}_{\pm}(y, \eta)$ with respect to η_0. □

Corollary 3.3 *After a linear symplectic change of coordinates around $\bar\rho$ the symbol $p(x, \xi)$ takes the form*

$$-p(x, \xi) = e(x, \xi)\{(\xi_0 - \phi(x, \xi))^2 + \psi(x, \xi')\}$$

with $e(\bar\rho) = \lambda$, where we have $\psi \geq 0$ near $\bar\rho'$, $\psi(x, \xi') - E(x, \xi') = O(|(x, \xi')|^3)$ and $\phi(x, \xi') = O(|(x, \xi')|^2)$ as $(x, \xi') \to \bar\rho'$.

Proof From the Malgrange preparation theorem (see [33, Theorem 7.5.5]) it follows that $-p(x, \xi)$ takes the desired form where $\phi(\bar\rho') = \psi(\bar\rho') = 0$. Comparing the quadratic parts of these two expressions of $-p(x, \xi)$ we obtain the required properties of ϕ and ψ since the non-negativity of ψ near $\bar\rho'$ follows from the hyperbolicity of $p(x, \cdot)$ with respect to ξ_0 near $x = 0$. □

As far as we are concerned with bicharacteristics, without restrictions we can assume that

$$-p(x, \xi) = (\xi_0 - \phi)^2 - \psi \tag{3.45}$$

since H_p and H_{ep} are proportional on Σ_1. In what follows we assume (3.45).

Lemma 3.13 *There is a constant $C > 0$ such that we have for $\rho \in \Sigma_1$ near $\bar\rho$*

$$|dp(\rho)| \leq C\left|\frac{\partial p}{\partial \xi_0}(\rho)\right|.$$

Proof Since $p(\rho) = 0$ for $\rho \in \Sigma_1$ one has

$$\psi(\rho)^{1/2} = |\xi_0 - \phi(\rho)| = \frac{1}{2}\left|\frac{\partial p}{\partial \xi_0}(\rho)\right|.$$

On the other hand, $\psi \geq 0$ implies $|d\psi| \leq C' \psi^{1/2}$ with some constant $C' > 0$ by Glaeser's inequality. Then the desired conclusion follows from

$$\frac{\partial p}{\partial x} = 2(\xi_0 - \phi)\frac{\partial \phi}{\partial x} + \frac{\partial \psi}{\partial x}, \qquad \frac{\partial p}{\partial \xi'} = 2(\xi_0 - \phi)\frac{\partial \phi}{\partial \xi'} + \frac{\partial \psi}{\partial \xi'}. \qquad □$$

Using Lemma 3.13 we give an estimate for an incoming or outgoing bicharacteristic $\gamma(s)$.

Lemma 3.14 *Let $\gamma(s)$ be an incoming or outgoing bicharacteristic. Then*

$$|\gamma(s)| \leq C|x_0(s)| \quad when \quad \gamma(s) \to \bar\rho.$$

Proof We first note that $\dot x_0(s) \neq 0$ as $\gamma(s) \to \bar\rho$ where the dot refers to differentiation with respect to s. Indeed, if $\dot x_0(s) = 0$ at some $\gamma(s) \in \Sigma_1$ near $\bar\rho$, then $\partial p(\gamma(s))/\partial \xi_0 = \dot x_0(s) = 0$, so that Lemma 3.13 implies $dp(\gamma(s)) = 0$, but this contradicts $\gamma(s) \in \Sigma_1$. Thus we may take x_0 as a new parameter of the curve $\gamma(s)$

near $\bar{\rho}$. Then Lemma 3.13 proves

$$\left|\frac{d\gamma}{dx_0}\right| \leq \frac{|dp|}{|\partial p/\partial \xi_0|} \leq C$$

on the curve $\gamma(s)$. This proves the assertion.   □
Thanks to Lemma 3.14 above, it can be understood that

$$O(|\rho|^l) = O(|x_0|^l) \quad \text{as} \quad \rho \to \bar{\rho}$$

as far as the limit is taken along a bicharacteristic, and then we simply denote this relation by O^l.

Lemma 3.15 *Let $\gamma(s)$ be an incoming or outgoing bicharacteristic. Then there exists $C > 0$ such that*

$$|x_0(s)| \leq C|\dot{x}_0(s)|$$

on $\gamma(s)$ near $\bar{\rho}$ where C is independent of the choice of the bicharacteristic $\gamma(s)$.

Proof We shall be working on the bicharacteristic $\gamma(s)$. Since $p(x, \xi) = 0$ on the trajectory, it follows that

$$|\dot{x}_0| = |\partial p/\partial \xi_0| = 2|\xi_0 - \phi| = 2\psi^{1/2}.$$

On the other hand, we have by Corollary 3.3 that $x_0^2 \leq E = \psi + O^3$. Since $|x_0|$ is small enough, we obtain the desired conclusion.   □

Proposition 3.8 *Let $\gamma(s) = (x(s), \xi(s))$ be an incoming or outgoing bicharacteristic. Then $v(s) = (x'(s), \xi'(s))$ satisfies*

$$|v(s)| \leq C |x_0(s)|^{3/2} \quad \text{on} \quad \gamma(s) \quad \text{near} \quad \bar{\rho}$$

where $C > 0$ is a constant independent of the choice of a bicharacteristic $\gamma(s)$.

Proof Setting $e_j(s) = (x_j(s)^2 + \xi_j(s)^2)/2$ for $1 \leq j \leq d$, we shall show that

$$e_j = \begin{cases} O^3 & \text{for} \quad 1 \leq j \leq l, \\ O^4 & \text{for} \quad l+1 \leq j \leq d. \end{cases}$$

We first consider the case $1 \leq j \leq k$. Since

$$\partial p/\partial x_j = 2\mu_j x_j + O^2, \qquad \partial p/\partial \xi_j = 2\mu_j \xi_j + O^2$$

it follows that $\dot{e}_j(s) = O^3$. This together with Lemma 3.15 implies that

$$de_j/dx_0 = O^2.$$

This gives the desired estimates. In case $l + 1 \leq j \leq d$, we have

$$\dot{x}_j = \partial p / \partial \xi_j = O^2, \qquad \dot{\xi}_j = -\partial p / \partial x_j = O^2$$

so that $e_j = O^4$ is obtained by the same argument as above. It remains to consider the case $k < j \leq l$. We first have as above $\dot{x}_j = \partial p / \partial \xi_j = O^2$ so that $x_j = O^2$. By using this we get $\dot{\xi}_j = -2\mu_0 x_j + O^2 = O^2$ so that $\xi_j = O^2$ and hence $e_j = O^4$. This completes the proof. $\square$

We now set, for $u = (y', \eta') \in \mathbb{R}^{2n}$ with $|u| < 1$:

$$\begin{cases} \Phi(x_0, u) = x_0^{-2} \phi(x_0, x_0 u), \\ \Psi(x_0, u) = x_0^{-2} \psi(x_0, x_0 u), \\ \pi^{\pm}(x_0, u) = \pm \Psi(x_0, u)^{1/2} + x_0 \Phi(x_0, u) \end{cases}$$

where the square root can be taken by virtue of the fact $\psi \geq 0$. Observe that Φ and Ψ are smooth as far as $|x_0|$ is small (i.e., C^{∞} or analytic, corresponding to the assumption on the principal symbol), that can be seen by inspecting the remainders of the Taylor expansions. Similarly $\Psi^{1/2}$ is smooth there, which is a consequence of the fact $\psi(x_0, x_0 u) = x_0^2 E(1, u) + \cdots$. Thus the functions $\pi^{\pm}$ are smooth whenever $|x_0|$ is small, say, in $\{|x_0| < a, \ |u| < 1\}$, and satisfy

$$\pm \pi^{\pm}(0, u) = E(1, u)^{1/2} = 1 + \frac{1}{2} E(0, u) + O(|u|^4) \quad \text{as } u \to 0.$$

Therefore, setting $v = (x', \xi')$ and $p^{\pm}(x, \xi) = \xi_0 - x_0 \pi^{\pm}(x_0, v/x_0)$ we have a factorization of the principal symbol

$$- p(x, \xi) = p^+(x, \xi) p^-(x, \xi) \quad \text{near } \bar{\rho} \tag{3.46}$$

which is valid in a certain cone (or, rather, a two-sided wedge having the ξ_0-axis as its edge):

$$|v| = |(x', \xi')| < a |x_0| \tag{3.47}$$

with some $a > 0$.

By virtue of Proposition 3.8 above, every incoming or outgoing bicharacteristic must stay, locally near $\bar{\rho}$, in the cone (3.47). Furthermore, we see, in view of the factorization (3.46), that it must be an integral curve of either H_{p+} or H_{p-} with the limit point $\bar{\rho}$, where the parameter must be changed from s to $t = t^{\pm}$. Then, setting

$$\gamma(t) = (x(t), \xi(t)), \qquad v(t) = (x'(t), \xi'(t))$$

we have $dx_0/dt = 1$ in either case, so that we may take x_0 to be the parameter $t = t^{\pm}$ simultaneously. Then

$$\frac{dx'}{dt}(t) = -\frac{\partial \pi^{\pm}}{\partial \eta'}(t, v/t), \qquad \frac{d\xi'}{dt}(t) = \frac{\partial \pi^{\pm}}{\partial y'}(t, v/t) \tag{3.48}$$

where $v/t \to 0$ as $t \to 0$ (by Proposition 3.8), and similarly for the scalar function $\xi_0 = \xi_0(t)$, that is

$$\frac{d\xi_0}{dt}(t) = \pi^{\pm}(t, v/t) - (v/t)\frac{\partial \pi^{\pm}}{\partial u}(t, v/t) \tag{3.49}$$

and $\xi_0(0) = 0$. Conversely, if we are given a solution of (3.48), then we can reproduce a bicharacteristic by integrating (3.49).

We are thus led to investigate the unique existence and the regularity for the solution of (3.48). Here, the uniqueness is for the one-sided problems, whereas the regularity is for the two-sided ones. Evidently, the regularity of $v(t)$ in (3.48) ensures that of $\xi_0 = \xi_0(t)$ in (3.49). Let us write (3.48) as follows:

$$\frac{dv}{dt} = G(t, v/t), \qquad v/t \to 0 \in \mathbb{R}^N \quad \text{as } t \to 0 \tag{3.50}$$

where $N = 2n$ and

$$G(x_0, u) = G^{\pm}(x_0, u) = \left(-\frac{\partial \pi^{\pm}}{\partial \eta'}(x_0, u), \frac{\partial \pi^{\pm}}{\partial y'}(x_0, u) \right).$$

It then follows that

$$G(0, 0) = 0, \qquad \sigma\big(G_u(0, 0)\big) \subset \{\lambda \in \mathbb{C} \mid \operatorname{Re}\lambda \le 0\} \tag{3.51}$$

where $\sigma\big(G_u(0, 0)\big)$ stands for the spectrum of the Jacobian matrix

$$G_u(0, 0) = \frac{\partial G}{\partial u}(0, 0).$$

Indeed we have

$$\pm \pi_u^{\pm}(0, u) = \frac{1}{2}E_u(0, u) + O(|u|^3)$$

and this implies that $\pm G_u^{\pm}(0, 0)$ is the Hamilton map of the quadratic form $E(0, u)$. Thus the eigenvalues of each one of $G_u^{\pm}(0, u)$ are 0 and $\pm i\mu_j$, for $1 \le j \le k$.

Here we recall a theorem of Briot and Bouquet in [11] and its C^∞ version given by de Hoog and Weiss [19] (see also Proposition 3.7 and its proof). We state it in a form which is convenient to our purpose:

Theorem 3.2 *Let $G = G(x_0, u) \in \mathbb{R}^N$ be a C^∞ (resp. analytic) function near $(x_0, u) = (0, 0) \in \mathbb{R} \times \mathbb{R}^N$ satisfying (3.51). Then the two-sided initial value problem (3.50) admits a C^∞ (resp. analytic) solution $v = v(t)$ near $t = 0$. Furthermore, the uniqueness is valid for each one of the one-sided problems, among solutions of C^1 class except for the end point $t = 0$.*

Proof of Proposition 3.1 The first part of the assertion follows from Theorem 3.2 and Proposition 3.8. In view of (3.48) and (3.49), we see that

$$\frac{dv}{dt}(0) = 0, \qquad \frac{d\xi_0}{dt}(0) = \pi^\pm(0,0) = \pm 1$$

so that the integral curves $\gamma = \gamma^\pm(t)$ of $H_{p\pm}$ with $\gamma^\pm(0) = \bar{\rho}$ satisfy, respectively $Fd\gamma^\pm(0)/dt = \pm d\gamma^\pm(0)/dt$. Since in (3.45) we made a normalization $p \mapsto e^{-1}p$ which implies $F \mapsto e^{-1}F$ therefore we have

$$e(\bar{\rho})F\frac{d\gamma^\pm}{dt}(0) = \pm\lambda\frac{d\gamma^\pm}{dt}(0) \qquad (3.52)$$

before normalization. Namely, the tangent lines at $\bar{\rho}$ of the two curves in Proposition 3.1 are spanned by eigenvectors of the Hamilton map F associated with the two non-vanishing real eigenvalues $\pm\lambda$. $\square$

Chapter 4
Microlocal Energy Estimates and Well-Posedness

Abstract Naturally the structure of the principal symbol $p(x, \xi)$ changes if (x, ξ) varies in the phase space and so does "microlocal" energy estimates. Having proved microlocal energy estimates, the usual next procedure would be to obtain "local" energy estimates by partition of unity. Then one must get rid of the errors caused by the partition of unity. Sometimes it happens that the microlocal energy estimates is too weak to control such errors. In this chapter we propose a new energy estimates for second order operators, much weaker than strictly hyperbolic ones, energy estimates with a gain of H^κ norm for a small $\kappa > 0$. We show that if for every $|\xi'| = 1$ one can find $P_{\xi'}$ which coincides with P in a small conic neighborhood of $(0, 0, \xi')$ for which the proposed energy estimates holds then the Cauchy problem for P is locally solvable in C^∞, which is crucial for our approach to the well-posedness of the Cauchy problem.

4.1 Parametrix with Finite Propagation Speed of Micro Supports

For notational convenience we first introduce

Definition 4.1 We denote by $S^m[\xi_0]$ the set of symbols, polynomial in ξ_0 with coefficients which are pseudodifferential symbols in (x, ξ'), of the form

$$P = \sum_{j=0}^{m} A_j(x, \xi') \xi_0^{m-j}, \quad A_j \in C^\infty(I; S_{phg}^j), \quad A_0 = 1 \text{ (or } - 1) \quad (4.1)$$

where I is some open interval on $\mathbb{R}$ containing 0, which is often omitted in the notation. We call m the order of P and the set of all pseudodifferential operators $\mathrm{Op}(P)$ with $P \in S^m[\xi_0]$ is denoted by $\Psi^m[D_0]$, which are actually differential operators in x_0 with coefficients which are pseudodifferential in x'. In applications it often occurs that $A_j(x, \xi')$ is not defined globally but only defined in an open conic set in $\mathbb{R}^{n+1} \times (\mathbb{R}^n \setminus \{0\})$. We use the same notation in this case also.

© Springer International Publishing AG 2017   71
T. Nishitani, *Cauchy Problem for Differential Operators with Double Characteristics*, Lecture Notes in Mathematics 2202,
DOI 10.1007/978-3-319-67612-8_4

Definition 4.2 Let I be an open interval containing the origin. We denote by $C^k(I; H^p)$ the set of all k-times continuously differentiable functions from I to the Sobolev space $H^p = H^p(\mathbb{R}^n)$ and denote by $C^k_+(I; H^p)$ the set of all $f \in C^k(I; H^p)$ vanishing in $x_0 \leq 0$.

We are concerned with the Cauchy problem for $P \in \Psi^m[D_0]$ with $A_0 = 1$

$$\begin{cases} Pu = f, & f = 0 \text{ in } x_0 \leq 0, \\ u = 0, & \text{in } x_0 \leq 0. \end{cases} \tag{4.2}$$

In this section we first introduce parametrix at $(0, 0, \xi')$, $|\xi'| \neq 0$, of the Cauchy problem for P, with finite propagation speed of micro supports following [71]. Then we show that if parametrix with finite propagation speed of micro supports exists at every $(0, 0, \xi')$ then the Cauchy problem for P is locally solvable near the origin $(0, 0) \in \mathbb{R} \times \mathbb{R}^n$. For two open conic sets W_i in $\mathbb{R}^n \times (\mathbb{R}^n \setminus \{0\})$ we denote $W_1 \Subset W_2$ if $W_1 \cap \{|\xi'| = 1\}$ is relatively compact in $W_2 \cap \{|\xi'| = 1\}$. We also denote by $\overline{W}_i$ the closure of W_i in $\mathbb{R}^n \times (\mathbb{R}^n \setminus \{0\})$.

Notation 4.1 We use the same letter a to denote both $a(x, \xi')$ and $a(x, D')$ if there is no confusion. According to this abbreviation $a_1 a_2 \cdots a_l$ denotes either $a_1(x, D')a_2(x, D') \cdots a_l(x, D')$ or $a_1(x, \xi')a_2(x, \xi') \cdots a_l(x, \xi')$. On the other hand $\mathrm{Op}(a_1 a_2 \cdots a_l)$ is abbreviated to $[a_1 a_2 \cdots a_l]$.

Definition 4.3 Let I be an open interval containing the origin. We say $R \in \mathscr{R}(I)$ if R is a linear operator which maps $\cap_{j=0}^\ell C^j_+(I; H^{q-j})$ into $\cap_{j=0}^\ell C^j_+(I; H^{p-j})$ for any $\ell \in \mathbb{N}$ and $p, q \in \mathbb{R}$ such that

$$\sum_{j=0}^\ell \|D_0^j Rf(t, \cdot)\|_{p-j}^2 \leq c_{\ell p q} \sum_{j=0}^\ell \int^t \|D_0^j f(\tau, \cdot)\|_{q-j}^2 d\tau, \quad t \in I.$$

We denote $\Psi^{-\infty}(I) = C^\infty(I; \mathrm{Op} S^{-\infty})$. We often write $\mathscr{R}$ and $\Psi^{-\infty}$ dropping I, which always assumed to be an interval containing the origin.

Definition 4.4 ([71]) We say that G is a parametrix of $P \in \Psi^m[D_0]$ at $(0, \rho')$, $\rho' = (\tilde{x}', \tilde{\xi}')$ with finite propagation speed of micro supports with loss of β ($\beta + m \geq 0$) derivatives (parametrix at $(0, \rho')$ with β loss, for short) if G is a linear operator from $C^0_+((-\tau, \tau); H^{s+\beta-j})$ into $\cap_{j=0}^{m-1} C^j_+((-\tau, \tau); H^{s-j})$ for any $s \in \mathbb{R}$ with some $\tau > 0$ and one can find a conic neighborhood Γ of ρ' and an open interval $0 \in I \subset (-\tau, \tau)$ such that;

(i) we have $PGh - h \in \Psi^{-\infty}(I) + \mathscr{R}(I)$ for any $h = h(x', \xi') \in S^0(\mathbb{R}^n \times \mathbb{R}^n)$ with $\mathrm{supp}\, h \subset \Gamma$,

(ii) for any $s \in \mathbb{R}$ there is $c_s > 0$ such that for any $f \in C^0_+(I; H^{s+\beta})$

$$\sum_{j=0}^{m-1} \|D_0^j Gf(t, \cdot)\|_{s-j}^2 \leq c_s \int^t \|f(\tau, \cdot)\|_{s+\beta}^2 d\tau, \quad t \in I,$$

(iii) for any conic neighborhoods $\Gamma_1 \Subset \Gamma_2 \Subset \Gamma$ of ρ' there exists an open interval $0 \in J \subset I$ such that for any $h_i(x', \xi') \in S^0, i = 1, 2$ supported in Γ_1 and $\Gamma \setminus \overline{\Gamma}_2$ respectively one has

$$D_0^j h_2 G h_1 \in \mathcal{R}(J), \quad j = 0, 1, \ldots, m - 1.$$

Lemma 4.1 *Assume that G verifies* (i) *and* (ii). *Then for any $\ell \in \mathbb{N}$ and $s \in \mathbb{R}$ and for any $f \in \cap_{j=0}^{\ell} C_+^j(I; H^{s+\beta-j})$ one has*

$$\sum_{j=0}^{\ell+m-1} \|D_0^j G h f(t, \cdot)\|_{s-j}^2 \le c_{s\ell} \sum_{j=0}^{\ell} \int_0^t \|D_0^j f(\tau, \cdot)\|_{s+\beta-j}^2 d\tau,$$

$$\sum_{j=0}^{\ell+m} \int_0^t \|D_0^j G h f(\tau, \cdot)\|_{s-j}^2 d\tau \le c_{s\ell} \sum_{j=0}^{\ell} \int_0^t \|D_0^j f(\tau, \cdot)\|_{s+\beta-j}^2 d\tau.$$

(4.3)

Proof We prove the first estimate. Assuming the estimate for $\ell = p$ we prove the corresponding estimate for $\ell = p+1$. It suffices to show that $\|D_0^{p+m} G h f(t)\|_{s-(p+m)}^2$ is bounded by the right-hand side with $\ell = p+1$. Since P is a monic polynomial in D_0 of degree m one can write

$$D_0^{p+m} = Q_1 P + Q_2, \quad Q_2 = \sum_{j=0}^{m-1} B_j D_0^j \qquad (4.4)$$

with $B_j \in S^{p+m-j}$ where $Q_1 \in \Psi^p[D_0]$. From (i) of Definition 4.4 it follows that $D_0^{p+m} G h f = Q_1(h + S + R)f + Q_2 G h f$ with $S \in \Psi^{-\infty}$ and $R \in \mathcal{R}$. Here note that from $D_0^j f(t) = i \int_0^t D_0^{j+1} f(\tau) d\tau$ it follows

$$\|D_0^j f(t)\|_s^2 \le C \int_0^t \|D_0^{j+1} f(\tau)\|_s^2 d\tau. \qquad (4.5)$$

Taking this into account we have

$$\|Q_1(h + S)f\|_{s-(p+m)}^2 \le C \sum_{j=0}^{p+1} \int_0^t \|D_0^j(h + S)f\|_{s-m-j}^2$$

which is bounded by the right-hand with $\ell = p + 1$ because of $\beta \ge -m$. Since the other terms $\|Q_1 R f\|_{s-(p+m)}^2$ and $\|Q_2 G h f\|_{s-(p+m)}^2$ are easily estimated then the assertion is proved. The proof of the second estimate is similar. $\qquad \square$

Applying Lemma 4.1 it is easy to check that if $S \in \Psi^{-\infty}$ and $R \in \mathcal{R}$ we have

$$D_0^k S G h, \quad D_0^k G h S, \quad D_0^k G h R \in \mathcal{R}, 0 \le k \le m - 1,$$

$$D_0^k R G h \in \mathcal{R}, 0 \le k \le m.$$

(4.6)

Let Γ_i and $h_i(x', \xi') \in S^0$ be as above then repeating the same arguments proving (4.3) we obtain for any $p \in \mathbb{R}$ and any $f \in \cap_{j=0}^{\ell} C_+^j(I; H^{q-j})$

$$\sum_{j=0}^{\ell+m-1} \|D_0^j h_2 G h_1 f(t, \cdot)\|_{p-j}^2 \le c_{\ell pq} \sum_{j=0}^{\ell} \int^t \|D_0^j f(\tau, \cdot)\|_{q-j}^2 d\tau,$$

$$\sum_{j=0}^{\ell+m} \int^t \|D_0^j h_2 G h_1 f(\tau, \cdot)\|_{p-j}^2 d\tau \le c_{\ell pq} \sum_{j=0}^{\ell} \int^t \|D_0^j f(\tau, \cdot)\|_{q-j}^2 d\tau.$$

(4.7)

Definition 4.5 Let $P, \tilde{P} \in \Psi^m[D_0]$. We say $P \equiv \tilde{P}$ at $(0, \rho')$ if one can write

$$P - \tilde{P} = \sum_{j=1}^{m} B_j(x, D') D_0^{m-j}$$

with $B_j \in S^j$ which are in $S^{-\infty}$ in a conic neighborhood of ρ' uniformly in x_0 for small $|x_0|$.

Lemma 4.2 *Let $\tilde{P} \equiv P$ at $(0, \rho')$ and G be a parametrix of P at $(0, \rho')$ with β loss. Then G is also a parametrix of $\tilde{P}$ at $(0, \rho')$ with β loss.*

Proof Let $h(x', \xi') \in S^0$ be supported in a small conic neighborhood of ρ' and write $\tilde{P} G h = (\tilde{P} - P) G h + P G h$. Since $(\tilde{P} - P) G h \in \mathscr{R}((-\tau, \tau))$ with some $\tau > 0$ by (4.6) the proof is immediate. $\square$

Proposition 4.1 *Let $P_i \in \Psi^{m_i}[D_0]$, $i = 1, 2$. If each P_i has a parametrix at $(0, \rho')$ with β_i loss then $P_1 P_2$ has a parametrix at $(0, \rho')$ with $\beta_1 + \beta_2$ loss.*

Proof Let G_i be parametrices of P_i at $(0, \rho')$ with β_i loss then by definition there is a conic neighborhood Γ of ρ' such that (i) and (iii) holds for both G_i. We fix a conic neighborhood $\tilde{\Gamma} \Subset \Gamma$ of ρ' and $\phi(x', \xi') \in S^0$ such that $\phi = 1$ on $\tilde{\Gamma}$ and supported in Γ. Then we show that $G = G_2 \phi G_1$ is a parametrix of $P_1 P_2$ at $(0, \rho')$ with $\beta_1 + \beta_2$ loss. We first check (i). Let $h(x', \xi') \in S^0$ be such that $\operatorname{supp} h \subset \tilde{\Gamma}$. By (i) for P_2 we have $P_1 P_2 G h = P_1(\phi + S + R) G_1 h$ with $S \in \Psi^{-\infty}$ and $R \in \mathscr{R}$ where $P_1 R G_1 h \in \mathscr{R}$ by (4.6). Write

$$P_1 S G_1 h = S(h + \tilde{S} + \tilde{R}) + [P_1, S] G_1 h$$

where the right-hand side belongs to $\Psi^{-\infty} + \mathscr{R}$ by (4.6). On the other hand writing $P_1 \phi G_1 h = \phi P_1 G_1 h + [P_1, \phi] G_1 h$ it is clear that $P_1 \phi G_1 h - h \in \Psi^{-\infty} + \mathscr{R}$ because of (4.6) and $\phi = 1$ on the support of h and this proves (i) for $P_1 P_2$.

We next examine (ii) with $\beta = \beta_1 + \beta_2$. Applying (4.4) with $P = P_2$ one can write $D_0^j G = Q_1 P_2 G_2 \phi G_1 + Q_2 G_2 \phi G_1$. From (ii) for G_i it is clear that

$$\|Q_2 G_2 \phi G_1 f(t)\|_{s-j}^2 \le C \sum_{v=0}^{m_2-1} \|D_0^v G_2 \phi G_1 f(t)\|_{s-v}^2 \le C' \int^t \|f(\tau)\|_{s+\beta_1+\beta_2}^2 d\tau.$$

Note $Q_1 P_2 G_2 \phi G_1 = Q_1(\phi + S + R)G_1$ with $S \in \Psi^{-\infty}$ and $R \in \mathcal{R}$. For $j \leq m_1 + m_2 - 1$ it follows from (ii) for G_1 that

$$\|Q_1(\phi + S + R)G_1 f(t)\|_{s-j}^2 \leq C \sum_{\nu=0}^{j-m_2} \|D_0^\nu(\phi + S + R)G_1 f(t)\|_{s+\beta_2-\nu}^2$$

$$\leq C' \int_0^t \|f(\tau)\|_{s+\beta_1+\beta_2}^2 d\tau.$$

Thus $G = G_2\phi G_1$ verifies (ii) with $m = m_1 + m_2$ and $\beta = \beta_1 + \beta_2$.

Finally we show (iii). Let Γ_i be conic neighborhoods of ρ' such that $\Gamma_1 \Subset \Gamma_2 \Subset \tilde{\Gamma}$ and let $h_i(x', \xi') \in S^0$, $i = 1, 2$ be supported in Γ_1 and $\tilde{\Gamma} \setminus \overline{\Gamma}_2$ respectively. Take conic neighborhoods Γ_i, $i = 3, 4, 5$ of ρ' so that $\Gamma_1 \Subset \Gamma_3 \Subset \Gamma_4 \Subset \Gamma_5 \Subset \Gamma_2$. Apply (4.4) with $P = P_2$ to get $h_2 D_0^{j+k} G h_1 = h_2 Q_1 P_2 G_2\phi G_1 h_1 + h_2 Q_2 G_2\phi G_1 h_1$ where $k \leq m_1 + m_2 - 1$. Take $\theta(x', \xi') \in S^0$ which is 1 on $\overline{\Gamma}_3$ and supported in Γ_4 and write

$$h_2 Q_2 G_2 \phi G_1 h_1 = h_2 Q_2 G_2 \theta \phi G_1 h_1 + h_2 Q_2 G_2(1 - \theta)\phi G_1 h_1$$

where we note that for any $N \in \mathbb{N}$ there is $\psi_N(x', \xi') \in S^0$ supported in $\Gamma \setminus \overline{\Gamma}_3$ such that $(1 - \theta)\#\phi - \psi_N \in S^{-N}$. Then taking (ii), (iii) for G_i into account we see that

$$\|h_2 Q_2 G_2(1 - \theta)\phi G_1 h_1 f(t)\|_{p-j}^2 \leq C \int_0^t \|f(\tau)\|_q^2 d\tau. \tag{4.8}$$

Let $\psi(x', \xi') \in S^0$ be such that $\psi = 1$ on $\overline{\Gamma}_5$ supported in Γ_2 and set $\hat{\theta} = (1 - \psi)\phi$. Write $h_2 Q_2 G_2 \theta\phi G_1 h_1 = h_2 Q_2 \hat{\theta} G_2 \theta\phi G_1 h_1 + h_2 Q_2(1 - \hat{\theta})G_2 \theta\phi G_1 h_1$ and note that

$$\operatorname{supp} \hat{\theta} \subset \Gamma \setminus \overline{\Gamma}_5, \quad \operatorname{supp} \theta \subset \Gamma_4, \quad 1 - \hat{\theta} = 0 \quad \text{on} \quad \operatorname{supp} h_2. \tag{4.9}$$

Then it follows from (ii) and (iii) for G_i that $\|h_2 Q_2 G_2 \theta\phi G_1 h_1 f(t)\|_{p-j}^2$ is bounded by the right-hand side of (4.8) again. We turn to consider $h_2 Q_1 P_2 G_2 \phi G_1 h_1 = h_2 Q_1(\phi + S + R)G_1 h_1$ with $S \in \Psi^{-\infty}$ and $R \in \mathcal{R}$. It is clear from (4.6) that

$$\|h_2 Q_1(S + R)G_1 h_1 f(t)\|_{p-j}^2 \leq C \sum_{\nu=0}^{j} \int_0^t \|D_0^\nu f(\tau)\|_{q-\nu}^2 d\tau. \tag{4.10}$$

Write $h_2 Q_1 \phi G_1 h_1 = h_2 Q_1 \phi \hat{\theta} G_1 h_1 + h_2 Q_1 \phi(1 - \hat{\theta})G_1 h_1$. Note (4.9) and $\operatorname{supp} h_1 \subset \Gamma_1$ then thanks to Lemma 4.1 and (4.7) we conclude that $\|h_2 Q_1 \phi G_1 h_1 f(t)\|_{p-j}^2$ is bounded by the right-hand side of (4.10). Therefore combining these estimates we get (iii) for $G = G_2\phi G_1$. $\qquad\square$

Corollary 4.1 *Let $P_i \in \Psi^{m_i}[D_0]$, $i = 1, \ldots, n$ have a parametrix at $(0, \rho')$ with β_i loss. Then $P_1 \cdots P_n$ has a parametrix at $(0, \rho')$ with $\beta_1 + \cdots + \beta_n$ loss.*

Proposition 4.2 *Let $T(x, \xi') \in S^0$ be elliptic at $(0, \rho')$ in the sense that $|T(x, \xi')| \geq c > 0$ in a conic neighborhood of $(0, \rho')$. Let P, $\tilde{P} \in \Psi^m[D_0]$ and assume $PT \equiv T\tilde{P}$ at $(0, \rho')$. If $\tilde{P}$ has a parametrix at $(0, \rho')$ with β loss then P has a parametrix at $(0, \rho')$ with β loss.*

Proof Let $\tilde{G}$ be a parametrix of $\tilde{P}$ at $(0, \rho')$ with β loss. It is well known that there is $\tilde{T} \in S^0$ such that $T \# \tilde{T} - 1 \in S^{-\infty}$ in a conic neighborhood of $(0, \rho')$ (see [34, Theorem 18.1.9]). Let $\phi(x', \xi') \in S^0$ be 1 in a conic neighborhood of ρ' and supported in another small conic neighborhood. Then one can easily check that $G = T\tilde{G}\phi\tilde{T}$ is a parametrix of P at $(0, \rho')$ with β loss. $\qquad\square$

Remark 4.1 Proposition 4.2 can be generalized as follows: assume that there exist $T(x, \xi') \in S^\ell_{1/2,1/2}$ and $\tilde{T}(x, \xi') \in S^{\tilde{\ell}}_{1/2,1/2}$ such that $T\tilde{T} = I$, $\tilde{T}T = I$ and $PT = T\tilde{P}$ where we refer to [33, Chapter VII] for the class $S^m_{1/2,1/2}$. Assume that $\tilde{P}$ has a parametrix at $(0, \rho')$ with β loss, where it is understood that $C^\infty(I; \mathrm{Op}S^{-\infty})$ is replaced by $C^\infty(I; \mathrm{Op}S^{-\infty}_{1/2,1/2})$ in the definition of parametrix, then P has a parametrix at $(0, \rho')$ with $\beta + \ell + \tilde{\ell}$ loss.

We next consider Fourier integral operators associated with local homogeneous canonical transformations preserving the x_0 coordinate. Let χ be a local homogeneous canonical transformation from a conic neighborhood of $(\hat{y}, \hat{\eta})$ to a conic neighborhood of $(\hat{x}, \hat{\xi})$ such that $\chi(\hat{y}, \hat{\eta}) = (\hat{x}, \hat{\xi})$ and $y_0 = x_0$. Since χ preserves the y_0 coordinate a generating function of χ has the form $x_0\eta_0 + g(x, \eta')$ and the amplitude is assumed to be independent of η_0 then the corresponding Fourier integral operator is represented as

$$Fu(x) = \int e^{i(g(x,\eta') - y'\eta')} a(x, y', \eta') \hat{u}(x_0, y') dy' d\eta'$$

in a convenient y' coordinates (see [20, 21, 29, 59]). Therefore one can regard x_0 as a parameter. We assume that F is elliptic near $(\hat{x}, \hat{\xi}, \hat{y}, \hat{\eta})$ and bounded from $H^k(\mathbb{R}^n_{y'})$ to $H^k(\mathbb{R}^n_{x'})$ for every $k \in \mathbb{R}$ uniformly in x_0 near $x_0 = 0$. Let F^* be the adjoint of F then $FF^* \in \mathrm{Op}S^0$ which is elliptic at $(0, \hat{x}', \hat{\xi}')$ so that one can choose $B(x, \xi') \in S^0$ such that $(FF^*)B - 1 \in S^{-\infty}$ in a conic neighborhood of $(\hat{x}', \hat{\xi}')$ uniformly in small x_0. We set $\tilde{F} = F^*B$. We say $K \in \mathscr{L}^{-\infty}$ if there is an open interval I containing 0 such that we have $K \in C^\infty(I; \mathscr{L}(H^p(\mathbb{R}^n_{y'}), H^q(\mathbb{R}^n_{x'})))$ for any $p, q \in \mathbb{R}$ where $\mathscr{L}(H^p, H^q)$ stands for the set of all bounded linear operators from H^p to H^q. We say $K \in \mathscr{L}^{-\infty}$ at $(0, \hat{y}', \hat{\eta}')$ if one can find a conic neighborhood Γ of $(\hat{y}', \hat{\eta}')$ such that for any $h \in S^0$ supported in Γ we have $Fh \in \mathscr{L}^{-\infty}$.

Proposition 4.3 *Notations being as above. Assume that $P(x, D), \tilde{P}(y, D) \in \Psi^m[D_0]$ satisfy $PF = F\tilde{P} + Q$ with $Q = \sum_{j=1}^m F_j D_0^{m-j}$ where $F_j \in \mathscr{L}^{-\infty}$ at $(0, \hat{y}', \hat{\eta}')$. If $\tilde{P}$ has a parametrix at $(0, \hat{y}', \hat{\eta}')$ with β loss then P has a parametrix at $(0, \hat{x}', \hat{\xi}')$ with the same loss.*

Proof Note that χ has the form $\chi(y, \eta) = (y_0, x'(y, \eta'), \eta_0 + \xi_0(y, \eta'), \xi'(y, \eta'))$. Set $\tilde{\chi}_s(y', \eta') = (x'(s, y', \eta'), \xi'(s, y', \eta'))$. Let $\tilde{G}$ be a parametrix of $\tilde{P}$ at $(0, \hat{y}', \hat{\eta}')$ with β loss verifying Definition 4.4 with $\tilde{\Gamma}$. Choose $\phi_i \in S^0$ which are 1 in a small conic neighborhood of $(\hat{y}', \hat{\eta}')$ supported in $\tilde{\Gamma}$ such that $\text{supp}\,\phi_1 \Subset \{\phi_2 = 1\}$ and take a conic neighborhood Γ of $(\hat{x}', \hat{\xi}')$ such that $\Gamma \Subset \tilde{\chi}_s(\{\phi_1 = 1\})$ for small $|s|$. We show that $G = F\phi_2 \tilde{G}\phi_1 \tilde{F}$ is a parametrix of P at $(0, \hat{x}', \hat{\xi}')$ with β loss. We start with proving (i). Let $h \in S^0$ be supported in Γ. Since $PF = F\tilde{P} + Q$ we have

$$PGh = F\tilde{P}\phi_2 \tilde{G}\phi_1 \tilde{F}h + Q\phi_2 \tilde{G}\phi_1 \tilde{F}h$$

where $Q\phi_2 \tilde{G}\phi_1 \tilde{F}h \in \mathscr{R}$ because $F_j\phi_2 \in \mathscr{L}^{-\infty}$ and (ii). The first term on the right-hand side is written $F\phi_2 \tilde{P}\tilde{G}\phi_1 \tilde{F}h + F[\tilde{P}, \phi_2]\tilde{G}\phi_1 \tilde{F}h$ where $F[\tilde{P}, \phi_2]\tilde{G}\phi_1 \tilde{F}h \in \mathscr{R}$ thanks to (4.6). Thus it suffices to consider $F\phi_2 \tilde{P}\tilde{G}\phi_1 \tilde{F}h = F\phi_2(\phi_1 + S + R)\tilde{F}h$ with some $S \in \Psi^{-\infty}$ and $R \in \mathscr{R}$. Therefore $F\phi_2 \tilde{P}\tilde{G}\phi_1 \tilde{F}h = F\phi_2 \phi_1 \tilde{F}h$ modulo $\Psi^{-\infty} + \mathscr{R}$. Since $\text{supp}\,h \subset \Gamma \Subset \tilde{\chi}_s(\{\phi_1 = 1\})$ for small $|s|$ we see $(\phi_2\phi_1 - 1)\tilde{F}h \in \mathscr{L}^{-\infty}$ (see [29, Theorem 10.1]). Thus we conclude that G verifies (i) for P. Since it is clear that G satisfies (ii) with β we turn to the condition (iii).

Let $\Gamma_1 \Subset \Gamma_2 \Subset \Gamma$ be conic neighborhoods of $(\hat{x}', \hat{\xi}')$ and let $h_i(x', \xi') \in S^0$, $i = 1, 2$ be supported in Γ_1 and $\Gamma \setminus \overline{\Gamma}_2$ respectively. Choose $\theta_i(y', \eta') \in S^0$ which are supported in a small conic neighborhood of $(\hat{y}', \hat{\eta}')$ so that

$$\Gamma_1 \Subset \tilde{\chi}_s(\{\theta_1 = 1\}), \quad \text{supp}\,\theta_1 \Subset \{\theta_2 = 1\}, \quad \tilde{\chi}_s(\text{supp}\,\theta_2) \Subset \Gamma_2 \qquad (4.11)$$

holds for small $|s|$. We show that $D_0^j h_2 G h_1 \in \mathscr{R}$ for $j = 0, \ldots, m - 1$. Write

$$h_2 G h_1 = h_2 F\phi_2 \tilde{G}\theta_1 \phi_1 \tilde{F}h_1 + h_2 F\phi_2 \tilde{G}[\phi_1, \theta_1]\tilde{F}h_1$$
$$+ h_2 F\phi_2 \tilde{G}\phi_1(1 - \theta_1)\tilde{F}h_1.$$

Since $(1 - \theta_1)\tilde{F}h_1, [\phi_1, \theta_1]\tilde{F}h_1 \in \mathscr{L}^{-\infty}$ it follows from (ii) that

$$D_0^j(h_2 F\phi_2 \tilde{G}\phi_1(1 - \theta_1)\tilde{F}h_1), \quad D_0^j(h_2 F\phi_2 \tilde{G}\phi_1[\phi_1, \theta_1]\tilde{F}h_1) \in \mathscr{R}$$

for $j = 0, \ldots, m - 1$. Thus it suffices to study $D_0^j(h_2 F\phi_2 \tilde{G}\theta_1 \phi_1 \tilde{F}h_1)$. Write

$$h_2 F\phi_2 \tilde{G}\theta_1 \phi_1 \tilde{F}h_1 = h_2 F\phi_2 \theta_2 \tilde{G}\theta_1 \phi_1 \tilde{F}h_1 + h_2 F\phi_2(1 - \theta_2)\tilde{G}\theta_1 \phi_1 \tilde{F}h_1.$$

Then by definition we have $D_0^j(1 - \theta_2)\tilde{G}\theta_1 \in \mathscr{R}$ for $0 \le j \le m - 1$ and hence $D_0^j(h_2 F\phi_2(1 - \theta_2)\tilde{G}\theta_1 \phi_1 \tilde{F}h_1) \in \mathscr{R}$ for $0 \le j \le m - 1$. On the other hand by (4.11) we have $h_2 F\phi_2 \theta_2 \in \mathscr{L}^{-\infty}$ and then from (ii) it follows that $D_0^j(h_2 F\phi_2 \theta_2 \tilde{G}\theta_1 \phi_1 \tilde{F}h_1) \in \mathscr{R}$ for $0 \le j \le m - 1$ which proves the assertion. $\qquad\square$

Theorem 4.1 ([71]) *Assume that $P \in \Psi^m[D_0]$ has a parametrix at $(0, 0, \xi')$ with $\beta(\xi')$ loss for every ξ' with $|\xi'| = 1$. Then the Cauchy problem for P is locally solvable in C^∞ near $(0, 0)$. More precisely there are an open neighborhood ω of the* *origin of* $\mathbb{R}^n$ *and an open interval* I *containing the origin such that for every* $s \in \mathbb{R}$ *with* $s + \beta \geq 0$ *and* $f \in C^0_+(I; H^{s+\beta})$ *there exists* $u \in \cap_{j=0}^{m-1} C^j_+(I; H^{s-j})$ *satisfying* $Pu = f$ *in* $I \times \omega$ *where* $\beta = \sup_{|\xi'|=1} \beta(\xi')$.

Proof By a compactness argument we can find a finite number of open conic neighborhoods W_i of $(0, \xi'_i) \in \mathbb{R}^n \times \mathbb{R}^n$, open intervals $0 \in I_i$ and G_i, β_i such that $\cup_i W_i \supset \{0\} \times (\mathbb{R}^n \setminus \{0\})$ and G_i, W_i, β_i and I_i verify Definition 4.4 with $G = G_i$, $\Gamma = W_i$, $\beta = \beta_i$ and $I = I_i$. Now we take another open conic coverings $\{U_i\}$, $\{V_i\}$ of $\{0\} \times (\mathbb{R}^n \setminus \{0\})$ with $U_i \Subset V_i \Subset W_i$ and a partition of unity $\{h_i(x', \xi')\}$ subordinate to $\{U_i\}$ such that

$$\sum_i h_i(x', \xi') = h(x')$$

where $h(x')$ is equal to 1 in a small neighborhood of the origin. Let us define

$$G = \sum_i G_i h_i$$

then we can conclude that there exist an open interval $0 \in J \subset \cap_i I_i$ and $S_i \in \Psi^{-\infty}(J)$, $R_i \in \mathscr{R}(J)$ such that

$$PG = \sum_i PG_i h_i = \sum h_i + S_i + R_i = h(x') + (S + R)$$

where $S = \sum S_i \in \Psi^{-\infty}(J)$ and $R = \sum R_i \in \mathscr{R}(J)$. Let $\chi_1(x') \in C^\infty_0(\mathbb{R}^n)$ be 1 near the origin such that supp $\chi_1 \Subset \{h = 1\}$. Then it is clear from the definition that there exists $c > 0$ such that

$$\int^t \|\chi_1 R f(\tau, \cdot)\|^2 d\tau \leq c t \int^t \|f(\tau, \cdot)\|^2 d\tau, \quad t \in J$$

for any $f \in C^0_+(J; L^2)$. From the Sobolev embedding theorem (Proposition 1.1) we have $\sup_{x'} |Sf(t, x')| \leq C \|Sf(t, \cdot)\|_{[n/2]+1} \leq C' \|f(t, \cdot)\|$ then it is easy to see

$$\|\chi_1 Sf(t, \cdot)\|^2 \leq C' \|f(t, \cdot)\|^2 \|\chi_1\|^2 \leq \|f(t, \cdot)\|^2/8, \quad t \in J$$

provided $\|\chi_1\|$ is sufficiently small. Thus with $T = -\chi_1(S + R)$ there is $\tau_1 > 0$ such that

$$\int^t \|Tf(\tau, \cdot)\|^2 d\tau \leq \frac{1}{2} \int^t \|f(\tau, \cdot)\|^2 d\tau, \quad t \leq \tau_1$$

for any $f \in C^0_+(J; L^2)$. Therefore $U = \sum_{k=0}^\infty T^k$ defines an operator on $L^2_+(\tilde{J}; L^2)$ where $\tilde{J} = \{x_0 < \tau_1\} \cap J$ such that

$$\int^t \|Uf(\tau, \cdot)\|^2 d\tau \leq C \int^t \|f(\tau, \cdot)\|^2 d\tau, \quad t \in \tilde{J}$$

and we have $(I - T)U = I$. Take $\chi(x') \in C_0^\infty(\mathbb{R}^n)$ which is 1 near the origin such that supp $\chi \Subset \{\chi_1 = 1\}$. Since $\chi PG = \chi(h + (S + R)) = \chi(I - T)$ it follows that

$$\chi PGUf = \chi f, \quad f \in C_+^0(\tilde{J}; L^2)$$

and hence GUf solves the equation $P(GUf) = f$ in a neighborhood of the origin of $\mathbb{R}^{n+1}$. Note that for $f \in C_+^0(\tilde{J}; L^2)$ we have $Uf = 0$ in $x_0 < 0$ and hence $GUf = 0$ in $x_0 < 0$. Denote $\beta = \max_i \beta_i$. Noting that for any $s \in \mathbb{R}$ there is $c_s > 0$ such that

$$\int^t \|Tf(\tau, \cdot)\|_s^2 d\tau \le c_s \int^t \|f(\tau, \cdot)\|^2 d\tau, \quad t \in \tilde{J}, \quad f \in C_+^0(\tilde{J}; L^2)$$

and $Uf = \sum_{k=0}^\infty T^k f = f + TUf$ one has from (ii) of Definition 4.4 that for any $s \in \mathbb{R}$ with $s + \beta \ge 0$

$$\|D_0^j GUf(t, \cdot)\|_{s-j}^2 \le c_s' \left\{ \|D_0^j Gf(t, \cdot)\|_{s-j}^2 + \int^t \|Uf(\tau, \cdot)\|^2 d\tau \right\}$$

$$\le c_s'' \left\{ \int^t \|f(\tau, \cdot)\|_{s+\beta}^2 d\tau + \int^t \|f(\tau, \cdot)\|^2 d\tau \right\}$$

for any $f \in C_+^0(\tilde{J}; H^{s+\beta})$ and $j = 0, 1, \ldots, m - 1$. This proves

$$\|D_0^j GUf(t, \cdot)\|_{s-j}^2 \le C_s \int^t \|f(\tau, \cdot)\|_{s+\beta}^2 d\tau, \quad t \in \tilde{J} \tag{4.12}$$

for $j = 0, 1, \ldots, m - 1$. From (4.12) it follows that $GUf \in \cap_{j=0}^{m-1} C_+^j(\tilde{J}; H^{s-j})$ and hence the assertion. $\square$

4.2   Energy Estimate (E) and Existence of Parametrix

In this section we work with operators $P \in \Psi^2[D_0]$;

$$P = -D_0^2 + A_1(x, D')D_0 + A_2(x, D') \tag{4.13}$$

where $A_j(x, \xi') \in S_{phg}^j$ depends smoothly on x_0. We propose energy estimates for P which ensures that P has a parametrix at $(0, 0, \hat{\xi}')$ with finite propagation speed of micro supports.

Definition 4.6 (E)   We say that $P \in \Psi^2[D_0]$ satisfies (E) if P can be written as

$$P = -\Lambda^2 + B\Lambda + \tilde{Q} \tag{4.14}$$

with

$$\begin{cases} \Lambda = D_0 - \lambda - \lambda_0, \quad \lambda \in S^1 \text{ is real and } \lambda_0 \in S^0, \\ B = b + b_0, \quad b \in S^1 \text{ is real and } b_0 \in S^0, \\ \tilde{Q} = Q + \hat{P}_1, \quad 0 \le Q \in S^2, \quad \hat{P}_1 \in S^1 \end{cases} \tag{4.15}$$

such that we have with some $\bar{\epsilon} > 0, C > 0$

$$(1 - \bar{\epsilon})(Qu, u) + \text{Re}(\hat{P}_1 u, u) \ge -C \|u\|^2 \tag{4.16}$$

and there exist $0 < \kappa < 1$ and $T > 0$ such that with the energy

$$N_s(u) = \|\Lambda u\|_s^2 + \text{Re}(\tilde{Q}u, u)_s + \|u\|_{s+\kappa}^2 + \theta^2 \|u\|_s^2$$

the following estimates holds: for every $s \in \mathbb{R}$ there are $a_s \in S^0$ with $\text{Re}\, a_s \in S^{-1}$ and $C_s > 0, \theta_s > 0$ such that for any $u \in \cap_{j=0}^2 C_+^j([-T, T]; H^{s+2-j})$ and $\theta \ge \theta_s$ and $|t| \le T$ one has

$$e^{-2\theta t} N_s(u(t)) + \theta \int^t e^{-2\theta x_0} N_s(u(x_0)) dx_0$$

$$\le C_s \, \text{Im} \int^t e^{-2\theta x_0} (\langle D' \rangle^s Pu, (\Lambda + a_s) \langle D' \rangle^s u) dx_0. \tag{4.17}$$

Remark 4.2 From (4.16) we have

$$\text{Re}\,(\tilde{Q}u, u) = \bar{\epsilon}(Qu, u) + (((1 - \bar{\epsilon})Q + \text{Re}\,\hat{P}_1)u, u) \ge \bar{\epsilon}\,(Qu, u) - C\|u\|^2$$

and hence $\text{Re}\,(\tilde{Q}u, u)_s + \theta^2 \|u\|_s^2 \ge \bar{\epsilon}\,\text{Re}\,(Qu, u)_s$ for $\theta \ge \theta_s$. Since $Q \ge 0$ we have $\text{Re}\,(Qu, u)_s \ge -c_s \|u\|_s^2$ from the Fefferman-Phong inequality and one can find $\theta_s > 0$ such that for $\theta \ge \theta_s$

$$\text{Re}\,(\tilde{Q}u, u)_s + \theta^2 \|u\|_s^2 \ge \theta^2 \|u\|_s^2 / 2.$$

Remark 4.3 Note that $|\text{Re}(\tilde{Q}u, u)_s - \text{Re}((Q + \text{Re}\,\hat{P}_1)u, u)_s| \le C_s'' \|u\|_s^2$ holds with some $C_s'' > 0$ and hence $N_s(u)$ in the definition (E) can be replaced by

$$\|\Lambda u\|_s^2 + \text{Re}\,((Q + \text{Re}\,\hat{P}_1)u, u)_s + \|u\|_{s+\kappa}^2 + \theta^2 \|u\|_s^2.$$

Theorem 4.2 ([85]) *Let $\hat{P} \in \Psi^2[D_0]$ and $0 \ne \hat{\xi}' \in \mathbb{R}^n$ be fixed. Assume that one can find $P \in \Psi^2[D_0]$ verifying (E) such that $P \equiv \hat{P}$ at $(0, 0, \hat{\xi}')$ and for any $s \in \mathbb{R}$ there is $C_s > 0$ such that there exists $u \in \cap_{j=0}^1 C_+^j(I; H^{s-j})$ verifying $Pu = f$ and*

$$N_s(u(t)) \le C_s \int^t \|f(\tau)\|_s^2 d\tau \tag{4.18}$$

for any $f \in C^0_+(I; H^s)$. Then $\hat{P}$ has a parametrix at $(0, 0, \hat{\xi}')$ with $-\kappa$ loss.

Proof of Theorem 4.2 Thanks to Lemma 4.2, defining $G : f \mapsto u$ by $Gf = u$, it suffices to show that G is a parametrix of P at $(0, 0, \hat{\xi}')$. By definition (i) is obviously satisfied. If $a_s \in S^0$ one has $|\,\|(\Lambda + a_s)\langle D'\rangle^s u\|^2 - \|\Lambda u\|^2_s| \leq C_s \|u\|^2_s$ then the inequality (4.18) implies that $\|\Lambda u(t)\|^2_s + \|u(t)\|^2_{s+\kappa} \leq C'_s \int^t \|Pu(x_0, \cdot)\|^2_s dx_0$ with some $C'_s > 0$. From this it follows that

$$\|D_0 u(t)\|^2_{s-1} + \|u(t)\|^2_s \leq C''_s \int^t \|Pu(x_0, \cdot)\|^2_{s-\kappa} dx_0 \tag{4.19}$$

which proves that G verifies (ii).

Thus it remains to show that G verifies (iii). In the rest of this section we prove a key proposition to the proof of (iii) and we finish the proof in the next section. We now assume that P verifies (E) and recall that the principal symbol p is

$$p(x, \xi) = -(\xi_0 - \lambda)^2 + b(\xi_0 - \lambda) + Q = -(\xi_0 - \lambda - b)(\xi_0 - \lambda) + Q.$$

Definition 4.7 ([40]) We say that $f(x, \xi') \in S^0$, depending on x_0 smoothly, is of spatial type if f satisfies with some $c_1 > 0$ and $0 < c < 1$

$$\{\xi_0 - \lambda, f\} \geq c_1 > 0,$$

$$\{\xi_0 - \lambda - b, f\}\{\xi_0 - \lambda, f\} \geq c_1 > 0, \tag{4.20}$$

$$\{f, Q\}^2 \leq 4c\,\{\xi_0 - \lambda - b, f\}\{\xi_0 - \lambda, f\}Q$$

uniformly in x_0 for small $|x_0|$.

Let $\chi(x') \in C^\infty_0(\mathbb{R}^n)$ be equal to 1 near $x' = 0$ and vanish in $|x'| \geq 1$. Set

$$d_\epsilon(x', \xi'; \bar{\rho}') = \left(\chi(x' - y')|x' - y'|^2 + |\xi'\langle\xi'\rangle^{-1} - \eta'\langle\eta'\rangle^{-1}|^2 + \epsilon^2\right)^{1/2}$$

with $\bar{\rho}' = (y', \eta')$ and denote for small $\nu > 0, \epsilon > 0$

$$f(x, \xi'; \bar{\rho}') = x_0 - 2\nu\epsilon + \nu d_\epsilon(x', \xi'; \bar{\rho}'). \tag{4.21}$$

Then it is easy to examine that f is a symbol of spatial type for $0 < \nu \leq \nu_0$ if ν_0 is small.

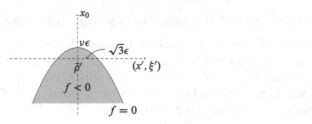

Indeed since $0 \leq Q \in S^2$ it follows from the Glaeser's inequality (see [57, Lemma 4.3.8], [33, Lemma 7.7.2]) that

$$\{Q, \nu d_\epsilon\}^2 \leq C\nu^2 Q \qquad (4.22)$$

with $C > 0$ independent of $\epsilon > 0$ and $\nu > 0$. On the other hand since it is clear that there is $C > 0$ independent of ν, ϵ such that

$$1 - C\nu \leq \{\xi_0 - \lambda, f\} \leq 1 + C\nu, \quad |\{b, f\}| \leq C\nu \qquad (4.23)$$

we get the assertion taking small ν_0, independent of $\bar{\rho}'$ and $\epsilon > 0$. Recall that one can write $P = -\Lambda^2 + B\Lambda + \tilde{Q}$ where

$$\Lambda = \xi_0 - \lambda - \lambda_0, \quad B = b + b_0, \quad \tilde{Q} = Q + \hat{P}_1, \quad \hat{P}_1 \in S^1, \quad \lambda_0, b_0 \in S^0.$$

Let $f(x, \xi')$ be of spatial type. We define ϕ following [40, 41] by

$$\phi(x, \xi') = \begin{cases} \exp(1/f(x, \xi')) & \text{if } f < 0, \\ 0 & \text{otherwise.} \end{cases} \qquad (4.24)$$

Denote

$$\phi_1 = f^{-1}\{\Lambda, f\}^{1/2}\phi, \quad m = f\{\Lambda, f\}^{-1/2}$$

then it is clear that

$$\phi, \; \phi_1 \in S^0, \; m = f\{\Lambda, f\}^{-1/2} \in S^0, \; \phi - m \# \phi_1 \in S^{-1}. \qquad (4.25)$$

Regularizing ϕ, ϕ_1 we consider

$$\Phi(x, \xi') = w_\delta^{-l}(\xi')\phi(x, \xi'), \quad \Phi_1(x, \xi') = w_\delta^{-l}(\xi')\phi_1(x, \xi')$$

where $w_\delta = \langle \delta \xi' \rangle$ with $0 \leq \delta \leq 1$ and $l \geq 0$. In what follows we simply write $a \in S^k$ for implying that a belongs to S^k uniformly in $\delta > 0$. We denote $\Psi = \langle \delta D' \rangle^{-l}\phi(x, D')$ and consider

$$\text{Im}(P\Psi u, \Lambda \Psi u)_s = \text{Im}([P, \Psi]u, \Lambda \Psi u)_s + \text{Im}(\Psi P u, \Lambda \Psi u)_s. \qquad (4.26)$$

To estimate $\mathrm{Im}([P, \Psi]u, \Lambda\Psi u)_s$ we follow the arguments in [41].

Definition 4.8 Let T, S be two real valued functionals on $C^1(\mathbb{R}; C_0^\infty(\mathbb{R}^n))$ depending on x_0. We say $T \sim S$ and $T \preceq S$ if there exists $C > 0$ independent of δ such that

$$|T(u) - S(u)| \leq C(\|\Lambda u\|^2_{s-\kappa/2} + \|u\|^2_{s+\kappa/2} + \|\Lambda\Psi u\|^2_s + \|\Psi u\|^2_{s+\kappa}),$$

$$T(u) \leq S(u) + C(\|\Lambda u\|^2_{s-\kappa/2} + \|u\|^2_{s+\kappa/2} + \|\Lambda\Psi u\|^2_s + \|\Psi u\|^2_{s+\kappa})$$

holds uniformly in x_0 for small $|x_0|$.

Consider $([\Lambda^2, \Psi]u, \Lambda\Psi u)_s = (\Lambda[\Lambda, \Psi]u, \Lambda\Psi u)_s + ([\Lambda, \Psi]\Lambda u, \Lambda\Psi u)_s$ where

$$(\Lambda[\Lambda, \Psi]u, \Psi\Lambda u)_s \sim -i\frac{d}{dx_0}([\Lambda, \Psi]u, \Psi\Lambda u)_s + ([\Lambda, \Psi]u, \Lambda\Psi\Lambda u)_s$$

because $\lambda \in S^1$ is real. Since it is clear that $([\Lambda, \Psi]u, [\Lambda, \Psi]\Lambda u)_s \sim 0$ we have

$$-\mathrm{Im}(\Lambda[\Lambda, \Psi]u, \Psi\Lambda u)_s \sim \frac{d}{dx_0}\mathrm{Re}([\Lambda, \Psi]u, \Psi\Lambda u)_s - \mathrm{Im}([\Lambda, \Psi]u, \Psi\Lambda^2 u)_s.$$

We next examine that

$$-\mathrm{Im}([\Lambda, \Psi]\Lambda u, \Lambda\Psi u)_s \sim -\|\Lambda\Phi_1 u\|^2_s. \tag{4.27}$$

Since $[\Lambda, \Psi] = [\Lambda, \langle\delta D'\rangle^{-l}]\langle\delta D'\rangle^l\Psi + i\mathrm{Op}(\Phi\{\Lambda, f\}f^{-2}) + T_1$ with $T_1 \in \mathrm{Op}S^{-1}$ we see

$$-\mathrm{Im}([\Lambda, \Psi]\Lambda u, \Lambda\Psi u)_s \sim -\mathrm{Re}([\Phi\{\Lambda, f\}f^{-2}]\Lambda u, \Lambda\Psi u)_s$$

$$\sim -\mathrm{Re}([\Phi\{\Lambda, f\}f^{-2}]\Lambda u, \Lambda\Phi u)_s$$

because $\Psi = \Phi(x, D') + R$ with $R \in \mathrm{Op}S^{-1}$. Note that from (4.25)

$$\mathrm{Re}([\{\Lambda, f\}f^{-2}\Phi]\Lambda u, \Lambda\Phi u)_s \sim \mathrm{Re}([\{\Lambda, f\}f^{-2}\Phi]\Lambda u, \Lambda m\Phi_1 u)_s$$

$$\sim \mathrm{Re}([\{\Lambda, f\}f^{-2}\Phi]\Lambda u, m\Lambda\Phi_1 u)_s.$$

Since we have $m\#\langle\xi'\rangle^{2s}\#(\{\Lambda, f\}f^{-2}\Phi) = \langle\xi'\rangle^{2s}\#\Phi_1 + T$ with $T \in S^{2s-1}$ and $(\Phi_1\Lambda u, \Lambda\Phi_1 u)_s \sim \|\Lambda\Phi_1 u\|^2_s$ one concludes the assertion (4.27) easily. Therefore we have

$$-\mathrm{Im}([\Lambda^2, \Psi]u, \Lambda\Psi u)_s \sim \frac{d}{dx_0}\mathrm{Re}([\Lambda, \Psi]u, \Psi\Lambda u)_s$$

$$- \mathrm{Im}([\Lambda, \Psi]u, \Psi\Lambda^2 u)_s - \|\Lambda\Phi_1 u\|^2_s. \tag{4.28}$$

We turn to consider $([B\Lambda, \Psi]u, \Lambda\Psi u)_s = (B[\Lambda, \Psi]u, \Lambda\Psi u)_s + ([B, \Psi]\Lambda u, \Lambda\Psi u)_s$. With $B_s = \langle D'\rangle^s B\langle D'\rangle^{-s}$ we write

$$(B[\Lambda, \Psi]u, \Lambda\Psi u)_s = 2i((\text{Im}B_s)\langle D'\rangle^s[\Lambda, \Psi]u, \langle D'\rangle^s\Lambda\Psi u)$$
$$+(B_s^*\langle D'\rangle^s[\Lambda, \Psi]u, \langle D'\rangle^s\Lambda\Psi u)$$
$$= 2i((\text{Im}B_s)\langle D'\rangle^s[\Lambda, \Psi]u, \langle D'\rangle^s\Lambda\Psi u) + ([\Lambda, \Psi]u, B\Lambda\Psi u)_s.$$

Since $\text{Im}B_s = \text{Im}B + r$ with $r \in S^0$ where $\text{Im}B \in S^0$ then we have

$$|((\text{Im}B_s)\langle D'\rangle^s[\Lambda, \Psi]u, \langle D'\rangle^s\Lambda\Psi u)| \leq C\|\Lambda u\|_{s-\kappa/2}^2 + C\|u\|_{s+\kappa/2}^2 \sim 0.$$

Since $|\text{Im}([\Lambda, \Psi]u, B[\Lambda, \Psi]u)_s| \leq C\|u\|_s^2 \sim 0$ we have

$$\text{Im}(B[\Lambda, \Psi]u, \Lambda\Psi u)_s \sim \text{Im}([\Lambda, \Psi]u, \Psi B\Lambda u)_s.$$

We turn to consider $\text{Im}([B, \Psi]\Lambda u, \Lambda\Psi u)_s$. Since $[B, \Psi] = T_1\Psi - i\,\text{Op}(\{b, \phi\}w_\delta^{-l}) + T_0$ with $T_1 \in S^0$ and $T_0 \in S^{-1}$ one sees

$$\text{Im}([B, \Psi]\Lambda u, \Lambda\Psi u)_s \sim -\text{Re}([\{b, \phi\}w_\delta^{-l}]\Lambda u, \Lambda\Psi u)_s$$
$$\sim \text{Re}([\{b, f\}f^{-2}\Phi]\Lambda u, \Lambda\Phi u)_s$$
$$\sim \text{Re}([\{b, f\}f^{-2}\Phi]\Lambda u, m\Lambda\Phi_1 u)_s.$$

Noting that $mf^{-2}\Phi = \{\Lambda, f\}^{-1}\Phi_1$ we get

$$\text{Im}([B, \Psi]\Lambda u, \Lambda\Psi u)_s \sim \text{Re}([\{\Lambda, f\}^{-1}\{b, f\}]\Lambda\Phi_1 u, \Lambda\Phi_1 u)_s$$

and hence

$$\begin{aligned}\text{Im}([B\Lambda, \Psi]u, \Lambda\Psi u)_s &\sim \text{Im}([\Lambda, \Psi]u, \Psi B\Lambda u)_s \\ &+ \text{Re}([\{\Lambda, f\}^{-1}\{b, f\}]\Lambda\Phi_1 u, \Lambda\Phi_1 u)_s.\end{aligned} \tag{4.29}$$

We finally consider $([\tilde{Q}, \Psi]u, \Lambda\Psi u)_s$. Noting that $\hat{P}_1 \in S^1$ we have

$$|([\hat{P}_1, \Psi]u, \Lambda\Psi u)_s| \leq C\|u\|_{s+\kappa/2}^2 + C\|\Lambda u\|_{s-\kappa/2}^2 \sim 0.$$

Since $[Q, \Psi] = [Q, \langle\delta D'\rangle^{-l}]\langle\delta D'\rangle^l\Psi - i\,\text{Op}(\{Q, \phi\}w_\delta^{-l}) + T_1 + T_0$ with real $T_1 \in S^0$ and $T_0 \in S^{-1}$ we see

$$\text{Im}([\tilde{Q}, \Psi]u, \Lambda\Psi u)_s \sim -\text{Re}([\{Q, \phi\}w_\delta^{-l}]u, \Lambda\Psi u)_s$$
$$\sim -\text{Re}([\{Q, \phi\}w_\delta^{-l}]u, \Lambda\Phi u)_s.$$

Indeed with $H = \text{Op}(\{Q, w_\delta^{-l}\}w_\delta^l) \in S^1$ it follows that $(H\Psi u, \Lambda\Psi u)_s \leq \|H\Psi u\|_s^2 + \|\Lambda\Psi u\|_s^2$. For $\|H\Psi u\|_s^2 = (H\Psi u, H\Psi u)_s$ we note

$$C\,\text{Re}(Q\Psi u, \Psi u)_s - \text{Re}(H\Psi u, H\Psi u)_s$$
$$= (\text{Op}(C\langle\xi'\rangle^{2s}Q - H^2\langle\xi'\rangle^{2s})\Psi u, \Psi u) + (T\Psi u, \Psi u)$$

with $T \in S^{2s}$. Choosing C so that $CQ - H^2 \geq 0$, which is possible by the Glaeser inequality, one has from the Fefferman-Phong inequality that

$$\text{Re}(H\Psi u, H\Psi u)_s \leq C\,\text{Re}(Q\Psi u, \Psi u)_s + C'\|\Psi u\|_s^2.$$

Taking $\text{Re}\,(\tilde{Q}u, u)_s + \theta^2\|u\|_s^2 \geq \bar{\epsilon}\,\text{Re}\,(Qu, u)_s$ into account we have $\|H\Psi u\|_s^2 \sim 0$. From (4.25) it follows

$$-\text{Re}([\{Q, \phi\}w_\delta^{-l}]u, \Lambda\Phi u)_s \sim -\text{Re}([\{Q, \phi\}w_\delta^{-l}]u, \Lambda m\Phi_1 u)_s$$
$$\sim -\text{Re}([\{Q, \phi\}w_\delta^{-l}]u, m\Lambda\Phi_1 u)_s$$
$$- \text{Re}([\{Q, \phi\}w_\delta^{-l}]u, [\Lambda, m]\Phi_1 u)_s.$$

Here we remark that with $K = \text{Op}(\{Q, \phi\}w_\delta^{-l})$

$$\text{Re}(Ku, [\Lambda, m]\Phi_1 u)_s = \text{Re}(\Phi_1[\Lambda, m]^*\langle D'\rangle^{2s}Ku, u) \sim 0$$

because $\Phi_1[\Lambda, m]^*\langle D'\rangle^{2s}K = -i\,\text{Op}(\Phi_1\{\Lambda, m\}\langle\xi'\rangle^{2s}K) + T$ with $T \in S^{2s}$. Thus we get

$$-\text{Re}(Ku, \Lambda\Phi u)_s \sim -\text{Re}(m\langle D'\rangle^{2s}Ku, \Lambda\Phi_1 u).$$

Noting that $m\#\langle\xi'\rangle^{2s}\#K = -\langle\xi'\rangle^{2s}\#(\{Q, f\}\{\Lambda, f\}^{-1})\#\Phi_1 + T$ with $T \in S^{2s}$ we obtain

$$\text{Im}([\tilde{Q}, \Phi]u, \Lambda\Phi u)_s \sim \text{Re}([\{Q, f\}\{\Lambda, f\}^{-1}]\Phi_1 u, \Lambda\Phi_1 u)_s. \qquad (4.30)$$

Here we summarize what we have proved in

Proposition 4.4 *We have*

$$\text{Im}([P, \Psi]u, \Lambda\Psi u)_s \sim \frac{d}{dx_0}\text{Re}([\Lambda, \Psi]u, \Psi\Lambda u)_s - \|\Lambda\Phi_1 u\|_s^2$$
$$- \text{Im}([\Lambda, \Psi]u, \Psi\Lambda^2 u)_s + \text{Im}([\Lambda, \Psi]u, \Psi B\Lambda u)_s$$
$$+ \text{Re}([\{\Lambda, f\}^{-1}\{b, f\}]\Lambda\Phi_1 u, \Lambda\Phi_1 u)_s \qquad (4.31)$$
$$+ \text{Re}([\{Q, f\}\{\Lambda, f\}^{-1}]\Phi_1 u, \Lambda\Phi_1 u)_s.$$

Noting that the sum of the third and fourth terms on the right-hand side of (4.31) is $\operatorname{Im}([\Lambda,\Psi]u,\Psi(-\Lambda^2+B\Lambda)u)_s$ and taking into account the identity $-\Lambda^2+B\Lambda = P-\tilde{Q}$ we study $-\operatorname{Im}([\Lambda,\Psi]u,\Psi\tilde{Q}u)_s$. Write $\tilde{Q}=Q+\operatorname{Re}\hat{P}_1+i\operatorname{Im}\hat{P}_1$ and note $\operatorname{Re}([\Lambda,\Psi]u,\Psi\operatorname{Im}\hat{P}_1u)_s\sim-\operatorname{Im}(\operatorname{Op}(\{\Lambda,\Phi\})u,\Phi\operatorname{Im}\hat{P}_1u)_s\sim0$. Hence one has

$$-\operatorname{Im}([\Lambda,\Psi]u,\Psi\tilde{Q}u)_s\sim-\operatorname{Im}([\Lambda,\Psi]u,\Psi(Q+\operatorname{Re}\hat{P}_1)u)_s$$
$$=-\operatorname{Im}(\Psi^*\langle D'\rangle^{2s}[\Lambda,\Psi]u,(Q+\operatorname{Re}\hat{P}_1)u).$$

Here note that $[\Lambda,\Psi]=[\Lambda,\langle\delta D'\rangle^{-l}]\langle\delta D'\rangle^l\Psi-i\operatorname{Op}(\{\Lambda,\phi\}w_\delta^{-l})+T_1+T_0$ and $\Psi^*=\Phi+iR_1+R_0$ with real $T_1,R_1\in S^{-1}$ and $T_0,R_0\in S^{-2}$. Then it follows that

$$\Psi^*\langle D'\rangle^{2s}[\Lambda,\Psi]=\Psi^*\langle D'\rangle^{2s}K\Psi-i\operatorname{Op}(\{\Lambda,\phi\}w_\delta^{-l}\Phi\langle\xi'\rangle^{2s})+A_1+A_0$$

with real $A_1\in S^{2s-1}$ and $A_0\in S^{2s-2}$ where $K=iK_0+K_{-1}$ with $K_j\in S^j$ and $K_0=-\{\Lambda,w_\delta^{-l}\}w_\delta^{-l}$ is real. Since $-\operatorname{Im}((A_1+A_0)u,(Q+\operatorname{Re}\hat{P}_1)u)\sim-\operatorname{Im}(A_0u,Qu)\sim0$ it follows that

$$-\operatorname{Im}([\Lambda,\Psi]u,\Psi\tilde{Q}u)_s\sim-\operatorname{Re}(\Phi_1u,\Phi_1(Q+\operatorname{Re}\hat{P}_1)u)_s$$
$$-\operatorname{Im}(K\Psi u,\Psi(Q+\operatorname{Re}\hat{P}_1)u)_s$$

because $\{\Lambda,\phi\}w_\delta^{-l}\Phi\langle\xi'\rangle^{2s}=\Phi_1\#\langle\xi'\rangle^{2s}\#\Phi_1+R$ with $R\in S^{2s-2}$. Note

$$(\Phi_1u,\Phi_1(Q+\operatorname{Re}\hat{P}_1)u)_s=(\Phi_1u,(Q+\operatorname{Re}\hat{P}_1)\Phi_1u)_s$$
$$+(\Phi_1u,[\Phi_1,Q+\operatorname{Re}\hat{P}_1]u)_s$$
$$\sim(\Phi_1u,(Q+\operatorname{Re}\hat{P}_1)\Phi_1u)_s+(\Phi_1u,[\Phi_1,Q]u)_s$$

where we have $\operatorname{Re}(\Phi_1u,[\Phi_1,Q]u)_s\sim0$ since $[\Phi_1,Q]+i\operatorname{Op}(\{\Phi_1,Q\})\in\operatorname{Op}S^{-1}$. Similarly we have

$$\operatorname{Im}(K\Psi u,\Psi(Q+\operatorname{Re}\hat{P}_1)u)_s\sim\operatorname{Re}(K_0\Psi u,(Q+\operatorname{Re}\hat{P}_1)\Psi u)_s.$$

Consider $C\operatorname{Re}((Q+\operatorname{Re}\hat{P}_1)v,v)_s-\operatorname{Re}((Q+\operatorname{Re}\hat{P}_1)v,K_0v)_s$. Note that

$$\operatorname{Re}((C-K_0)\langle D'\rangle^{2s}(Q+\operatorname{Re}\hat{P}_1))$$
$$=\operatorname{Op}(\sqrt{C-K_0}\langle\xi'\rangle^s\#(Q+\operatorname{Re}\hat{P}_1)\#\sqrt{C-K_0}\langle\xi'\rangle^s)+R$$

with $R\in S^{2s-2}$ where $C>0$ is chosen so that $C-K_0$ is positive. Taking into account $\operatorname{Re}((Q+\operatorname{Re}\hat{P}_1)v,v)\geq-C'\|v\|^2$ we see that

$$C\operatorname{Re}((Q+\operatorname{Re}\hat{P}_1)\Psi u,\Psi u)_s-\operatorname{Re}((Q+\operatorname{Re}\hat{P}_1)\Psi u,K_0\Psi u)_s\geq-C\|\Psi u\|_s^2$$

and noting $|\mathsf{Re}(\tilde{Q}u,u)_s - \mathsf{Re}((Q + \mathsf{Re}\,\hat{P}_1)u,u)_s| \leq C''_s\|u\|_s^2$ one has $\mathsf{Im}(K\Psi u,\Psi(Q + \mathsf{Re}\,\hat{P}_1)u)_s \sim 0$. Thus we have

$$\begin{aligned}
\mathsf{Im}([\Lambda,\Psi]u,\Psi(-\Lambda^2 + B\Lambda)u)_s &= \mathsf{Im}([\Lambda,\Psi]u,\Psi Pu)_s - \mathsf{Im}([\Lambda,\Psi]u,\Psi\tilde{Q}u)_s \\
&\sim \mathsf{Im}([\Lambda,\Psi]u,\Psi Pu)_s - \mathsf{Re}(\Phi_1 u,(Q + \mathsf{Re}\,\hat{P}_1)\Phi_1 u)_s.
\end{aligned} \tag{4.32}$$

From (4.28)–(4.30) and (4.32) we conclude that

$$\begin{aligned}
\mathsf{Im}([P,\Psi]u,\Lambda\Psi u)_s \sim &\frac{d}{dx_0}\mathsf{Re}([\Lambda,\Psi]u,\Psi\Lambda u)_s - \|\Lambda\Phi_1 u\|_s^2 \\
&- \mathsf{Re}((Q + \mathsf{Re}\,\hat{P}_1)\Phi_1 u,\Phi_1 u)_s \\
&+ \mathsf{Re}\big([\{\Lambda,f\}^{-1}\{b,f\}]\Lambda\Phi_1 u,\Lambda\Phi_1 u\big)_s \\
&+ \mathsf{Re}\big([\{\Lambda,f\}^{-1}\{Q,f\}]\Phi_1 u,\Lambda\Phi_1 u\big)_s + \mathsf{Im}([\Lambda,\Psi]u,\Psi Pu)_s.
\end{aligned}$$

Remark that with $\alpha = (1 - \{\Lambda,f\}^{-1}\{b,f\})^{1/2}$ and $\beta = \alpha^{-1}\{\Lambda,f\}^{-1}\{Q,f\}$ one has

$$-\|\Lambda\Phi_1 u\|_s^2 + \mathsf{Re}\big([\{\Lambda,f\}^{-1}\{b,f\}]\Lambda\Phi_1 u,\Lambda\Phi_1 u\big)_s \sim -\|\alpha\Lambda\Phi_1 u\|_s^2,$$

$$-\|\alpha\Lambda\Phi_1 u\|_s^2 - \mathsf{Re}((Q + \mathsf{Re}\,\hat{P}_1)\Phi_1 u,\Phi_1 u)_s + \mathsf{Re}([\{\Lambda,f\}^{-1}\{Q,f\}]\Phi_1 u,\Lambda\Phi_1 u)_s$$

$$\sim -\|(\alpha\Lambda - \beta/2)\Phi_1 u\|_s^2 - \mathsf{Re}\big([Q + \mathsf{Re}\,\hat{P}_1 - \beta^2/4]\Phi_1 u,\Phi_1 u\big)_s$$

because $\alpha\#\alpha - \alpha^2 \in S^{-2}$, $\beta\#\beta - \beta^2 \in S^0$ and $\alpha\#\beta - \alpha\beta \in S^0$. Here we note that f is of spatial type implies that $Q - \beta^2/4 \geq 0$;

$$\begin{aligned}
Q - \frac{1}{4}\beta^2 &= \frac{1}{4}\{\Lambda,f\}^{-2}\big(4\{\Lambda,f\}^2 Q - \alpha^{-2}\{Q,f\}^2\big) \\
&= \frac{1}{4}\{\Lambda,f\}^{-2}\alpha^{-2}\big(4\{\Lambda,f\}^2\alpha^2 Q - \{Q,f\}^2\big) \\
&= \frac{1}{4}\{\Lambda,f\}^{-2}\alpha^{-2}\big(4\{\Lambda - b,f\}\{\Lambda,f\} - \{Q,f\}^2\big) \geq 0.
\end{aligned}$$

Our aim is to prove that micro supports of solutions propagate with *finite* speed and no need of precise estimate for the speed so that we can use a small space $\bar{\epsilon} > 0$ assumed in the inequality (4.16). To do so we specify our ϕ. Let ϕ be defined with f in (4.21). From (4.22) and (4.23) it is clear that one can choose $\nu > 0$ in (4.21) small so that

$$\frac{1}{4}\beta^2 = \frac{1}{4}\left(\frac{\{\Lambda,f\}^{-2}\{Q,f\}^2}{1 - \{\Lambda,f\}^{-1}\{b,f\}}\right) \leq \bar{\epsilon}Q$$

where $\bar{\epsilon}$ is given in (4.16). Then there are $C, C' > 0$ such that

$$\mathsf{Re}\big((Q + \mathsf{Re}\,\hat{P}_1 - \beta^2/4)\Phi_1 u, \Phi_1 u\big)_s = \big(((1 - \bar{\epsilon})Q + \mathsf{Re}\,\hat{P}_1)\Phi_1 u, \Phi_1 u\big)_s$$
$$+ \big((\bar{\epsilon}Q - \beta^2/4)\Phi_1 u, \Phi_1 u\big)_s$$
$$\geq -C\|\Phi_1 u\|_s^2 \geq -C'\|u\|_s^2.$$

Remark 4.4 If there is no space in the inequality (4.16), that is if $\bar{\epsilon} = 0$, it seems to be hard to control the term

$$\mathsf{Re}\big((Q + \mathsf{Re}\,\hat{P}_1 - \beta^2/4)\Phi_1 u, \Phi_1 u\big)_s$$

which is one of the main troubles when we study the well-posedness of the Cauchy problem under the "non strict" IPH condition.

Remark 4.5 Assume that $\lambda = 0$ and $b = 0$. Assume also that one can find $0 \leq d(x', \xi') \in S^0$ such that $\{d, Q\} = 0$. Then (4.20) holds for $f = x_0 - 2\nu\epsilon + \nu M d$ with any $M > 0$. In some cases, using this fact, we can obtain a sharp estimate of the propagation of micro supports (see [76] for example).
We summarize what we have proved in

Lemma 4.3 *Let ϕ be defined by f in (4.21) and let $\Psi = \langle\delta D'\rangle^{-l}\phi(x, D')$. Then there exists $\nu_0 > 0$ such that for any $0 < \nu \leq \nu_0$ and $l \geq 0$ we have*

$$\mathsf{Im}([P, \Psi]u, \Lambda\Psi u)_s \preceq \frac{d}{dx_0}\mathsf{Re}([\Lambda, \Psi]u, \Psi\Lambda u)_s + \mathsf{Im}([\Lambda, \Psi]u, \Psi P u)_s$$

uniformly in δ.

We turn to $\mathsf{Im}(P\Psi u, \Lambda\Psi u)_s$. Let $\tilde{\Lambda} = \Lambda + a$ with $a \in S^0$ where $\mathsf{Re}\,a \in S^{-1}$. Then repeating similar arguments as above we see

$$\mathsf{Im}(\langle D'\rangle^s[P, \Psi]u, a\langle D'\rangle^s\Psi u) \sim \mathsf{Im}(\langle D'\rangle^s[P, \Phi]u, a\langle D'\rangle^s\Phi u) \sim 0$$

because $[P, \Phi] = -i\,\mathsf{Op}(\{P, \Phi\}) + R, R \in \mathsf{Op}S^0$ and hence

$$\mathsf{Im}(\langle D'\rangle^s P\Psi u, a\langle D'\rangle^s\Psi u) \sim \mathsf{Im}(\langle D'\rangle^s\Psi P u, a\langle D'\rangle^s\Psi u)$$
$$\geq -C\|\Psi P u\|_s^2 - C\|\Psi u\|_s^2$$

so that $\mathsf{Im}(\langle D'\rangle^s P\Psi u, \Lambda\langle D'\rangle^s\Psi u) \succeq \mathsf{Im}(\langle D'\rangle^s P\Psi u, \tilde{\Lambda}\langle D'\rangle^s\Psi u) - C\|\Psi P u\|_s^2$. Write

$$\mathsf{Im}(\langle D'\rangle^s P\Psi u, \Lambda\langle D'\rangle^s\Psi u) = \mathsf{Im}(\langle D'\rangle^s P\Psi u, \langle D'\rangle^s\Lambda\Psi u) + \mathsf{Im}(\langle D'\rangle^s\Psi P u, [\Lambda, \langle D'\rangle^s]\Psi u)$$
$$+ \mathsf{Im}(\langle D'\rangle^s[P, \Psi]u, [\Lambda, \langle D'\rangle^s]\Psi u).$$

Noting $[\Lambda, \langle D'\rangle^s] + i\,\mathrm{Op}\{\Lambda, \langle \xi'\rangle^s\} \in \mathrm{Op}S^{s-2}$ the same reasoning as above shows

$$\mathrm{Im}(\langle D'\rangle^s[P, \Psi]u, [\Lambda, \langle D'\rangle^s]\Psi u) \sim \mathrm{Im}(\langle D'\rangle^s[P, \Phi]u, [\Lambda, \langle D'\rangle^s]\Phi u) \sim 0$$

and then we conclude $\mathrm{Im}(P\Psi u, \Lambda\Psi u)_s \geq \mathrm{Im}(\langle D'\rangle^s P\Psi u, \tilde{\Lambda}\langle D'\rangle^s\Psi u) - C\|\Psi Pu\|_s^2$. Recalling $\mathrm{Im}(P\Psi u, \Lambda\Psi u)_s = \mathrm{Im}([P, \Psi]u, \Lambda\Psi u)_s + \mathrm{Im}(\Psi Pu, \Lambda\Psi u)_s$ from Lemma 4.3 we have

$$\mathrm{Im}(\langle D'\rangle^s P\Psi u, \tilde{\Lambda}\langle D'\rangle^s\Psi u) \preceq \frac{d}{dx_0}\mathrm{Re}([\Lambda, \Psi]u, \Psi\Lambda u)_s + C\|\Psi Pu\|_s^2. \qquad (4.33)$$

Proposition 4.5 *Assume that P verifies Definition 4.6. Let ϕ be as in Lemma 4.3 and let $u \in \cap_{j=0}^1 C_+^j((-T, T); H^{\ell-j})$ and $Pu \in C_+^0((-T, T); H^{\ell'})$ with some ℓ, ℓ'. Assume that $u \in L^\infty((-T, T); H^{s+\kappa/2})$ and $\Lambda u \in L^\infty((-T, T); H^{s-\kappa/2})$ and $\phi Pu \in L^\infty((-T, T); H^s)$. Then we have $\phi u \in L^\infty((-T, T); H^{s+\kappa})$ and $\Lambda\phi u \in L^\infty((-T, T); H^s)$. Moreover we have for any $|t| \leq T$*

$$N_s(\phi u(t)) + \int^t N_s(\phi u)dx_0 \leq C_s\left(N_{s-\kappa/2}(u(t)) + \int^t \left(\|\phi Pu\|_s^2 + N_{s-\kappa/2}(u)\right)dx_0\right).$$

Proof Take $l \geq 0$ so that $l + \min\{\ell - 2, \ell'\} \geq s$ then it is clear that

$$\Psi u = \langle \delta D'\rangle^{-l}\phi(x, D')u \in \cap_{j=0}^2 C^j((-T, T); H^{s+2-j}).$$

We multiply (4.33) by $e^{-2\theta x_0}$ and integrate the resulting inequality in x_0. Note that the modulo term yields

$$C_1 \int^t e^{-2\theta x_0}N_{s-\kappa/2}(u)dx_0 + C_2 \int^t e^{-2\theta x_0}N_s(\Psi u)dx_0.$$

The second term can be cancelled against the left-hand side of (4.17) taking θ large. Then fixing such a θ we have from (4.17) for any $|t| \leq T$

$$N_s(\langle \delta D'\rangle^{-l}\phi(x, D')u(t)) + \int^t N_s(\langle \delta D'\rangle^{-l}\phi(x, D')u)dx_0$$

$$\leq C_s\left(N_{s-\kappa/2}(u(t)) + \int^t \left(\|\langle \delta D'\rangle^{-l}\phi(x, D')Pu\|_s^2 + N_{s-\kappa/2}(u)\right)dx_0\right).$$

From the assumption the right-hand side is bounded uniformly both in $t \in (-T, T)$ and in $\delta > 0$. Then letting $\delta \to 0$ we obtain the assertion. $\square$

4.3   Energy Estimate (E) and Finite Propagation Speed of Micro Supports

In this section we prove Theorem 4.2 completing the proof of (iii) for G. Let ϕ_ϵ be defined with $f_\epsilon = x_0 - 2\nu\epsilon + \nu d_\epsilon(x', \xi'; \bar\rho')$ by (4.21). Note that if $\epsilon_1 < \epsilon_2$ then we have

$$\{f_{\epsilon_1}(x, \xi'; \bar\rho') < 0\} \cap \{x_0 \geq 0\} \Subset \{f_{\epsilon_2}(x, \xi'; \bar\rho') < 0\} \cap \{x_0 \geq 0\}.$$

We start with

Lemma 4.4 *Assume that* $u \in \cap_{j=0}^1 C_+^j(I; H^{\ell+1-j})$ *and* $Pu \in C^0(I; H^{\ell'})$, $\phi_{\epsilon_0} Pu \in C^0(I; H^{s_0})$ *with some* $\ell, \ell', s_0 \in \mathbb{R}$. *Then for every* $0 < \epsilon < \epsilon_0$ *we have*

$$\phi_\epsilon u \in \cap_{j=0}^1 C_+^j(I; H^{s+1-j})$$

for all $s \leq s_0 + \kappa/2 - 2$. *Moreover for* $s \leq s_0 - \kappa/2$ *we have*

$$N_s(\phi_\epsilon u(t)) \leq c_s\Big(N_\ell(u(t)) + \int^t \big(\|\phi_{\epsilon_0} Pu(x_0, \cdot)\|_{s_0}^2 + \|Pu(x_0, \cdot)\|_{\ell'}^2\big) dx_0\Big).$$

Proof Take $\epsilon < \epsilon_j < \epsilon_0$ so that $\epsilon < \cdots < \epsilon_{j+1} < \epsilon_j < \cdots < \epsilon_0$ and $\epsilon_j \to \epsilon$ as $j \to \infty$. We write $\phi_j = \phi_{\epsilon_j}$ and $f_j = f_{\epsilon_j}$ in this proof. Let s' be $s' \leq \ell$. By induction on j we show that

$$\begin{aligned} N_{s'+j\kappa/2}(\phi_j u(t)) \leq c_s\Big(N_\ell(u(t)) + \int^t \big(\|\phi_{\epsilon_0} Pu(x_0, \cdot)\|_{s_0}^2 + \|Pu(x_0, \cdot)\|_{\ell'}^2 \\ + N_\ell(u(x_0))\big)dx_0\Big). \end{aligned} \tag{4.34}$$

We choose $g_j(x, \xi') \in S^0$, depends on x_0 smoothly, so that $\operatorname{supp} g_j \subset \{f_j < 0\}$ and $\{f_{j+1} < 0\} \subset \{g_j = 1\}$. Noting $\phi_{j+1} P g_j u = \phi_{j+1} g_j Pu + \phi_{j+1}[P, g_j]u$ we apply Proposition 4.5 with $s = s' + (j+1)\kappa/2$, $\phi = \phi_{j+1}$ and $u = g_j u$ to get

$$N_{s'+(j+1)\kappa/2}(\phi_{j+1} g_j u) + \int^t N_{s'+(j+1)\kappa/2}(\phi_{j+1} g_j u) dx_0$$

$$\leq C(s', j)\Big(N_{s'+j\kappa/2}(g_j u) + \int^t \big(\|\phi_{j+1} P g_j u\|_{s'+(j+1)\kappa/2}^2 + N_{s'+j\kappa/2}(g_j u)\big) dx_0\Big).$$

Since $\phi_{j+1} g_j \equiv \phi_{j+1}$ modulo $S^{-\infty}$ and $\|\phi_{j+1}[P, g_j]u\|_{s'+(j+1)\kappa/2}^2 \leq CN_\ell(u)$ it is clear that

$$\|\phi_{j+1} P g_j u\|_{s'+(j+1)\kappa/2}^2 \leq C\|\phi_{j+1} Pu\|_{s'+(j+1)\kappa/2}^2 + C(s', j)\big(N_\ell(u) + \|Pu\|_{\ell'}^2\big)$$

and hence the right-hand side is bounded by

$$\tilde{C}(s',j)\left(\|\phi_{\epsilon_0}Pu\|^2_{s'+(j+1)\kappa/2} + \|Pu\|^2_{\ell'} + N_\ell(u)\right)$$

since $\phi_{j+1} \equiv k_j\phi_{\epsilon_0}$ modulo $S^{-\infty}$ with some $k_j \in S^0$. We next consider $N_{s'+j\kappa/2}(g_ju)$ and $N_{s'+(j+1)\kappa/2}(\phi_{j+1}g_ju)$. Since $g_j \equiv k'_j\phi_j$ modulo $S^{-\infty}$ with some $k'_j \in S^0$ it follows that

$$N_{s'+j\kappa/2}(g_ju) \le c(s',j)\left(N_{s'+j\kappa/2}(\phi_ju) + N_\ell(u)\right),$$

$$N_{s'+(j+1)\kappa/2}(\phi_{j+1}u) \le c(s',j)\left(N_{s'+(j+1)\kappa/2}(\phi_{j+1}g_ju) + N_\ell(u)\right).$$

This proves (4.34) for $j+1$. Taking the largest j_0 with $s' + j\kappa/2 \le s_0$ we get

$$\phi_{j_0}u \in L^\infty(I;H^{s_0+\kappa/2}), \quad \Lambda\phi_{j_0}u \in L^\infty(I;H^{s_0-\kappa/2}).$$

Since $\phi_\epsilon Pu \in L^\infty(I;H^{s_0})$ and there is $k \in S^0$ such that $k\phi_{j_0} \equiv 1$ on the support of ϕ_ϵ we see that

$$D_0^2\phi_\epsilon u \in L^\infty(I;H^{s_0+\kappa/2-2}), \quad D_0\phi_\epsilon u \in L^\infty(I;H^{s_0+\kappa/2-1})$$

which proves $\phi_\epsilon u \in \cap^1_{j=0}C^j_+(I;H^{s_0+\kappa/2-1-j})$ and hence the assertion.   □

Let Γ_i $(i = 0,1,2)$ be open conic sets in $\mathbb{R}^n \times (\mathbb{R}^n \setminus \{0\})$ such that $\Gamma_2 \cap \{|\xi'| = 1\}$ is relatively compact and $\Gamma_0 \Subset \Gamma_1 \Subset \Gamma_2$. Let us take $h_i(x',\xi') \in S^0$ with supp $h_1 \subset \Gamma_0$ and supp $h_2 \subset \Gamma_2 \setminus \overline{\Gamma}_1$.

Proposition 4.6 *Assume that P verifies Definition 4.6 and let Γ_i be as above. Then there is $\bar{t} = \bar{t}(\Gamma_i) > 0$ such that for any $h_i(x',\xi') \in S^0$ being as above and for any $q \in \mathbb{R}$, $c > 0$ and $p \in \mathbb{R}$ one can find $C > 0$ such that for any $u \in \cap^1_{j=0}C^j_+(I;H^{q-j})$ and $f \in C^0_+(I;H^q)$ satisfying $Pu = h_1f$ and $N_q(u(t)) \le c \int^t \|f(x_0,\cdot)\|^2_q dx_0$ we have*

$$\sum_{j=0}^1 \|D_0^jh_2u(t)\|^2_{p-j} \le C\int^t \|f(x_0,\cdot)\|^2_q dx_0, \quad t \le \bar{t}. \tag{4.35}$$

Proof Recall $f_\epsilon = x_0 - 2\nu\epsilon + \nu d_\epsilon(x',\xi';\bar{\rho}')$. It is clear that one can choose a small $\hat{\epsilon} > 0$ so that $\{x_0 \ge 0\} \cap \{f_{\hat{\epsilon}} \le 0\} \cap (\mathbb{R} \times \Gamma_0) = \emptyset$ for every $\bar{\rho}' = (y',\eta') \in \mathbb{R}^{2n} \setminus \Gamma_1$, $\eta' \ne 0$. We take $0 < \epsilon < \hat{\epsilon}$ so close to $\hat{\epsilon}$. From the compactness arguments we can find finitely many $\bar{\rho}'_i = (y'_i,\eta'_i) \in \overline{\Gamma}_2 \setminus \Gamma_1$, $i = 1,\dots,M$ such that with $\bar{t} = \nu\epsilon/2$ we have

$$\overline{\Gamma}_2 \setminus \Gamma_1 \Subset \bigcup_{i=1}^M \{(x',\xi') \mid f_\epsilon(\bar{t},x',\xi';\bar{\rho}'_i) \le 0\}.$$

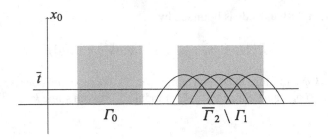

Let $\phi_{i,\epsilon}$ be defined by (4.24) with $f_\epsilon(x, \xi'; \bar{\rho}_i')$. Then we have $h_2 \equiv k \sum \phi_{i,\epsilon}$ modulo $S^{-\infty}$ for $0 \le x_0 \le \bar{t}$ with some $k \in S^0$ since $\sum \phi_{i,\epsilon} > 0$ on $[0, \bar{t}] \times \text{supp}\, h_2$. Note that $\phi_{i,\epsilon} h_1 f \in C^0([0, \bar{t}], H^{p+\kappa/2})$ for all $p \in \mathbb{R}$. Now we apply Lemma 4.4 with $\phi_{\epsilon_0} = \phi_{i,\hat{\epsilon}}, \phi_\epsilon = \phi_{i,\epsilon}, s_0 = p + \kappa/2$ so that we obtain for $t \le \bar{t}$

$$N_p(\phi_{i,\epsilon} u(t)) \le c_p\Big(N_q(u(t)) + \int^t \big(\|\phi_{i,\hat{\epsilon}} h_1 f\|_{p+\kappa/2}^2 + \|h_1 f\|_q^2\big) dx_0\Big).$$

Since $\phi_{i,\hat{\epsilon}} h_1 \in S^{-\infty}$ and $N_q(u(t)) \le c \int^t \|f(x_0, \cdot)\|_q^2 dx_0$ by assumption we have for $t \le \bar{t}$

$$N_p(\phi_{i,\epsilon} u(t)) \le c_p\Big(N_q(u(t)) + \int^t \|f(x_0, \cdot)\|_q^2 dx_0\Big) \le c_{pq} \int^t \|f(x_0, \cdot)\|_q^2 dx_0.$$

Summing up over i we have the following estimate

$$\|D_0 h_2 u(t)\|_{p-1}^2 + \|h_2 u(t)\|_p^2 \le C_{pq} \int^t \|f(x_0, \cdot)\|_q^2 dx_0, \quad j = 0, 1, \quad t \le \bar{t}$$

which is (4.35).   □

Completion of the Proof of Theorem 4.2 We show that G verifies (iii). Let Γ_i and $h_i(x', \xi')$ be as above. Let $Pu = h_1 f$ where $f \in \cap_{j=0}^\ell C_+^j(I; H^{q-j})$. Then it is enough to prove

$$\sum_{j=0}^{\ell+1} \|D_0^j h_2 u(t)\|_{p-j}^2 \le C \sum_{j=0}^\ell \int^t \|D_0^j f(x_0, \cdot)\|_{q-j}^2 dx_0, \quad t \le \bar{t}. \tag{4.36}$$

Thanks to Proposition 4.6 it follows that (4.36) holds for $\ell = 0$. We now assume that (4.36) holds for $\ell = r$. From $Pu = h_1 f$ one can write $D_0^{r+2} h_2 u = \sum_{j=0}^{r+1} B_j D_0^j u + D_0^r h_2 h_1 f$ where $B_j \in \text{Op}S^{r+2-j}$. Choose $\theta(x', \xi') \in S^0$ supported in $\Gamma_2 \setminus \overline{\Gamma}_1$ such that $\theta = 1$ on the support of h_2. From the induction hypothesis it follows that

$$\sum_{j=0}^{r+1} \|B_j D_0^j \theta u(t)\|_{p-(r+2)}^2 \le C \sum_{j=0}^{r+1} \|D_0^j \theta u(t)\|_{p-j}^2 \le C_{rpq} \sum_{j=0}^r \int^t \|D_0^j f\|_{q-j}^2 dx_0$$

for $t \leq \bar{t}$. On the other hand noting $B_j(1 - \theta) \in \mathrm{Op}S^{-\infty}$ and assuming

$$\sum_{j=0}^{\ell+1} \|D_0^j u(t)\|_{s-j}^2 \leq \sum_{j=0}^{\ell} C_{\ell s} \int^t \|f(\tau)\|_{s-\kappa-j}^2 d\tau \tag{4.37}$$

for any $f \in \cap_{j=0}^{\ell} C_+^0 (I; H^{s-\kappa-j})$ which will be checked below, we conclude that

$$\sum_{j=0}^{r+1} \|B_j(1 - \theta)D_0^j u\|_{p-(r+2)}^2 \leq C \sum_{j=0}^{r+1} \|D_0^j u\|_{q+\kappa-j}^2 \leq C_{rpq} \sum_{j=0}^{r} \int^t \|D_0^j f\|_{q-j}^2.$$

Noting $h_2 h_1 \in \mathrm{Op}S^{-\infty}$ we have from (4.5)

$$\|h_2 h_1 D_0^r f(t)\|_{p-(r+2)}^2 \leq C \int^t \|D_0^{r+1} f(\tau)\|_{q-r}^2 d\tau$$

and therefore we conclude that (4.35) holds for $\ell = r + 1$.

Finally we check (4.37) which holds for $\ell = 0$ by (4.19). We now assume that (4.37) holds for $\ell = r$. As before one can write $D_0^{r+2} u = \sum_{j=0}^{r+1} B_j D_0^j u + D_0^r f$ where $B_j \in \mathrm{Op}S^{r+2-j}$. From the induction hypothesis it follows that

$$\sum_{j=0}^{r+1} \|B_j D_0^j u(t)\|_{s-(r+2)}^2 \leq C \sum_{j=0}^{r+1} \|D_0^j u(t)\|_{s-j}^2 \leq C_{rs} \sum_{j=0}^{r} \int^t \|D_0^j f(\tau)\|_{s-j}^2 dx_0.$$

Note that the term $\|D_0^r f(t)\|_{s-(r+2)}^2 \leq \|D_0^r f(t)\|_{s-\kappa-(r+1)}^2$ can be estimated by the right-hand side of (4.37) with $\ell = r + 1$ thanks to (4.5). $\qquad\square$

Chapter 5
Cauchy Problem: No Tangent Bicharacteristics

Abstract The main purpose of this chapter is to prove two new results on C^∞ well-posedness mentioned in the end of Sect. 1.4. If there is no transition of spectral type and no tangent bicharacteristics then the Cauchy problem is C^∞ well-posed for P of order m under the strict IPH condition. If the positive trace is zero, the IPH condition is reduced to the Levi condition. In this case we prove that when p is of spectral type 2 on Σ and there is no tangent bicharacteristics, the Levi condition is necessary and sufficient in order that the Cauchy problem is C^∞ well-posed for P of order m. The same result holds for second order differential operators of spectral type 1 on Σ with 0 positive trace. To prove these assertions using the results in Chap. 2 we first derive microlocal energy estimates. Then making use of the idea developed in Chap. 4 we prove the C^∞ well-posedness of the Cauchy problem.

5.1   Main Results on Well-Posedness

We study the Cauchy problem for $P = \mathrm{Op}\big(\sum_{j=0}^m A_j(x, \xi')\xi_0^{m-j}\big) \in \Psi^m[D_0]$ with principal symbol $p(x, \xi)$;

$$p(x, \xi) = \sum_{j=0}^m A_{j0}(x, \xi')\xi_0^{m-j}$$

where $A_j(x, \xi') \sim A_{j0} + A_{j1} + \cdots$ and $A_0 = 1$.

Lemma 5.1 *Let Γ be a conic neighborhood of $\rho' = (\hat{x}', \hat{\xi}')$, $\hat{\xi}' \neq 0$ and I be an open interval containing 0. Assume that p has a factorization $p(x, \xi) = p^1(x, \xi)p^2(x, \xi)$ when $(x_0, x', \xi') \in I \times \Gamma$ where*

$$p^\nu(x, \xi) = \sum_{j=0}^{m_\nu} a_j^\nu(x, \xi')\xi_0^{m_\nu - j}, \quad \nu = 1, 2, \quad a_0^\nu = 1$$

© Springer International Publishing AG 2017

T. Nishitani, *Cauchy Problem for Differential Operators with Double Characteristics*, Lecture Notes in Mathematics 2202, DOI 10.1007/978-3-319-67612-8_5

with $a_j^\nu(x, \xi') \in C^\infty(I \times \Gamma)$ which are homogeneous of degree j in ξ'. Assume that p^1 and p^2 have no common zero ξ_0 for $(x_0, x', \xi') \in I \times \Gamma$. Then one can find $P^i \in \Psi^{m_i}[D_0]$, $i = 1, 2$ such that $P \equiv P^1 P^2$ at $(0, \rho')$.

Proof Write

$$P^\nu = \mathrm{Op}\Big(\sum_{j=0}^{m_\nu} A_j^\nu \xi_0^{m_\nu - j} \Big)$$

where $A_0^\nu = 1$ and $A_j^\nu \sim \sum a_{jk}^\nu$ for $j \neq 0$. In the symbol $\sigma(P^1 P^2)$ the coefficient a_{jk}^ν appears for the first time in terms of order $m - k$ where $m = m_1 + m_2$. To make the terms of order $m - k$ in $\sigma(P - P^1 P^2)$ vanish near $(0, \rho')$ we must solve an equation of the form

$$p^2 \sum_{j=0}^{m_1} a_{jk}^1 \xi_0^{m_1 - j} + p^1 \sum_{j=0}^{m_2} a_{jk}^2 \xi_0^{m_2 - j} = q$$

where q is a polynomial of degree less than m in ξ_0 which is homogeneous of degree $m - k$. Denote by $H(x, \xi')$ the coefficient matrix of this equation for the unknowns $\{a_{jk}^\nu\}$, $\nu = 1, 2$, $j = 0, \ldots, m_\nu$ then $\det H(x, \xi')$ is the resultant of p^1 and p^2 as polynomials in ξ_0 which is different from 0 by assumption. Checking the homogeneity in ξ' of entries of the inverse matrix of H we get the assertion. $\square$

Corollary 5.1 *Let $p(x, \xi) \in S^m[\xi_0]$ be the principal symbol of $P \in \Psi^m[D_0]$ and assume that $p(x, \xi_0, \xi') = 0$ has at most double roots at $(0, \rho')$. Then we can find $P^i \in \Psi^2[D_0]$ and $Q^j \in \Psi^1[D_0]$ such that $P \equiv P^1 \cdots P^r \cdot Q^1 \cdots Q^k$ at $(0, \rho')$.*

Proof Let λ_i, $i = 1, \ldots, r$ be the different double roots. Then one can find a factorization

$$p(x, \xi) = \prod_{i=1}^{r} (\xi_0^2 + a_1^i(x, \xi)\xi_0 + a_2^i(x, \xi')) \prod_{j=1}^{k} (\xi_0 - c_j(x, \xi')) = \prod p^i \prod q^j$$

where p^i, $i = 1, \ldots, r$, q^j, $j = 1, \ldots, k$ have no common zero ξ_0 in a conic neighborhood of $(0, \rho')$. Then repeated applications of Lemma 5.1 proves the assertion. $\square$

We recall

Definition 5.1 Let $P \in \Psi^m[D_0]$. Then we define $P_{sub}(x, \xi) \in S^{m-1}(\mathbb{R}^{n+1} \times \mathbb{R}^{n+1})$ which is homogeneous of degree $m - 1$ in ξ such that

$$P - \mathrm{Op}\,(p + P_{sub}) \in \mathrm{Op}S^{m-2}.$$

Let j be fixed and denote $\lambda = \lambda_j$. Then in a conic neighborhood of $(0, \rho')$ one can find a factorization

$$p(x, \xi) = p^1(x, \xi)p^2(x, \xi)$$

where $p^1 = \xi_0^2 + a_1^1(x, \xi')\xi_0 + a_2^1(x, \xi')$ and $p^2(x, \xi) = \sum_{j=0}^{m-2} a_j^2(x, \xi')\xi_0^{m-2-j}$ such that $p^2(0, \hat{x}', \lambda, \hat{\xi}') \neq 0$ and $p^1(0, \hat{x}', \lambda, \hat{\xi}') = 0$. Then by Lemma 5.1 we can find $P^1 \in \Psi^2[D_0]$ and $P^2 \in \Psi^{m-2}[D_0]$ such that $P \equiv P^1 P^2$ at $(0, \rho')$.

Lemma 5.2 *Notations being as above and let* $\rho = (0, \hat{x}', \lambda, \hat{\xi}')$. *Then*

$$P_{sub}(\rho) = p^2(\rho)P_{sub}^1(\rho), \quad \mathrm{Tr}^+ F_p(\rho) = p^2(\rho)\mathrm{Tr}^+ F_{p^1}(\rho).$$

Proof Write $P^1 P^2 = \mathrm{Op}(p^1 + P_{sub}^1 + R^1)\mathrm{Op}(p^2 + P_{sub}^2 + R^2)$ where $R^1 \in S^0$ and $R^2 \in S^{m-2}$. The right-hand side is

$$\mathrm{Op}((p^1 + P_{sub}^1)\#(p^2 + P_{sub}^2) + R)$$

with $R \in S^{m-2}$. This proves that $P_{sub}(x, \xi) = \{p^1, p^2\}/(2i) + p^1 P_{sub}^2 + p^2 P_{sub}^1$. Since p^1 vanishes at ρ of order 2 hence $\{p^1, p^2\}$ vanishes at ρ then the first assertion is clear. Since $\partial_x^\alpha \partial_\xi^\beta p^1(\rho) = 0$ for $|\alpha + \beta| \leq 1$ the second assertion follows immediately. □

Definition 5.2 We say that P satisfies the Levi condition on Σ if

$$P_{sub}(\rho) = 0, \quad \forall \rho \in \Sigma \tag{5.1}$$

and we say that P satisfies the Levi condition near ρ if one can find a conic neighborhood V of ρ such that (5.1) is verified on $V \cap \Sigma$.
Here we recall Theorem 1.12.

Definition 5.3 We say that P satisfies the IPH condition on Σ if

$$\mathrm{Im}\, P_{sub}(\rho) = 0, \quad -\mathrm{Tr}^+ F_p(\rho) \leq P_{sub}(\rho) \leq \mathrm{Tr}^+ F_p(\rho), \quad \forall \rho \in \Sigma \tag{5.2}$$

and say that P satisfies the strict IHP condition on Σ if

$$\mathrm{Im}\, P_{sub}(\rho) = 0, \quad -\mathrm{Tr}^+ F_p(\rho) < P_{sub}(\rho) < \mathrm{Tr}^+ F_p(\rho), \quad \forall \rho \in \Sigma. \tag{5.3}$$

We also say that P satisfies the IPH (resp. strict IPH) condition near ρ if there is a conic neighborhood V such that (5.2) (resp. (5.3)) is satisfied in $V \cap \Sigma$.

Corollary 5.2 *Notations being as above. Then P satisfies the IPH (resp. strict IHP) condition near ρ if and only if P^1 satisfies the IPH (resp. strict IHP) condition near ρ.*

Lemma 5.3 *If P satisfies the IPH or the strict IPH or the Levi condition then P**
satisfies the same condition.

Proof Since $P = \text{Op}(p + P_{sub}) + R$ with $R \in \text{Op}S^{m-2}$ and hence

$$P^* = \text{Op}(p + \bar{P}_{sub}) + \tilde{R}, \quad \tilde{R} \in \text{Op}S^{m-2}. \tag{5.4}$$

Thus we have $(P^*)_{sub} = \bar{P}_{sub}$ and hence $\text{Re}\,P^*_{sub} = \text{Re}\,P_{sub}$ and $\text{Im}\,P^*_{sub} = -\text{Im}\,P_{sub}$
from which the assertion follows immediately. $\qquad\square$

Theorem 5.1 *Assume* (1.5) *and that there is no transition and no tangent bichar-*
acteristics. Then the Cauchy problem for P is C^∞ well-posed near the origin if the
strict IPH condition is satisfied on Σ.

Remark 5.1 If p is of spectral type 1 on Σ then there is no tangent bicharacteristics
by Proposition 3.3. In this case Theorem 5.1 was first proved in [32] and [40]. The
(non strict) IPH condition is not sufficient in general for the C^∞ well-posedness of
the Cauchy problem. An example will be given in the last section of Chap. 8.
Recall that if $\text{Tr}^+F_p = 0$ then the IPH condition reduces to the Levi condition.

Theorem 5.2 *Assume* (1.5) *and that p is of spectral type 2 on Σ and has no tangent*
bicharacteristics. Then the Cauchy problem for P is C^∞ well-posed near the origin
if the Levi condition $P_{sub} = 0$ is satisfied on Σ. If $\text{Tr}^+F_p = 0$ on Σ then the Levi
condition is also necessary.

Theorem 5.3 ([32]) *Let $m = 2$. Assume* (1.5) *and p is of spectral type 1 on*
Σ. Then the Cauchy problem for P is C^∞ well-posed near the origin if the Levi
condition $P_{sub} = 0$ is satisfied on Σ. If $\text{Tr}^+F_p = 0$ on Σ then the Levi condition is
also necessary.

Remark 5.2 Assume that p is of spectral type 1 on Σ and $\text{Tr}^+F_p = 0$. Then from
Lemma 1.7 the quadratic form p_ρ takes the form $p_\rho = -\xi_0^2 + \sum_{j=1}^r \xi_j^2$ in a suitable
symplectic coordinates system. Therefore Σ is an involutive manifold since

$$(T_\rho\Sigma)^\sigma = \langle H_{\xi_0}, H_{\xi_1}, \dots, H_{\xi_r}\rangle \subset \langle H_{\xi_0}, H_{\xi_1}, \dots, H_{\xi_r}\rangle^\sigma = T_\rho\Sigma, \quad \rho \in \Sigma.$$

On the other hand Σ is neither involutive nor symplectic if p is of spectral type 2 on
Σ even though $\text{Tr}^+F_p = 0$ is verified.

Remark 5.3 It is not clear whether one can extend Theorem 5.3 for general m. See
Remark 5.6 where we give some comments on this question.

5.2   Energy Identity

In this section we derive an energy identity for

$$P = -M\Lambda + B\Lambda + Q(x, D')$$

with $M = D_0 - m(x, D')$, $\Lambda = D_0 - \lambda(x, D')$ where $\lambda(x, \xi') \in S^1$, $m(x, \xi') \in S^1$, $B(x, \xi') \in S^1$, $Q(x, \xi') \in S^2$. Here and in what follows by $a(x, \xi') \in S^m$ we denote symbols a with values in $S^m(\mathbb{R}^n \times \mathbb{R}^n)$ depending smoothly on $x_0 \in \mathbb{R}$ or $x_0 \in I$ where I is an interval containing 0. Define P_θ by

$$P(e^{\theta x_0} u) = e^{\theta x_0} P_\theta u$$

with a large positive parameter $\theta > 0$. We put $M_\theta = M - i\theta$, $\Lambda_\theta = \Lambda - i\theta$ such that $P_\theta = -M_\theta \Lambda_\theta + B\Lambda_\theta + Q$.

Proposition 5.1 *We have the following identity.*

$$2\mathsf{Im}(P_\theta u, \Lambda_\theta u) = \frac{d}{dx_0}(\|\Lambda_\theta u\|^2 + \mathsf{Re}(Qu, u)) + 2\theta\|\Lambda_\theta u\|^2$$

$$+2((\mathsf{Im}\,B)\Lambda_\theta u, \Lambda_\theta u) + 2((\mathsf{Im}\,m)\Lambda_\theta u, \Lambda_\theta u) + 2\mathsf{Re}(\Lambda_\theta u, (\mathsf{Im}\,Q)u)$$

$$+2\theta\mathsf{Re}(Qu, u) + \mathsf{Im}([D_0 - \mathsf{Re}\,\lambda, \mathsf{Re}\,Q]u, u) + 2\mathsf{Re}((\mathsf{Re}\,Q)u, (\mathsf{Im}\,\lambda)u).$$

Proof Since $2\mathsf{Im}(B\Lambda_\theta u, \Lambda_\theta u) = 2((\mathsf{Im}\,B)\Lambda_\theta u, \Lambda_\theta u)$ is clear we compute

$$-2\mathsf{Im}(M_\theta \Lambda_\theta u, \Lambda_\theta u) + 2\mathsf{Im}(Qu, \Lambda_\theta u) = I_1 + I_2.$$

Noting $d/dx_0 = iM_\theta + im - \theta$ it is easy to see

$$I_1 = \frac{d}{dx_0}\|\Lambda_\theta u\|^2 + 2\theta\|\Lambda_\theta u\|^2 + 2((\mathsf{Im}\,m)\Lambda_\theta u, \Lambda_\theta u).$$

We now consider $I_2 = 2\mathsf{Im}(Qu, D_0 u) + 2\theta\mathsf{Re}(Qu, u) + 2\mathsf{Im}(Qu, -\lambda u)$ where we see

$$2\mathsf{Im}(Qu, D_0 u) = 2\mathsf{Im}\{-D_0(Qu, u) + (QD_0 u, u) + ([D_0, Q]u, u)\}$$

$$= 2\mathsf{Re}\frac{d}{dx_0}(Qu, u) + 2\mathsf{Im}(D_0 u, Q^* u) + 2\mathsf{Im}([D_0, Q]u, u)$$

$$= 2\frac{d}{dx_0}((\mathsf{Re}\,Q)u, u) + 2\mathsf{Im}(D_0 u, Qu) + 2\mathsf{Im}(D_0 u, (Q^* - Q)u)$$

$$+2\mathsf{Im}([D_0, Q]u, u).$$

Therefore we get

$$2\mathsf{Im}(Qu, D_0 u) = \frac{d}{dx_0}((\mathsf{Re}\,Q)u, u) + \mathsf{Im}(D_0 u, (Q^* - Q)u) + \mathsf{Im}([D_0, Q]u, u).$$

Noting $([d/dx_0, \text{Im}\, Q]u, u)$ is real and $D_0 = \Lambda_\theta + i\theta + \lambda$ we have

$$\text{Im}([D_0, Q]u, u) = \text{Im}([D_0, \text{Re}\, Q]u, u),$$
$$\text{Im}(D_0 u, (Q^* - Q)u) = 2\text{Re}(D_0 u, (\text{Im}\, Q)u)$$

and hence

$$I_2 = \frac{d}{dx_0}((\text{Re}\, Q)u, u) + \text{Im}([D_0, \text{Re}\, Q]u, u) + 2\text{Re}(\Lambda_\theta u, (\text{Im}\, Q)u)$$
$$+ 2\text{Re}(\lambda u, (\text{Im}\, Q)u) + 2\theta \text{Re}(Qu, u) + 2\text{Im}(Qu, -\lambda u).$$

Since

$$2\text{Re}(\lambda u, (\text{Im}\, Q)u) + 2\text{Im}(Qu, -\lambda u) = -2\text{Im}((\text{Re}\, Q)u, \lambda u),$$
$$-2\text{Im}((\text{Re}\, Q)u, \lambda u) = -2\text{Im}((\text{Re}\, Q)u, (\text{Re}\, \lambda)u)$$
$$+ 2\text{Re}((\text{Re}\, Q)u, (\text{Im}\, \lambda)u)$$
$$= -\text{Im}([\text{Re}\, \lambda, \text{Re}\, Q]u, u)$$
$$+ 2\text{Re}((\text{Re}\, Q)u, (\text{Im}\, \lambda)u)$$

we have

$$I_2 = \frac{d}{dx_0}((\text{Re}\, Q)u, u) + \text{Im}([D_0 - \text{Re}\, \lambda, \text{Re}\, Q]u, u)$$
$$+ 2\text{Re}(\Lambda_\theta u, (\text{Im}\, Q)u) + 2\theta \text{Re}(Qu, u) + 2\text{Re}((\text{Re}\, Q)u, (\text{Im}\, \lambda)u)$$

and hence the result.   □
  Note that from

$$-2\text{Im}(\Lambda_\theta u, u) = 2\theta \|u\|^2 + \frac{d}{dx_0}\|u\|^2 + 2\text{Im}(\lambda u, u) \tag{5.5}$$

one obtains

$$\|\Lambda_\theta u\|^2 \geq \theta^2 \|u\|^2 + \theta \frac{d}{dx_0}\|u\|^2 + 2\theta((\text{Im}\, \lambda)u, u). \tag{5.6}$$

Replacing $\|\Lambda_\theta u\|^2$ in Proposition 5.1 by the estimate (5.6) we get

Proposition 5.2 *We have*

$$2\text{Im}(P_\theta u, \Lambda_\theta u) \geq \frac{d}{dx_0}(\|\Lambda_\theta u\|^2 + ((\text{Re}\, Q)u, u) + \theta^2 \|u\|^2) + \theta \|\Lambda_\theta u\|^2$$
$$+ 2\theta \text{Re}(Qu, u) + 2((\text{Im}\, B)\Lambda_\theta u, \Lambda_\theta u) + 2((\text{Im}\, m)\Lambda_\theta u, \Lambda_\theta u)$$

$$+2\mathsf{Re}(\Lambda_\theta u, (\mathsf{Im}\, Q)u) + \mathsf{Im}([D_0 - \mathsf{Re}\lambda, \mathsf{Re}\, Q]u, u)$$
$$+2\mathsf{Re}((\mathsf{Re}\, Q)u, (\mathsf{Im}\,\lambda)u) + \theta^3\|u\|^2 + 2\theta^2((\mathsf{Im}\,\lambda)u, u).$$

Before closing this section we note that

Lemma 5.4 *Assume that* p *admits a microlocal elementary factorization* $p = -M\Lambda + Q$ *at* $\rho \in \Sigma$. *Then we have* $\mathrm{Tr}^+ F_p(\rho) = \mathrm{Tr}^+ Q_\rho$.

Proof From the assumption one can write $p(x, \xi) = -(\xi_0 + \lambda)(\xi_0 - \lambda) + Q(x, \xi')$ where $|\{\xi_0 - \lambda, Q\}| \le CQ$. At ρ it is clear

$$p_\rho(x, \xi) = -(\xi_0 + \lambda_\rho)(\xi_0 - \lambda_\rho) + Q_\rho(x, \xi').$$

By a linear symplectic change of coordinates one may assume that

$$p_\rho = \xi_0(\xi_0 - \ell) + Q_\rho(x, \xi').$$

Since $|\{\xi_0, Q_\rho\}| \le CQ_\rho$ one concludes that Q_ρ is independent of x_0 and hence $Q_\rho = Q_\rho(x', \xi')$. By a linear symplectic change of coordinates again we may assume that $\ell = \xi_1$ or $\ell = 0$ according to $\ell \ne 0$ and $\ell = 0$. Now it is easy to see that

$$\det(tI - F_{p_\rho}) = t^2 \det(tI - F_{Q_\rho})$$

which proves that non-zero eigenvalues of F_{p_ρ} coincides with those of F_{Q_ρ} counting the multiplicity.   □

5.3 Microlocal Energy Estimates, Spectral Type 1

We fix any $\rho = (0, \hat{x}', \hat{\xi}_0, \hat{\xi}') \in \Sigma$. In this section we assume that p is of spectral type 1 near ρ and the strict IPH condition is satisfied there. Thanks to Proposition 2.1 p admits a microlocal elementary factorization verifying the conditions stated there. We extend these ϕ_j (given in Proposition 2.5) to be 0 outside a conic neighborhood of $\rho' = (\hat{x}', \hat{\xi}')$ so that they belong to S^1 for small $|x_0|$. Using such extended ϕ_j we define λ by the same formula in Proposition 2.5;

$$\lambda = \sum_{j=1}^{r} \gamma_j(x, \xi')\phi_j(x, \xi')$$

where the coefficients $\gamma_j(x, \xi')$ are extended outside a conic neighborhood of ρ' for small $|x_0|$ so that $\gamma_j \in S^0$ and $|\gamma| < 1$. Let us write $p = -(\xi_0 + \lambda)(\xi_0 - \lambda) + Q$.

Recall

$$Q = \sum_{j=1}^{r} \phi_j^2 - (\sum_{j=1}^{r} \gamma_j \phi_j)^2 \geq c \sum_{j=1}^{r} \phi_j^2$$

with some $c > 0$.

Lemma 5.5 *Let V be a small conic neighborhood of ρ' on which Proposition 2.1 holds. Then there exist $\tau > 0$ and $f(x, \xi') \in S^0$ such that $f(x, \xi') = 0$ in a conic neighborhood of ρ' and $f(x, \xi') \geq c\langle \xi' \rangle$ outside V for $|x_0| \leq \tau$ and satisfies*

$$\{\xi_0 - \lambda, f\} = 0, \quad |x_0| \leq \tau.$$

Proof Choose conic neighborhoods $V_1 \Subset V \Subset V_3$ of ρ' such that extended ϕ_i coincides with the original one in V and vanishes outside V_3. Take $0 \leq \chi(x', \xi') \leq 1$, homogeneous of degree 0 in ξ' ($|\xi'| \geq 1$), which are 1 near ρ' supported in V_1. We now define $f(x, \xi')$ solving the following Cauchy problem for a first order differential equation

$$\{\xi_0 - \lambda, f\} = 0, \quad f(0, x', \xi') = (1 - \chi(x', \xi'))\langle \xi' \rangle. \tag{5.7}$$

Since $f(x, \xi)$ is constant along characteristic curve of $\xi_0 - \lambda$ one can find, by a compactness argument, positive constants $c > 0$ and $\tau > 0$ such that $f(x, \xi') \geq c\langle \xi' \rangle$ for $(x', \xi') \in V_3 \setminus V$ and $f(x, \xi') = 0$ in a small conic neighborhood of ρ' for $|x_0| \leq \tau$. Since $\lambda = 0$ outside V_3 so that $\{\xi_0 - \lambda, f\} = \partial f / \partial x_0 = 0$ the assertion follows. $\square$

Lemma 5.6 *Let $f(x, \xi')$ be as above. Taking $k > 0$ large and $\tau > 0$ small we put*

$$p = -(\xi_0 + \lambda)(\xi_0 - \lambda) + \hat{Q}, \quad \hat{Q} = Q + k^2 f(x, \xi')^2$$

which coincides with the original p in a conic neighborhood of ρ' and we have

$$\left| \{\xi_0 - \lambda, \hat{Q}\} \right| \leq C\hat{Q}, \quad \left| \{\xi_0 + \lambda, \xi_0 - \lambda\} \right| \leq C(\hat{Q}^{1/2} + |\lambda|), \quad |\lambda| \leq C\hat{Q}^{1/2}$$

for $|x_0| < \tau$.

Proof Since $\hat{Q}$ coincides with Q in a conic neighborhood of ρ' for small $|x_0|$ by Lemma 5.5 and hence p coincides with the original one there clearly. Let us consider $|\{\xi_0 - \lambda, \hat{Q}\}|$ which is bounded by CQ on a conic neighborhood V of ρ' for small $|x_0|$ by Proposition 2.1. Since $f(x, \xi') \geq c\langle \xi' \rangle$ outside V, taking k large one can assume that $\hat{Q} \geq c'\langle \xi' \rangle^2$ there with some $c' > 0$. Therefore $|\{\xi_0 - \lambda, \hat{Q}\}|$ is bounded by $C\hat{Q}$ outside V. Noting that $\{\xi_0 + \lambda, \xi_0 - \lambda\} = 2\{\lambda, \xi_0 - \lambda\}$ and $\{\phi_j, \xi_0 - \lambda\}$ is a linear combination of $\phi_j, j = 1, \ldots, r$ in V and then repeating the same arguments we conclude that

$$\left|\{\xi_0 + \lambda, \xi_0 - \lambda\}\right| \le C(\hat{Q}^{1/2} + |\lambda|)$$

which is the second assertion. The proof of the third assertion is clear.   □
   Consider

$$\tilde{P} = p + P_1 + P_0, \quad p = -(\xi_0 + \lambda)(\xi_0 - \lambda) + \hat{Q} \qquad (5.8)$$

which coincides with the original P near ρ. In what follows to simplify notations we denote $\tilde{P}$ by P and $\hat{Q}$ by Q again:

$$P \text{ stands for } \tilde{P} \quad \text{and} \quad Q \text{ stands for } \hat{Q}. \qquad (5.9)$$

We denote

$$\phi_{r+1}(x, \xi') = kf(x, \xi'). \qquad (5.10)$$

Here we make a general remark.

Lemma 5.7 *Assume that $a(x, \xi') \in S^1$ vanishes near $(0, \rho')$ on $\Sigma' = \{\phi_j = 0, j = 1, \ldots, r\}$. Then one can write*

$$a(x, \xi') = \sum_{j=1}^{r+1} c_j \phi_j(x, \xi'), \quad c_j \in S^0. \qquad (5.11)$$

Proof It suffices to repeat the same arguments proving Lemma 5.6.   □

Proposition 5.3 *Let P be as above. Choosing k large one can write*

$$P = -M\Lambda + Q + \hat{P}_1 + B_0\Lambda + \hat{P}_0, \quad M = \xi_0 + \lambda, \ \Lambda = \xi_0 - \lambda$$

where the factorization $p = -(\xi_0 + \lambda)(\xi_0 - \lambda) + Q$ satisfies Lemma 5.6 and $\hat{P}_1 \in S^1$ and $B_0, \hat{P}_0 \in S^0$. Moreover there exist $c > 0$, $C > 0$ such that

$$\left|\operatorname{Im}\hat{P}_1\right| \le C\sqrt{Q}, \quad \operatorname{Tr}^+ Q_\rho + \operatorname{Re}\hat{P}_1(\rho) \ge c\langle\xi'\rangle \ \text{if} \ Q(\rho) = 0.$$

Proof Since P satisfies the strict IPH condition near ρ it follows that $\operatorname{Im} P_{sub} = 0$ on Σ near ρ. Write $P_{sub} = P_s(x, \xi') + b(x, \xi')(\xi_0 - \lambda)$. Since $p = -\xi_0^2 + q$ and $|\lambda| \le C\sqrt{q}$ and hence $\Sigma = \{\xi_0 = 0, q = 0\}$ it is clear that

$$\operatorname{Im} P_{sub}\big|_\Sigma = \operatorname{Im} P_s\big|_{q=0}$$

near ρ so that $\mathrm{Im}\, P_s = 0$ if $q = 0$. Let $\{\phi_i\}$ be as in Proposition 2.1. Since $\mathrm{Im}\, P_s$ is a linear combination of $\{\phi_i\}$ near $(0, \rho')$ and hence

$$|\mathrm{Im}\, P_s(x, \xi')| \leq C \sum_i |\phi_i| \leq C\sqrt{Q}. \tag{5.12}$$

Outside a conic neighborhood of $(0, \rho')$, applying the same arguments proving Lemma 5.6 we get (5.12). We return to P. Recalling that

$$p(x, \xi) = -(\xi_0 + \lambda)\#(\xi_0 - \lambda) + Q - i\{\xi_0 + \lambda, \xi_0 - \lambda\}/2 + R \tag{5.13}$$

with $R \in S^0$ and $|\{\xi_0 + \lambda, \xi_0 - \lambda\}| \leq C\sqrt{Q}$ near $(0, \rho')$ hence repeating the same arguments for outside a conic neighborhood of $(0, \rho')$ we get the first assertion because

$$\mathrm{Im}\hat{P}_1 = \mathrm{Im}\, P_s - \{\xi_0 + \lambda, \xi_0 - \lambda\}/2. \tag{5.14}$$

To check the last assertion note that $\mathrm{Re}\,\hat{P}_1 = \mathrm{Re}\, P_s$. Since $\mathrm{Re}\, P_s = \mathrm{Re}\, P_{sub}$ on Σ near ρ therefore Lemma 5.4 and the strict IPH condition shows that $\mathrm{Tr}^+ Q_\rho + \mathrm{Re}\,\hat{P}_1(\rho) > 0$ on Σ. On the other hand, as checked in the proof of Lemma 5.6 one can assume $Q \neq 0$ outside Σ for small $|x_0|$. Thus we get the second assertion. $\square$

Let $\hat{P}_0 = 0$ and we apply Proposition 5.2. Since λ and Q are real we have

$$\begin{aligned}
2\mathrm{Im}(P_\theta u, \Lambda_\theta u) \geq \frac{d}{dx_0}(\|\Lambda_\theta u\|^2 &+ ((Q + \mathrm{Re}\,\hat{P}_1)u, u) \\
&+ \theta^2\|u\|^2) + \theta\|\Lambda_\theta u\|^2 + 2((\mathrm{Im}\, B_0)\Lambda_\theta u, \Lambda_\theta u) \\
&+ 2\theta((Q + \mathrm{Re}\,\hat{P}_1)u, u) + 2\mathrm{Re}(\Lambda_\theta u, (\mathrm{Im}\,\hat{P}_1)u) \\
&+ \mathrm{Im}([D_0 - \lambda, Q + \mathrm{Re}\,\hat{P}_1]u, u) + \theta^3\|u\|^2.
\end{aligned} \tag{5.15}$$

Choosing $\epsilon > 0$ sufficiently small we have $(1 - \epsilon)\mathrm{Tr}^+ Q_\rho + \mathrm{Re}\,\hat{P}_1(\rho) > 0$ if $Q(\rho) = 0$. Then from Theorem 1.8 there exist $c > 0, C > 0$ such that

$$((Q + \mathrm{Re}\,\hat{P}_1)u, u) \geq c\|u\|_{1/2} + \epsilon(Qu, u) - C\|u\|^2. \tag{5.16}$$

Note that $\mathrm{Im}([D_0 - \lambda, Q]u, u) \geq -\mathrm{Re}((\mathrm{Op}\{\xi_0 - \lambda, Q\})u, u) - C\|u\|^2$. Since $CQ - \{\xi_0 - \lambda, Q\} \geq 0$ with some $C > 0$ then the Fefferman-Phong inequality shows

$$C(Qu, u) + \mathrm{Im}([D_0 - \lambda, Q]u, u) \geq -C_1\|u\|^2. \tag{5.17}$$

We turn to $|\mathrm{Im}([D_0 - \lambda, \mathrm{Re}\,\hat{P}_1]u, u)|$. Noting $[D_0 - \lambda, \mathrm{Re}\,\hat{P}_1] \in \mathrm{Op}S^1$ it is clear that

$$|\mathrm{Im}([D_0 - \lambda, \mathrm{Re}\,\hat{P}_1]u, u)| \leq C\|u\|_{1/2}^2. \tag{5.18}$$

Thus from (5.16) to (5.18) we conclude that for $\theta \geq \theta_0$

$$\theta((Q + \operatorname{Im}\hat{P}_1)u, u) + \operatorname{Im}([D_0 - \lambda, Q + \operatorname{Re}\hat{P}_1]u, u)$$
$$\geq c\,\theta\|u\|_{1/2}^2 + c'\theta(Qu, u) - C\theta\|u\|^2. \tag{5.19}$$

Since $\|(\operatorname{Im}\hat{P}_1)u\|^2 \leq C(Qu, u) + C\|u\|^2$ we get

$$2|\operatorname{Re}(\Lambda_\theta u, (\operatorname{Im}\hat{P}_1)u)| \leq \|\Lambda_\theta u\|^2 + C(Qu, u) + C'\|u\|^2. \tag{5.20}$$

Noting $\hat{P}_0 \in S^0$ and $B_0 \in S^0$ we obtain

$$2\operatorname{Im}(P_\theta u, \Lambda_\theta u) \geq \frac{d}{dx_0}(\|\Lambda_\theta u\|^2 + ((Q + \operatorname{Re}(\hat{P}_1 + \hat{P}_0))u, u)$$
$$+ \theta^2\|u\|^2) + \theta((Q + \operatorname{Re}\hat{P}_1)u, u)$$
$$+ c\,\theta(\|\Lambda_\theta u\|^2 + \|u\|_{1/2}^2 + \theta^2\|u\|^2).$$

Integrating this inequality in x_0 yields

Proposition 5.4 *There exist $T > 0$, $c > 0$, $C > 0$, $\theta_0 > 0$ such that for any $\theta \geq \theta_0$ and any $u \in C^2([-T, T]; C_0^\infty(\mathbb{R}^n))$ vanishing in $x_0 \leq \tau$, $|\tau| \leq T$ one has*

$$\{\|\Lambda_\theta u(t)\|^2 + ((Q + \operatorname{Re}\hat{P}_1)u(t), u(t)) + \|u(t)\|_{1/2}^2 + c\,\theta^2\|u(t)\|^2\}$$

$$+ c\,\theta \int_\tau^t \{\|\Lambda_\theta u(s)\|^2 + ((Q + \operatorname{Re}\hat{P}_1)u(s), u(s)) \tag{5.21}$$

$$+ \|u(s)\|_{1/2} + \theta^2\|u(s)\|^2\}ds \leq C\operatorname{Im}\int_\tau^t (P_\theta u(s), \Lambda_\theta u(s))ds.$$

Lemma 5.8 *One can write*

$$\langle D'\rangle^s P = (-M\Lambda + \tilde{B}\Lambda + Q + \tilde{P}_1 + \tilde{P}_0)\langle D'\rangle^s, \quad M = \xi_0 + \lambda, \quad \Lambda = \xi_0 - \lambda$$

where $\tilde{B}$, $\tilde{P}_0 \in S^0$ and $\tilde{P}_1$ verifies the same conditions that $\hat{P}_1$ satisfies in Proposition 5.3.

Proof Write $P = -\Lambda^2 + B\Lambda + \tilde{Q}$ with $\Lambda = \xi_0 - \lambda$ and $B = B_0 - 2\lambda$, $\tilde{Q} = Q + \hat{P}_1 + \hat{P}_0$. Noting $[\Lambda, \langle D'\rangle^s] \in \operatorname{Op}S^s$, $[\Lambda, [\Lambda, \langle D'\rangle^s]] \in \operatorname{Op}S^s$ it is easy to check that

$$[\Lambda^2, \langle D'\rangle^s] = R_1\Lambda\langle D'\rangle^s + R_2\langle D'\rangle^s$$

with some $R_i \in S^0$. We turn to consider $[B\Lambda, \langle D'\rangle^s]$. Let us write $[B\Lambda, \langle D'\rangle^s] = B[\Lambda, \langle D'\rangle^s] + [B, \langle D'\rangle^s]\Lambda$ and note

$$B[\Lambda, \langle D'\rangle^s] = (ia_1\lambda + a_2)\langle D'\rangle^s \tag{5.22}$$

where $a_i \in S^0$ and $a_1 = -2\{\lambda, \langle \xi' \rangle^s\}\langle \xi' \rangle^{-s}$ is real. It is clear that we can write

$$[B, \langle D' \rangle^s]\Lambda = R_1 \Lambda \langle D' \rangle^s + R_2 \langle D' \rangle^s$$

with $R_i \in S^0$. We finally check $[\tilde{Q}, \langle D' \rangle^s]$. Since it is clear that $[\tilde{Q}, \langle D' \rangle^s]\langle D' \rangle^{-s} - [Q, \langle D' \rangle^s]\langle D' \rangle^{-s} \in \text{Op}S^0$ it suffices to consider $[Q, \langle D' \rangle^s]\langle D' \rangle^{-s}$. Then noticing that $[Q, \langle D' \rangle^s]\langle D' \rangle^{-s} - (1/i)\text{Op}(\{Q, \langle \xi' \rangle^s\}\langle \xi' \rangle^{-s}) \in \text{Op}S^0$ and that one can write

$$\{Q, \langle \xi' \rangle^s\}\langle \xi' \rangle^{-s} = \sum_{j=1}^{r+1} c_j \phi_j$$

with real $c_j \in S^0$ thanks to Lemma 5.7 we conclude that

$$[Q, \langle D' \rangle^s] = -\text{Op}\left(i\left(\sum_{j=1}^{r+1} c_j \phi_j\right) + r\right)\langle D' \rangle^s \tag{5.23}$$

with some $r \in S^0$. From this and (5.22) it is clear that $\text{Re}\,\tilde{P}_1 = \text{Re}\,\hat{P}_1$. Since $|\lambda| \leq C\sqrt{Q}$ and $|\phi_j| \leq C\sqrt{Q}$ the assertion follows.   □

Repeating the same arguments proving (5.21) for $\text{Im}\,(\langle D' \rangle^s Pu, \Lambda \langle D' \rangle^s u)$ we obtain energy estimates of $\langle D' \rangle^s u$. To formulate thus obtained estimate denote

$$N_s(u) = \|\Lambda u\|_s^2 + \text{Re}((Q + \text{Re}\hat{P}_1)u, u)_s + \|u\|_{s+1/2}^2 \tag{5.24}$$

where $\Lambda = D_0 - \lambda$ again. Remark that $CN_s(u) \geq \|D_0 u\|_{s-1/2} + \|u\|_{s+1/2}$. Here we note that

$$\langle \xi' \rangle^s \# Q \# \langle \xi' \rangle^{-s} - Q + i\{\langle \xi' \rangle^s, Q\}\langle \xi' \rangle^{-s} \in S^0 \tag{5.25}$$

so that $\left|\text{Re}(\langle D' \rangle^s Qu, \langle D' \rangle^s u) - (Q\langle D' \rangle^s u, \langle D' \rangle^s u)\right| \leq C\|u\|_s^2$. We also note $\Lambda \langle D' \rangle^s = \langle D' \rangle^s \Lambda + r\langle D' \rangle^s$ with $r \in S^0$ and hence

$$\|\Lambda u\|_s^2 \leq C\|\Lambda \langle D' \rangle^s u\|^2 + C\|u\|_s^2.$$

Taking $e^{\theta x_0} P_\theta e^{-\theta x_0} = P$, $e^{\theta x_0} \Lambda_\theta e^{-\theta x_0} = \Lambda$ into account we obtain

Proposition 5.5 *There exists $T > 0$ such that for any $s \in \mathbb{R}$ there are $\theta_s > 0$ and $C_s > 0$ such that for any $u \in \cap_{j=0}^2 C^j([-T, T]; H^{s+2-j}(\mathbb{R}^n))$ vanishing in $x_0 \leq \tau$, $|\tau| \leq T$ and $\theta \geq \theta_s$ one has*

$$e^{-2\theta t} N_s(u(t)) + \theta \int_\tau^t e^{-2\theta x_0} N_s(u(x_0)) dx_0$$

$$\leq C_s \text{Im} \int_\tau^t e^{-2\theta x_0} (\langle D' \rangle^s Pu, \Lambda \langle D' \rangle^s u) dx_0.$$

Corollary 5.3 *For any $s \in \mathbb{R}$ there exist $C_s, C'_s > 0$ such that we have*

$$\|D_0 u(t)\|^2_{s-1/2} + \|u(t)\|^2_{s+1/2} \leq C_s N_s(u(t)) \leq C'_s \int_0^t \|Pu(x_0)\|^2_s dx_0$$

for any $u \in \cap_{j=0}^2 C^j_+([-T,T]; H^{s+2-j}(\mathbb{R}^n))$.

From Lemma 5.3 it follows that $P^*(x, D)$ satisfies the strict IPH condition. Then repeating the same arguments as proving Proposition 5.5 and Corollary 5.3 we conclude that Corollary 5.3 holds also for P^*. Reversing the time direction we get

Proposition 5.6 *There exists $T > 0$ such that for any $s \in \mathbb{R}$ there is $C > 0$ such that for any $u \in \cap_{j=0}^2 C^j([-T,T]; H^{s+2-j}(\mathbb{R}^n))$ vanishing in $x_0 \geq \tau$, $|\tau| \leq T$ one has*

$$N_s(u(t)) + \int_t^\tau N_s(u(x_0))dx_0 \leq C \int_t^\tau \|P^* u\|^2_s dx_0. \tag{5.26}$$

5.4 Microlocal Energy Estimates, Spectral Type 2

We fix any $\rho \in \Sigma$ and we study the case that p is of spectral type 2 near ρ assuming the strict IPH condition. Thanks to Proposition 2.5 we have a microlocal factorization of $p = -\xi_0^2 + \sum_{j=1}^r \phi_j^2$ near ρ such that

$$p = -(\xi_0 + \lambda)(\xi_0 - \lambda) + Q$$

where $\lambda = \phi_1 + O(\sum_{j=1}^r \phi_j^2)$. The main difference from the case that p is of spectral type 1 near ρ is that we have no control of ϕ_1^2 by Q and what is the best we can expect is the inequality

$$CQ \geq \sum_{j=2}^r \phi_j^2 + \phi_1^4 |\xi'|^{-2}.$$

As in the previous section we extend ϕ_j, given in Proposition 2.5, outside a conic neighborhood of ρ' to be 0 so that they belong to S^1 for small $|x_0|$. Using such extended ϕ_j we define λ by the same formula in Proposition 2.5;

$$\lambda = \phi_1 + L(\phi')\phi_1 + \gamma \phi_1^3 \langle \xi' \rangle^{-2}$$

where the coefficients of L are extended outside a conic neighborhood of ρ' for small $|x_0|$. Choosing a conic neighborhood of ρ' enough small we may assume that

$$\lambda = b\phi_1 \tag{5.27}$$

with $b \in S^0$ which satisfies $c_1 \leq b(x, \xi') \leq c_2$ with some $c_i > 0$ for small $|x_0|$. Recall

$$Q = \sum_{j=2}^{r} \phi_j^2 + a(\phi)\phi_1^4 \langle \xi' \rangle^{-2} + b(\phi')L(\phi')\phi_1^2 \geq c(|\phi'|^2 + \phi_1^4 \langle \xi' \rangle^{-2})$$

with some $c > 0$ where $\phi' = (\phi_2, \ldots, \phi_r)$. We define $f(x, \xi')$ solving (5.7) as before.

Lemma 5.9 *Let $f(x, \xi')$ be as above. Taking $k > 0$ large we put*

$$p = -(\xi_0 + \lambda)(\xi_0 - \lambda) + \hat{Q}, \quad \hat{Q} = Q + k^2 f(x, \xi')^2$$

which coincides with the original p in a conic neighborhood of ρ' and we have

$$\left| \{\xi_0 - \lambda, \hat{Q}\} \right| \leq C\hat{Q}, \quad \left| \{\xi_0 + \lambda, \xi_0 - \lambda\} \right| \leq C(\hat{Q}^{1/2} + |\lambda|)$$

for $|x_0| < \tau$ with small $\tau > 0$.

Proof Note that $\{\xi_0 + \lambda, \xi_0 - \lambda\} = 2\{\lambda, \xi_0 - \lambda\}$ and $\{\phi_j, \xi_0 - \lambda\}$ is a linear combination of ϕ_j, $j = 1, \ldots, r$ and $\lambda = \phi_1 + L(\phi')\phi_1 + \gamma \phi_1^3 \langle \xi' \rangle^{-2}$ on a conic neighborhood V of ρ'. Thus to prove the lemma it suffices to repeat the same arguments proving Lemma 5.6.   □

Define $\tilde{P}$ by (5.8) and use the same convention (5.9) and (5.10).

Proposition 5.7 *There exists $a \in S^0$ such that we can write*

$$P = -\tilde{M}\tilde{\Lambda} + Q + \hat{P}_1 + B\tilde{\Lambda} + \hat{P}_0$$

where $\tilde{\Lambda} = \xi_0 - \lambda - a$, $\tilde{M} = \xi_0 + \lambda + a$ and B, $\hat{P}_0 \in S^0$, $\hat{P}_1 \in S^1$. Moreover one has

$$\mathrm{Im}\,\hat{P}_1 = \sum_{j=2}^{r+1} c_j\phi_j, \quad c_j \in S^0, \quad \mathrm{Tr}^+ Q_\rho + \mathrm{Re}\,\hat{P}_1(\rho) \geq c\,\langle \xi' \rangle, \quad c > 0, \quad \text{if } Q(\rho) = 0.$$

Proof Write $P_{sub} = P_s + b(\xi_0 - \lambda)$. As we see in the proof of Proposition 5.3 we have $\mathrm{Im}\,P_s = 0$ on $q = 0$ near $(0, \rho')$. Note (5.13) and that both $\{\xi_0 + \lambda, \xi_0 - \lambda\}$ and $\mathrm{Im}\,P_s$ are linear combinations of ϕ_j, $j = 1, \ldots, r$ near $(0, \rho')$ then by Lemma 5.7 we can write

$$\mathrm{Im}\,\hat{P}_1 = \mathrm{Im}\,P_s - \{\xi_0 + \lambda, \xi_0 - \lambda\}/2 = \sum_{j=1}^{r+1} c_j\phi_j \qquad (5.28)$$

with some real $c_j \in S^0$. Recalling $b\phi_1 = ((\xi_0 + \lambda) - (\xi_0 - \lambda))/2$ and (5.27) one can write

$$-(\xi_0 + \lambda)\#(\xi_0 - \lambda) + ic_1\phi_1$$
$$= -(\xi_0 + \lambda + ib^{-1}c_1/2)\#(\xi_0 - \lambda - ib^{-1}c_1/2) + r$$

with some $r \in S^0$. Since it is clear that $B\#(\xi_0 - \lambda) = B\#(\xi_0 - \lambda - ib^{-1}c_1/2) + r'$ with $r' \in S^0$ we get the assertion for $\mathsf{Im}\,\hat{P}_1$. The rest of the proof is just a repetition of that of Proposition 5.3.   □

From Proposition 5.7 we can write $P = -\tilde{M}\tilde{\Lambda} + B\tilde{\Lambda} + \tilde{Q}$ where

$$\begin{cases} \tilde{M} = \xi_0 + \lambda + a = \xi_0 + \tilde{\lambda}, \\ \tilde{\Lambda} = \xi_0 - \lambda - a = \xi_0 - \tilde{\lambda}, \\ \tilde{Q} = Q + \hat{P}_1 + \hat{P}_0. \end{cases} \tag{5.29}$$

Recall that Proposition 5.2 gives

$$2\mathsf{Im}(P_\theta u, \tilde{\Lambda}_\theta u) \geq \frac{d}{dx_0}(\|\tilde{\Lambda}_\theta u\|^2 + ((\mathsf{Re}\,\tilde{Q})u, u) + \theta^2\|u\|^2)$$
$$+ \theta\|\tilde{\Lambda}_\theta u\|^2 + 2\theta\mathsf{Re}(\tilde{Q}u, u) + 2((\mathsf{Im}\,B)\tilde{\Lambda}_\theta u, \tilde{\Lambda}_\theta u)$$
$$- 2((\mathsf{Im}\,\tilde{\lambda})\tilde{\Lambda}_\theta u, \tilde{\Lambda}_\theta u) + 2\mathsf{Re}(\tilde{\Lambda}_\theta u, (\mathsf{Im}\,\tilde{Q})u) \tag{5.30}$$
$$+ \mathsf{Im}([D_0 - \mathsf{Re}\,\tilde{\lambda}, \mathsf{Re}\,\tilde{Q}]u, u) + 2\mathsf{Re}((\mathsf{Re}\,\tilde{Q})u, (\mathsf{Im}\,\tilde{\lambda})u)$$
$$+ \theta^3\|u\|^2 + 2\theta^2((\mathsf{Im}\,\tilde{\lambda})u, u)$$

where

$$\tilde{\Lambda}_\theta = \tilde{\Lambda} - i\theta, \quad \tilde{M}_\theta = \tilde{M} - i\theta, \quad P_\theta = -\tilde{M}_\theta\tilde{\Lambda}_\theta + B\tilde{\Lambda}_\theta + \tilde{Q}.$$

Since $\mathsf{Im}\,\tilde{\lambda} \in S^0$ it is clear that

$$|((\mathsf{Im}\,\tilde{\lambda})\tilde{\Lambda}_\theta u, \tilde{\Lambda}_\theta u)| \leq C\|\tilde{\Lambda}_\theta u\|^2, \quad |((\mathsf{Im}\,\tilde{\lambda})u, u)| \leq C\|u\|^2. \tag{5.31}$$

It is also clear

$$((\mathsf{Im}\,B)\tilde{\Lambda}_\theta u, \tilde{\Lambda}_\theta u) \geq -C\|\tilde{\Lambda}_\theta u\|^2 \tag{5.32}$$

with some $C > 0$ because $\mathsf{Im}\,B \in S^0$. To simplify notation we denote

$$\psi = \phi_1^2\langle\xi'\rangle^{-1}.$$

Lemma 5.10 *There exist $C_1, C_2 > 0$ such that we have*

$$\|\psi u\|^2 + \sum_{j=2}^{r+1} \|\phi_j u\|^2 \le C_1(Qu, u) + C_2\|u\|^2.$$

Proof Take $C_1 > 0$ so that $C_1 Q - \psi^2 - \sum_{j=2}^{r+1} \phi_j^2 \ge 0$. Then from the Fefferman-Phong inequality it follows that

$$C_1(Qu, u) \ge (\mathrm{Op}(\psi^2 + \sum_{j=2}^{r+1} \phi_j^2)u, u) - C_2\|u\|^2.$$

Noting that $\psi^2 = \psi \# \psi + R_1$ and $\phi_j^2 = \phi_j \# \phi_j + R_j$ with $R_j \in S^0$ the proof is immediate. $\square$

We now study $\mathrm{Re}\, \tilde{Q} = Q + \mathrm{Re}\, \hat{P}_1 + \mathrm{Re}\, \hat{P}_0$ where $\mathrm{Re}\, \hat{P}_1 \in S^1$. From Proposition 5.7 taking sufficiently small $\epsilon_0 > 0$ we have

$$(1 - \epsilon_0)\mathrm{Tr}^+ Q_\rho + \mathrm{Re}\, \hat{P}_1(\rho) \ge c\langle \xi' \rangle \quad \text{if} \quad Q(\rho) = 0$$

with some $c > 0$ and then from Theorem 1.8 (Melin's inequality) it follows that

$$\mathrm{Re}((Q + \mathrm{Re}\, \hat{P}_1)u, u) \ge \epsilon_0 \mathrm{Re}(Qu, u) + c'\|u\|_{1/2}^2 - C\|u\|^2 \tag{5.33}$$

with some $c' > 0$. Thus we conclude with some $c > 0$

$$\mathrm{Re}(\tilde{Q}u, u) \ge \epsilon_0(Qu, u) + c\|u\|_{1/2}^2 - C\|u\|^2. \tag{5.34}$$

We estimate the term $\mathrm{Re}((\mathrm{Re}\, \tilde{Q})u, (\mathrm{Im}\, \tilde{\lambda})u)$. We have $\mathrm{Re}(\mathrm{Im}\, \tilde{\lambda} \# Q) = (\mathrm{Im}\, \tilde{\lambda})Q + R$ with $R \in S^0$ since $\mathrm{Im}\, \tilde{\lambda} \in S^0$ and hence

$$\mathrm{Re}(Qu, (\mathrm{Im}\, \tilde{\lambda})u) \le (\mathrm{Op}((\mathrm{Im}\, \tilde{\lambda})Q)u, u) + C'\|u\|^2.$$

Take $C > 0$ so that $C - \mathrm{Im}\, \tilde{\lambda} \ge 0$ then $C(Qu, u) - (\mathrm{Op}((\mathrm{Im}\, \tilde{\lambda})Q)u, u) \ge -C_1\|u\|^2$ by the Fefferman-Phong inequality since $0 \le (C - \mathrm{Im}\, \tilde{\lambda})Q \in S^2$. Thus we have

$$C(Qu, u) \ge \mathrm{Re}(Qu, (\mathrm{Im}\, \tilde{\lambda})u) - C_2\|u\|^2.$$

Noting $|((\mathrm{Re}\, \hat{P}_1)u, (\mathrm{Im}\, \tilde{\lambda})u)| \le C\|u\|_{1/2}^2$ for $\mathrm{Re}\, \hat{P}_1 \in S^1$ it follows from (5.34) that with some $C_3 > 0$

$$C_3 \mathrm{Re}(\tilde{Q}u, u) + 2\mathrm{Re}((\mathrm{Re}\, \tilde{Q})u, (\mathrm{Im}\, \tilde{\lambda})u) \ge -C\|u\|^2. \tag{5.35}$$

Recall that $\operatorname{Im} \tilde{Q} = \operatorname{Im} \hat{P}_1 + \operatorname{Im} \hat{P}_0$ and note $\operatorname{Im} \hat{P}_1 = \sum_{j=2}^{r+1} c_j \# \phi_j + r$ with $c_j, r \in S^0$ by (5.28). Thus in view of Lemma 5.10 it is easy to see

$$|(\tilde{\Lambda}_\theta u, (\operatorname{Im} \hat{P}_1)u)| \leq C\|\tilde{\Lambda}_\theta u\|^2 + C\sum_{j=2}^{r+1} \|\phi_j u\|^2 + C\|u\|^2$$

$$\leq C\|\tilde{\Lambda}_\theta u\|^2 + C'(Qu, u) + C'\|u\|^2.$$

Therefore we get

$$|(\tilde{\Lambda}_\theta u, (\operatorname{Im} \tilde{Q})u)| \leq C\|\tilde{\Lambda}_\theta u\|^2 + C(Qu, u) + C\|u\|^2. \tag{5.36}$$

We consider $\operatorname{Im}([D_0 - \operatorname{Re} \tilde{\lambda}, \operatorname{Re} \tilde{Q}]u, u)$. Recall $\xi_0 - \operatorname{Re} \tilde{\lambda} = \xi_0 - \lambda + R$ where $R \in S^0$. Since

$$[D_0 - \lambda, Q] + i\operatorname{Op}(\{\xi_0 - \lambda, Q\}) \in \operatorname{Op}S^0$$

and $|\{\xi_0 - \lambda, Q\}| \leq CQ$ by Lemma 5.9 it follows from the Fefferman-Phong inequality that

$$|([D_0 - \lambda, Q]u, u)| \leq C(Qu, u) + C\|u\|^2. \tag{5.37}$$

Since $[D_0 - \lambda, \operatorname{Re} \hat{P}_1 + \operatorname{Re} \hat{P}_0] \in \operatorname{Op}S^1$ we get

$$\operatorname{Im}([D_0 - \operatorname{Re} \tilde{\lambda}, \operatorname{Re} \tilde{Q}]u, u) \leq C(Qu, u) + C\|u\|_{1/2}^2. \tag{5.38}$$

Taking $\|\Lambda_\theta u\|^2 \leq C\|\tilde{\Lambda}_\theta u\|^2 + C\|u\|^2$ into account from (5.34), (5.35), (5.32), (5.36) and (5.38) we have

Proposition 5.8 *There exist $T > 0$, $c > 0$, $C > 0$ and $\theta_0 > 0$ such that for any $\theta \geq \theta_0$ and any $u \in C^2([-T, T]; C_0^\infty(\mathbb{R}^n))$ vanishing in $x_0 \leq \tau$ we have*

$$c(\|\Lambda_\theta u(t)\|^2 + \|u(t)\|_{1/2}^2 + \theta^2\|u(t)\|^2)$$

$$+c\theta \int_\tau^t \left(\|\Lambda_\theta u(x_0)\|^2 + \operatorname{Re}(Qu, u) + \|u(x_0)\|_{1/2}^2 + \theta^2\|u(x_0)\|^2\right)dx_0$$

$$+c\int_\tau^t \|\Lambda_\theta u(x_0)\|^2 dx_0 \leq C\int_\tau^t \|P_\theta u(x_0)\|^2 dx_0.$$

We now derive estimates for higher order derivatives of u.

Lemma 5.11 *We can write*

$$\langle D'\rangle^s P = (-\tilde{M}\tilde{\Lambda} + \tilde{B}\tilde{\Lambda} + Q + \tilde{P}_1 + \tilde{P}_0)\langle D'\rangle^s$$

with $\tilde{\Lambda} = \xi_0 - \lambda - \tilde{a}$, $\tilde{M} = \xi_0 + \lambda + \tilde{a}$ where $\tilde{a} \in S^0$ is pure imaginary and $\tilde{B}$, $\tilde{P}_0 \in S^0$. Moreover $\tilde{P}_1$ verifies the same conditions that $\hat{P}_1$ satisfies in Proposition 5.7.

Proof Write $P = -\tilde{\Lambda}^2 + \tilde{B}\tilde{\Lambda} + \tilde{Q}$ with $\tilde{\Lambda} = \xi_0 - \lambda - a$ and $\tilde{B} = B - 2\lambda - 2a$, $\tilde{Q} = Q + P_1 + P_0$. Repeating the same arguments proving Lemma 5.8 one has (5.22) and (5.23); $\tilde{B}[\tilde{\Lambda}, \langle D' \rangle^s] \langle D' \rangle^{-s} = (ia_1\lambda + a_2)\langle D' \rangle^s$ where $a_i \in S^0$ and $a_1 = -2\{\lambda, \langle \xi' \rangle^s\} \langle \xi' \rangle^{-s}$ is real. It is clear that $\mathsf{Re}\,\tilde{P}_1 = \mathsf{Re}\,\hat{P}_1$. Repeating the same arguments in the proof of Proposition 5.7 we move the term

$$i(a_1\lambda - c_1\phi_1)$$

into Λ, which yields $\tilde{a} = i(a_1 - b^{-1}c_1)/2$ and we get the desired assertion.   □
Note (5.25) and that $\tilde{\Lambda}\langle D' \rangle^s = \langle D' \rangle^s \Lambda + r\langle D' \rangle^s$ with $r \in S^0$ so that

$$\|\Lambda u\|_s^2 \leq C\|\tilde{\Lambda}\langle D' \rangle^s u\|^2 + C\|u\|_s^2.$$

Repeating the same arguments for $\mathsf{Im}\,(\langle D' \rangle^s Pu, \tilde{\Lambda}\langle D' \rangle^s u)$ proving Proposition 5.8 we have

Proposition 5.9 *The same assertion as Proposition 5.5 holds.*

Corollary 5.4 *The same assertion as Corollary 5.3 holds.*
Repeating the same arguments proving Proposition 5.6 we obtain

Proposition 5.10 *The same assertion as Proposition 5.6 holds.*

5.5 Case of Spectral Type 2 with Zero Positive Trace

In this section we continue to study energy estimates for $P = \mathrm{Op}(p + P_{sub}) + R$ with $R \in S^0$ where p is of spectral type 2 on Σ. As in Sect. 5.4 we fix any $\rho \in \Sigma$. Let p be the same extended symbol defined in Sect. 5.4. Since the IPH condition is reduced to the Levi condition if the positive trace is zero, then we assume that the Levi condition is satisfied on Σ in this section;

$$P_{sub} = 0 \quad \text{on } \Sigma. \tag{5.39}$$

Proposition 5.11 *There exists $a \in S^0$ such that we can write*

$$P = -\tilde{M}\tilde{\Lambda} + Q + \hat{P}_1 + B\tilde{\Lambda} + \hat{P}_0$$

where $\tilde{\Lambda} = \xi_0 - \lambda - a$, $\tilde{M} = \xi_0 + \lambda + a$ and B, $\hat{P}_0 \in S^0$. Moreover we have

$$\hat{P}_1 = \sum_{j=2}^{r+1} c_j\phi_j, \quad c_j \in S^0.$$

Proof Proof is clear from that of Proposition 5.7.   □

Write $P = -\check{M}\tilde{\Lambda} + B\tilde{\Lambda} + \tilde{Q}$ and recall (5.30). We now study $\operatorname{Re}\tilde{Q} = Q + \operatorname{Re}\hat{P}_1 + \operatorname{Re}\hat{P}_0$ where $\operatorname{Re}\hat{P}_1 \in S^1$. From Proposition 5.11 one can write

$$\operatorname{Re}\hat{P}_1 = \sum_{j=2}^{r+1} c_j \phi_j \tag{5.40}$$

with $c_j \in S^0$. Since $c_j\phi_j = c_j\#\phi_j + R_j$, $R_j \in S^0$ and hence for any $\epsilon > 0$

$$(\operatorname{Re}\hat{P}_1 u, u) \geq -\epsilon \sum_{j=2}^{r+1} \|\phi_j u\|^2 - C_\epsilon \|u\|^2.$$

Thanks to Lemma 5.10 we have

$$(((1 - \bar{\epsilon})Q + \operatorname{Re}\hat{P}_1)u, u) \geq -C\|u\|^2 \tag{5.41}$$

with any fixed $\bar{\epsilon} > 0$. Thus we conclude

$$\operatorname{Re}(\tilde{Q}u, u) \geq \bar{\epsilon}\,(Qu, u) - C\|u\|^2. \tag{5.42}$$

Lemma 5.12 *Let $a \in S^0$. Then there exist $C, C' > 0$ such that*

$$|(\operatorname{Op}(a\phi_1)u, u)| \leq C(\|\psi u\|^2 + \|\phi_2 u\|^2 + \|\phi_{r+1}u\|^2) + C'\|u\|^2, \quad u \in C_0^\infty(\mathbb{R}^n).$$

Proof Since $\{\phi_1, \phi_2\} \geq c\langle\xi'\rangle$ near $(0, \rho')$ then there are $c_j \in S^0$ such that $a\phi_1 - c_1\{\psi, \phi_2\} - c_2\psi$ vanishes near $(0, \rho')$ and hence one can write with some $c_3 \in S^0$

$$a\phi_1 = c_1\{\psi, \phi_2\} + c_2\psi + c_3\phi_{r+1}$$

choosing V in Lemma 5.5 small if necessary indeed we have $\phi_{r+1} = f(x, \xi') \geq c'\langle\xi'\rangle$ outside V. Note $c_1\{\psi, \phi_2\} = \{\psi, c_1\phi_2\} - \{\psi, c_1\}\phi_2$ and

$$|(\operatorname{Op}(\{\psi, c_1\phi_2\})u, u)| \leq C(\|\psi u\|^2 + \|\phi_2 u\|^2 + \|u\|^2)$$

which follows from $\operatorname{Op}(\{\psi, c_1\phi_2\}) = i\,[\psi, \operatorname{Op}(c_1\phi_2)] + R$ with $R \in \operatorname{Op}S^0$. Thus we conclude the assertion.   □

Recall that $\{\xi_0 - \lambda, \phi_j\}$, $j = 2, \ldots, r$ is a linear combination of $\phi_1, \phi_2, \ldots, \phi_r$ near $(0, \rho')$ and then thanks to Lemma 5.7 one can write $\{\xi_0 - \lambda, \phi_j\} = \sum_{k=1}^{r+1} c_{jk}\phi_k$ for $j = 2, \ldots, r$. From Lemma 5.12 and (5.40) it follows that

$$|([D_0 - \lambda, \operatorname{Re}\hat{P}_1]u, u)| \leq C(\|\psi u\|^2 + \sum_{j=2}^{r+1} \|\phi_j u\|^2) + C'\|u\|^2$$

which together with (5.37) proves $\left|([D_0-\lambda,\mathsf{Re}\,\tilde{Q}]u,u)\right| \le C(Qu,u)+C\|u\|^2$. Note that $[R,Q]-\mathsf{Op}(\{R,Q\})/i \in \mathsf{Op}S^0$ and $|\{R,Q\}|^2 \le CQ$ with some $C>0$ by the Glaeser's inequality. Then applying the Fefferman-Phong inequality again we get

$$|(\mathsf{Op}(\{R,Q\})u,u)| \le C(Qu,u) + C'\|u\|^2$$

and hence we have $|([R,Q]u,u)| \le C(Qu,u)+C\|u\|^2$ with some $C>0$. Collecting these estimates we get

$$\mathsf{Im}([D_0-\mathsf{Re}\,\tilde{\lambda},\mathsf{Re}\,\tilde{Q}]u,u) \le C(Qu,u)+C\|u\|. \tag{5.43}$$

Lemma 5.13 *There is $C>0$ such that*

$$\|\langle D'\rangle^{1/3}u\|^2 \le C(\|\psi u\|^2 + \|\phi_2\|^2 + \|\phi_{r+1}u\|^2 + \|u\|^2), \quad u \in C_0^\infty(\mathbb{R}^n).$$

Proof Note that $\{\phi_1\langle\xi'\rangle^{-1/3},\phi_2\} = \{\phi_1,\phi_2\}\langle\xi'\rangle^{-1/3} + \phi_1\langle\xi'\rangle^{-1/3}A$ with $A \in S^0$ which proves

$$\mathsf{Re}\,(i\,[\mathsf{Op}(\phi_1\langle\xi'\rangle^{-1/3}),\phi_2]u,u) \ge (\mathsf{Op}(\{\phi_1,\phi_2\}\langle\xi'\rangle^{-1/3})u,u)$$
$$-C\|\mathsf{Op}(\phi_1\langle\xi'\rangle^{-1/3})u\|^2 - C\|u\|^2.$$

Since $\{\phi_1,\phi_2\}\langle\xi'\rangle^{-1/3} \ge c'\langle\xi'\rangle^{2/3}$ near $(0,\rho')$ with some $c'>0$ then from the same arguments as before we have

$$\{\phi_1,\phi_2\}\langle\xi'\rangle^{-1/3} + C_1\phi_{r+1}\langle\xi'\rangle^{-1/3} \ge c\langle\xi'\rangle^{2/3}$$

with some $c>0$, $C_1>0$ which proves

$$(\mathsf{Op}(\{\phi_1,\phi_2\}\langle\xi'\rangle^{-1/3})u,u) + C_1(\mathsf{Op}(\phi_{r+1}\langle\xi'\rangle^{-1/3})u,u)$$
$$\ge c\|\langle D'\rangle^{1/3}u\|^2 - C\|u\|^2.$$

This shows

$$\|\langle D'\rangle^{1/3}u\|^2 \le C_2\big(\|\phi_2u\|^2 + \|\mathsf{Op}(\phi_1\langle\xi'\rangle^{-1/3})u\|^2 + \|\phi_{r+1}u\|^2\big) + C'\|u\|^2. \tag{5.44}$$

Since $(\phi_1\langle\xi'\rangle^{-1/3})\#(\phi_1\langle\xi'\rangle^{-1/3}) = \mathsf{Re}\,((\phi_1^2\langle\xi'\rangle^{-1})\#\langle\xi'\rangle^{1/3}) + R$ with $R \in S^0$ we get

$$C_2\|\mathsf{Op}(\phi_1\langle\xi'\rangle^{-1/3})u\|^2 \le C_3\|\psi u\|^2 + 2^{-1}\|\langle D'\rangle^{1/3}u\|^2 + C_4\|u\|^2. \tag{5.45}$$

Therefore the assertion follows from (5.44) and (5.45) immediately. $\square$

Taking $\|\Lambda_\theta u\|^2 \le C\|\tilde{\Lambda}_\theta u\|^2 + C\|u\|^2$ into account we have from the estimates (5.31), (5.32), (5.42), (5.35), (5.36), (5.43) and Lemma 5.13 that

Proposition 5.12 *There exist $T > 0$, $c > 0$, $C > 0$ and $\theta_0 > 0$ such that for any $\theta \geq \theta_0$ and any $u \in C^2([-T, T]; C_0^\infty(\mathbb{R}^n))$ vanishing in $x_0 \leq \tau$, $|\tau| \leq T$ we have*

$$c \left(\|\Lambda_\theta u(t)\|^2 + \|u(t)\|_{1/3}^2 + \theta^2 \|u(t)\|^2 \right)$$

$$+ c\,\theta \int_\tau^t \left(\|\Lambda_\theta u\|^2 + \mathsf{Re}(\tilde{Q}u, u) + \|u\|_{1/3}^2 + \theta^2 \|u\|^2 \right) dx_0 \leq C \int_\tau^t \|P_\theta u\|^2 dx_0.$$

Recall $P_\theta = e^{-\theta x_0} P e^{\theta x_0}$ and $\Lambda_\theta = e^{-\theta x_0} \tilde{\Lambda} e^{\theta x_0}$. Then from the inequality (5.30) and the estimates (5.31), (5.32), (5.42), (5.35), (5.36), (5.43) together with Lemma 5.13, where u is replaced by $e^{-\theta x_0} u$, we have

Proposition 5.13 *There exist $\theta_0 > 0$, $c > 0$ such that we have*

$$c\,e^{-2\theta t} \left(\|\tilde{\Lambda} u(t)\|^2 + \|u(t)\|_{1/3}^2 + \theta^2 \|u(t)\|^2 \right)$$

$$+ c\,\theta \int_\tau^t e^{-2\theta x_0} \left(\|\tilde{\Lambda} u\|^2 + \mathsf{Re}(\tilde{Q}u, u) + \|u\|_{1/3}^2 + \theta^2 \|u\|^2 \right) dx_0$$

$$\leq \mathsf{Im} \int_\tau^t e^{-2\theta x_0} (Pu, \tilde{\Lambda} u) dx_0, \quad |t| \leq T$$

for any $u \in C_+^2([-T, T]; C_0^\infty(\mathbb{R}^n))$ and any $\theta \geq \theta_0$.

We now derive energy estimates for higher order derivatives of u.

Lemma 5.14 *Denote $\lambda + a$ in Proposition 5.7 still by λ. Then one can write*

$$\langle D' \rangle^s P = (-\tilde{M}\tilde{\Lambda} + \tilde{B}\tilde{\Lambda} + Q + \tilde{P}_1 + \tilde{P}_0) \langle D' \rangle^s$$

where $\tilde{\Lambda} = \xi_0 - \lambda - \tilde{a}$ and $\tilde{M} = \xi_0 + \lambda + \tilde{a}$ with $\tilde{a} \in S^0$ which is pure imaginary and $\tilde{B}$, $\tilde{P}_0 \in S^0$. Moreover $\tilde{P}_1$ verifies the same conditions that $\hat{P}_1$ verifies in Proposition 5.7 and $\mathsf{Re}\,\tilde{P}_1 = \mathsf{Re}\,\hat{P}_1$.

Proof It suffices to repeat the proof of Lemma 5.11. □

Taking Lemma 5.14 into account we repeat the same arguments proving Proposition 5.13 for

$$\mathsf{Im}(\langle D' \rangle^s Pu, \tilde{\Lambda} \langle D' \rangle^s u)$$

to obtain an energy estimate of $\langle D' \rangle^s u$. To formulate thus obtained estimate we define with $\tilde{Q} = Q + \tilde{P}_1$

$$N_s(u) = \|\Lambda u\|_s^2 + \mathsf{Re}(\tilde{Q}u, u)_s + \|u\|_{s+1/3}^2 + \theta^2 \|u\|_s^2 \tag{5.46}$$

where $\Lambda = D_0 - \lambda$ again and recall that $\operatorname{Re} \tilde{Q} = Q + \operatorname{Re} \hat{P}_1$. Here we remark that $\operatorname{Re}(\langle \xi' \rangle^s \# \tilde{Q} \# \langle \xi' \rangle^{-s}) - \operatorname{Re} \tilde{Q} \in S^0$ so that

$$\left| \operatorname{Re}(\langle D' \rangle^s \tilde{Q}u, \langle D' \rangle^s u) - ((\operatorname{Re} \tilde{Q}) \langle D' \rangle^s u, \langle D' \rangle^s u) \right| \leq C \|u\|_s^2.$$

Note $\|\Lambda u\|_s^2 \leq C \|\tilde{\Lambda} \langle D' \rangle^s u\|^2 + C_s \|u\|_s^2$ since $\tilde{\Lambda} \langle D' \rangle^s = \langle D' \rangle^s \Lambda + r \langle D' \rangle^s$ with $r \in S^0$. Then we have

Proposition 5.14 *With $N_s(u)$ defined by (5.46) the same assertion as Proposition 5.5 holds.*

Corollary 5.5 *Let $N_s(u)$ be defined by (5.46). For any $s \in \mathbb{R}$ there exist $C_s, C_s' > 0$ such that we have*

$$\|D_0 u(t)\|_{s-2/3}^2 + \|u(t)\|_{s+1/3}^2 \leq C_s N_s(u(t)) \leq C_s' \int_0^t \|Pu(x_0)\|_s^2 dx_0$$

for any $u \in \cap_{j=0}^2 C_+^j([-T, T]; H^{s+2-j}(\mathbb{R}^n))$.

Note that P^* satisfies the Levi condition thanks to Lemma 5.3. Repeating the same arguments as proving Proposition 5.6 we get

Proposition 5.15 *With $N_s(u)$ defined by (5.46) the same assertion as Proposition 5.6 holds.*

5.6 Case of Spectral Type 1 with Zero Positive Trace

In this section we study the case that p is of spectral type 1 on Σ and the Levi condition (5.39) is satisfied. In view of Proposition 2.2 we see that p admits a local elementary factorization $p(x, \xi) = -(\xi_0 + \lambda)(\xi_0 - \lambda) + Q$. Write

$$P_{sub} = P_s(x, \xi') + b(x, \xi')(\xi_0 - \lambda)$$

then we have

Lemma 5.15 *Assume that P satisfies the Levi condition on Σ. Then there is $C > 0$ such that*

$$|P_s(x, \xi')| \leq C \sqrt{Q(x, \xi')}, \quad |\{\xi_0 - \lambda, P_s\}| \leq C \sqrt{Q(x, \xi')}.$$

Proof Write $p = -\xi_0^2 + q$ then it is clear that $P_s = 0$ on $\Sigma' = \{q = 0\}$ because $\Sigma = \{\xi_0 = 0, q = 0\}$ and $q = \lambda^2 + Q = 0$ implies $\lambda = 0$. Let $\{V_i\}, \{\chi_i\}$ be as in the proof of Proposition 2.2. Since $V_i \cap \Sigma' = \{\phi_{i\alpha} = 0\}$ and hence P_s is a linear combination of $\{\phi_{i\alpha}\}$ in V_i then it is clear that for $(x, \xi') \in V_i$

$$|P_s(x, \xi')| \leq C \sum_\alpha |\phi_{i\alpha}| \leq C\sqrt{q_i} \leq C'\sqrt{Q}$$

which proves the first assertion. We turn to $\{\xi_0 - \lambda, P_s\}$. Arguing as in the proof of Proposition 2.2 we see

$$\{\xi_0 - \lambda, P_s\} = \sum \chi_i\{\xi_0 - \lambda_i, P_s\} - \sum \lambda_i\{\chi_i, P_s\}.$$

Since P_s is a linear combination of $\{\phi_{i\alpha}\}$ on the support of χ_i and $\{\xi_0 - \lambda_i, \phi_{i\alpha}\}$ is also a linear combination of $\{\phi_{i\alpha}\}$ in V_i we see easily

$$\sum |\chi_i\{\xi_0 - \lambda_i, P_s\}| \leq C \sum \chi_i|\phi_{i\alpha}| \leq C' \sum \chi_i\sqrt{q_i} \leq C''\sqrt{Q}.$$

Together with the estimate $|\lambda_i| \leq C\sqrt{Q}$ proves the second assertion. □
Noting (5.14) and recalling $|\{\xi_0 + \lambda, \xi_0 - \lambda\}| \leq C\sqrt{Q}$ we have from Lemma 5.15

Proposition 5.16 *Assume that p satisfies (1.5) and of spectral type 1 on Σ and that P verifies the Levi condition on Σ. Then one can write*

$$P = -M\Lambda + Q + \hat{P}_1 + B_0\Lambda + \hat{P}_0, \quad M = (\xi_0 + \lambda), \quad \Lambda = (\xi_0 - \lambda)$$

where $B_0, \hat{P}_0 \in S^1$ and $p = -(\xi_0 + \lambda)(\xi_0 - \lambda) + Q$ is a local elementary factorization of p and $\hat{P}_1$ verifies with some $C > 0$

$$|\hat{P}_1| \leq C\sqrt{Q}, \quad |\{\xi_0 - \lambda, \operatorname{Re}\hat{P}_1\}| \leq C\sqrt{Q}.$$

Recall (5.15). We first check that we have with some $c > 0$ and $C > 0$

$$((Q + \operatorname{Re}\hat{P}_1)u, u) \geq c(Qu, u) - C\|u\|^2.$$

In fact note that $2((\operatorname{Re}\hat{P}_1)u, u) \leq \epsilon\|(\operatorname{Re}\hat{P}_1)u\|^2 + \epsilon^{-1}\|u\|^2$ for any small $\epsilon > 0$ and $(\operatorname{Re}\hat{P}_1)\#(\operatorname{Re}\hat{P}_1) - (\operatorname{Re}\hat{P}_1)^2 \in S^0$. Since $Q/2 - \epsilon(\operatorname{Re}\hat{P}_1)^2 \geq 0$ choosing $\epsilon > 0$ small we get

$$(Qu, u)/2 - \epsilon\|(\operatorname{Re}\hat{P}_1)u\|^2 \geq -C\|u\|^2$$

by the Fefferman-Phong inequality and hence the assertion.

For $\operatorname{Im}([D_0 - \lambda, Q]u, u)$ one has the estimate (5.17) because $|\{\xi_0 - \lambda, Q\}| \leq CQ$. We turn to estimate $|\operatorname{Im}([D_0 - \lambda, \operatorname{Re}\hat{P}_1]u, u)|$. Note $2|\operatorname{Im}([D_0 - \lambda, \operatorname{Re}\hat{P}_1]u, u)| \leq \|[D_0 - \lambda, \operatorname{Re}\hat{P}_1]u\|^2 + \|u\|^2$. Since $CQ - \{\xi_0 - \lambda, \operatorname{Re}\hat{P}_1\}^2 \geq 0$ with some $C > 0$, the same argument as above gives

$$C(Qu, u) \geq \|[D_0 - \lambda, \operatorname{Re}\hat{P}_1]u\|^2 - C_1\|u\|^2$$

with some $C_1 > 0$. Thus we conclude that

$$2C(Qu, u) + \text{Im}([D_0 - \lambda, Q + \text{Re}\,\hat{P}_1]u, u) \geq -(2C_1 + 1)\|u\|^2. \tag{5.47}$$

Consider $|\text{Re}(\Lambda_\theta u, (\text{Im}\,\hat{P}_1)u)|$. Note $2|\text{Re}(\Lambda_\theta u, (\text{Im}\,\hat{P}_1)u)| \leq \|\Lambda_\theta u\|^2 + \|(\text{Im}\,\hat{P}_1)u\|^2$ and $CQ - (\text{Im}\,\hat{P}_1)^2 \geq 0$ with some $C > 0$ we obtain

$$2|\text{Re}(\Lambda_\theta u, (\text{Im}\,\hat{P}_1)u)| \leq \|\Lambda_\theta u\|^2 + C(Qu, u) + C_1\|u\|^2 \tag{5.48}$$

with some $C_1 > 0$. With $E(u) = \|\Lambda_\theta u\|^2 + (Qu, u) + \theta^2\|u\|^2$ we get

$$\theta^{-1}\|P_\theta u\|^2 \geq c\frac{d}{dx_0}E(u) + c\,\theta E(u)$$

with some $c > 0$ for $\theta \geq \theta_0$. Integrating this inequality in x_0 we get

Lemma 5.16 *Assume that P satisfies the Levi condition on Σ. Then there exists $T > 0$ such that for any $u \in C^2([-T, T]; C_0^\infty(\mathbb{R}^n))$ vanishing in $x_0 \leq \tau$, $|\tau| \leq T$ we have*

$$\{\|\Lambda_\theta u(t, \cdot)\|^2 + (Qu(t), u(t)) + \theta^2\|u(t, \cdot)\|^2\}$$

$$+\theta\int_\tau^t \{\|\Lambda_\theta u(x_0, \cdot)\|^2 + (Qu, u) + \theta^2\|u(x_0, \cdot)\|^2\}dx_0 \tag{5.49}$$

$$\leq C\theta^{-1}\int_\tau^t \|P_\theta u(x_0, \cdot)\|^2 dx_0.$$

We estimate higher order derivatives of u. Consider $\langle D'\rangle^s P_\theta = P_\theta\langle D'\rangle^s - [P_\theta, \langle D'\rangle^s]$. With $R_s = [P_\theta, \langle D'\rangle]\langle D'\rangle^{-s}$ one can write

$$\langle D'\rangle^s P_\theta = (P_\theta - R_s)\langle D'\rangle^s.$$

Lemma 5.17 *Assume that $|M - \Lambda| \leq C\sqrt{Q}$ with some $C > 0$. Then we have*

$$|(R_s u, \Lambda_\theta u)| \leq C_s(\|\Lambda_\theta u\|^2 + \text{Re}(Qu, u) + \|u\|^2).$$

Proof Recall that $P_\theta = -M_\theta\Lambda_\theta + Q + \hat{P}_1 + B_0\Lambda_\theta + \hat{P}_0$. Then it is easy to see

$$[M_\theta\Lambda_\theta, \langle D'\rangle^s]\langle D'\rangle^{-s} = a\Lambda_\theta + bM_\theta + R_1$$

$$= (a + b)\Lambda_\theta + b(M - \Lambda) + R_2$$

with some $a, b, R_i \in \text{Op}S^0$. On the other hand we have $[Q, \langle D'\rangle^s]\langle D'\rangle^{-s} + i\,\text{Op}(T) \in \text{Op}S^0$ with $T = \{Q, \langle \xi'\rangle^s\}\langle \xi'\rangle^{-s}$. From the non-negativity of Q one has $|T^2| \leq CQ$. Noting that $\|Tu\|^2 = (\text{Op}(T\#T)u, u)$ and $T\#T - T^2 \in S^0$ the Fefferman-Phong inequality shows that $C\mathrm{Re}(Qu, u) - \|Tu\|^2 \geq -C\|u\|^2$. Since $|M - \Lambda|^2 \leq CQ$ the Fefferman-Phong inequality again shows

$$C\,\mathrm{Re}(Qu, u) - \|(M - \Lambda)u\|^2 \geq -C\|u\|^2.$$

Since $[\hat{P}_1 + \hat{P}_0, \langle D' \rangle^s]\langle D' \rangle^{-s} \in \mathrm{Op}S^0$ and $[B_0\Lambda_\theta, \langle D' \rangle^s]\langle D' \rangle^{-s} = c_0\Lambda_\theta + c_1$ with $c_j \in \mathrm{Op}S^{j-1}$ one has

$$|(R_s u, \Lambda_\theta)| \leq C\{\|\Lambda_\theta u\|^2 + \mathrm{Re}(Qu, u) + \|u\|^2\}$$

which is the desired assertion. $\qquad\square$

Thanks to Lemmas 5.16 and 5.17 we have

Proposition 5.17 *For any $s \in \mathbb{R}$ and any $u \in \cap_{j=0}^2 C^j([-T, T]; H^{s+2-j}(\mathbb{R}^n))$ vanishing in $x_0 \leq \tau$, $|\tau| \leq T$ we have*

$$\|\Lambda_\theta u(t)\|_s^2 + \theta^2\|u(t)\|_s^2 + \theta\int_\tau^t (\|\Lambda_\theta u(x_0)\|_s^2 + \theta^2\|u(x_0)\|_s^2)dx_0$$

$$\leq C_s\theta^{-1}\int_\tau^t \|P_\theta u(x_0)\|_s^2 dx_0. \tag{5.50}$$

Corollary 5.6 *There exists $T > 0$ such that for any $s \in \mathbb{R}$ there is $C_s > 0$ such that for any $u \in \cap_{j=0}^2 C_+^j([-T, T]; H^{s+2-j})$ we have*

$$\|D_0 u(t)\|_{s-1}^2 + \|u(t)\|_s^2 \leq C_s\int_0^t \|Pu(x_0)\|_s^2 dx_0.$$

By Lemma 5.3 it follows that P^* satisfies the Levi condition. Thus the energy estimates in Proposition 5.17 holds for P^*. Reversing the time direction we have

Proposition 5.18 *There exists $T > 0$ such that for any $s \in \mathbb{R}$ there are $\theta_s > 0$ and $C_s > 0$ such that for any $\theta \geq \theta_s$ and any $u \in \cap_{j=0}^2 C^j([-T, T]; H^{s+2-j}(\mathbb{R}^n))$ vanishing in $x_0 \geq \tau$, $|\tau| \leq T$ one has*

$$\|\Lambda_\theta u(t)\|_s^2 + \theta^2\|u(t)\|_s^2 + \theta\int_t^\tau (\|\Lambda_\theta u(x_0)\|_s^2 + \theta^2\|u(x_0)\|_s^2)dx_0$$

$$\leq C_s\theta^{-1}\int_t^\tau \|P_\theta^* u(x_0)\|_s^2 dx_0, \quad -T \leq t \leq \tau.$$

Remark 5.4 The proof of Proposition 5.17 is based on the "local elementary factorization" and not on "microlocal elementary factorization". One can find some related discussions about local and microlocal elementary factorizations in [90].

5.7   Proof of Main Results

Proof of Theorems 5.1 and 5.2: In order to prove the existence part of theorems, in
view of Theorem 4.1, it is enough to show that P has a parametrix at $(0, 0, \xi')$ for
any $|\xi'| = 1$. From Corollaries 4.1, 5.1 and 5.2 it is enough to prove that a second
order operator $P \in \Psi^2[D_0]$ satisfying the assumption in Theorems 5.1 or 5.2 near
$(0, 0, \xi')$ has a parametrix at $(0, 0, \xi')$. By virtue of Propositions 5.3, 5.7 and 5.11
one can write

$$P = -\tilde{\Lambda}^2 + (-2\lambda - 2a + B)\tilde{\Lambda} + Q + \hat{P}_1 + \hat{P}_0$$

where $\tilde{\Lambda} = D_0 - \lambda - a$ with real $\lambda \in S^1$ and $a, B \in S^0$. Here $a = 0$ if Proposition 5.3
holds. From (5.16), (5.33) and (5.41) we have

$$(((1 - \bar{\epsilon})Q + \text{Re}\,\hat{P}_1)u, u) \geq -C\|u\|^2.$$

Thus (E) is verified thanks to Propositions 5.5, 5.9 and 5.14. Note that from
Proposition 5.6, 5.10 and 5.15 one can find $T > 0$ such that the following inequality
holds for any $s \in \mathbb{R}$ and any $u \in C^2([-T, T]; H^\infty(\mathbb{R}^n))$ vanishing in $x_0 \geq T$

$$\int_t^T \|u(x_0)\|_s^2 dx_0 \leq C(s, T) \int_t^T \|P^* u(x_0)\|_s^2 dx_0. \tag{5.51}$$

Proposition 5.19 *For any* $f \in C_+^0([-T, T]; H^s)$ *there is* $u \in \cap_{j=0}^1 C_+^j((-T, T); H^{s-j})$
such that $Pu = f$ *and*

$$N_s(u(t)) \leq C_s \int^t \|f(\tau)\|_s^2 d\tau, \quad -T \leq t \leq T.$$

Proof Let $f \in C_+^0(I; H^s)$ and then $f \in L^2((0, T); H^s)$. If $v \in C_0^\infty((-T, T) \times \mathbb{R}^n)$
then (5.51) implies

$$\left| \int_0^T (f, v)dt \right| \leq C \left(\int_0^T \|f\|_s^2 dt \right)^{1/2} \left(\int_0^T \|P^* v\|_{-s}^2 dt \right)^{1/2}.$$

Applying the Hahn-Banach theorem to the anti-linear form: $P^* v \mapsto \int_0^T (f, v)dt$ we
conclude that there exists $u \in L^2((0, T); H^s)$ such that

$$\int_0^T (f, v)dt = \int_0^T (u, P^* v)dt \tag{5.52}$$

holds for any $v \in C_0^\infty((-T, T) \times \mathbb{R}^n)$. This proves that $Pu = f$ in $(0, T) \times \mathbb{R}^n$.
We now take $f_n \in \mathscr{S}(\mathbb{R}^{n+1})$ vanishing in $x_0 < 0$ such that $f_n \to f$ in $C_+^0(I; H^s)$
as $n \to \infty$. Since $f_n \in L^2((0, T); H^{s+3})$ we obtain the corresponding solution

$u_n \in L^2((0, T); H^{s+3})$. One can conclude that $D_0^j u_n \in L^2((0, T); H^{s+3-j})$ from the equation $P u_n = f_n$ (see [34, Appendix B]). Thus we have $D_0^j u_n \in C^j([0, T]; H^{s+2-j})$ for $j = 0, 1, 2$. Since (5.52) holds for any $v \in C_0^\infty((-T, T) \times \mathbb{R}^n)$ we conclude that $D_0^j u_n(0, x') = 0$ for $j = 0, 1$. Thus defining $u_n = 0$ in $x_0 < 0$ we conclude $P u_n = f_n$ in $(-T, T) \times \mathbb{R}^n$ and $u_n \in \cap_{j=0}^2 C_+^j((-T, T); H^{s+2-j})$. Thanks to Corollaries 5.3–5.5 the sequence u_n converges to some $u \in \cap_{j=0}^1 C_+^j((-T, T); H^{s-j})$. Since $N_s(u_n)$ is convergent we conclude the assertion. $\qquad \square$

Since P verifies Definition 4.6 taking Proposition 5.19 into account one can apply Theorem 4.2 to conclude that P has a parametrix at $(0, 0, \xi')$. Therefore the existence of solutions is proved.

Proof of Theorem 5.3 In case of Theorem 5.3, the operator P in Proposition 5.17 and Corollary 5.6 is nothing but the original operator then to prove the existence of solution it suffices to repeat the same arguments proving Proposition 5.19.

In Theorems 5.2 and 5.3, the necessary part follows from Theorem 1.12 because $\mathrm{Tr}^+ F_p = 0$. $\qquad \square$

Remark 5.5 It is clear that if $f \in C_+^1(I; H^\infty(\mathbb{R}^n))$ then we obtain C^m solution u near the origin.

Remark 5.6 We have only a control of H^s norm but no control of $H^{s+\kappa}$ norm in (5.50), no matter how small $\kappa > 0$. Therefore the estimates (5.50) is too weak to apply Theorem 4.2.

We turn to the uniqueness of solution. We choose a new system of local coordinates y so that

$$y_0 = x_0 + \epsilon \sum_{j=1}^n x_j^2, \quad y_j = x_j, \quad j = 1, 2, \ldots, n \qquad (5.53)$$

which is so called Holmgren transform where $\epsilon > 0$ is a small positive constant. Denote by $\tilde{p}(y, \eta)$ the principal symbol of P in these coordinates. It is clear that

$$\tilde{p}(y, \eta) = p(y_0 + \epsilon|y'|^2, y', \eta_0, \eta' - 2\epsilon\eta_0 y'). \qquad (5.54)$$

The geometrical formulation of the assumptions, assumption (1.5), spectral type 1(2), $\mathrm{Tr}^+ F_p$, (no) existence of tangent bicharacteristics, no transition of spectral type, implies that *they are independent of the choice of local coordinates system.* We also note that P_{sub} is invariantly defined at double characteristics (see [34], for example).

Lemma 5.18 *Assume that $p(x, \xi)$ is a hyperbolic polynomial in the direction ξ_0. For any $R > 0$ one can find $\epsilon_0 > 0$ such that $\tilde{p}(y, \eta)$ is a hyperbolic polynomial in the direction η_0 for $|y| \leq R$ if $|\epsilon| < \epsilon_0$.*

Proof Write $q(s, \tau, \eta', y, \epsilon) = p(y_0 + \epsilon|y'|^2, y', \tau + s, \eta' - 2\epsilon s y')$ which is a polynomial in s of degree at most m with the leading term $a_0(y, \epsilon)s^m$. It is clear that $a_0(y,0) = 1$ and one can find $\epsilon_0 > 0$ so that $a_0(y,\epsilon) \geq 1/2$ for $|\epsilon| \leq \epsilon_0$ and $|y| \leq R$. Therefore the roots s of

$$q(s,\tau,\eta',y,\epsilon) = 0$$

are continuous in (τ,η',y,ϵ) for $|y| \leq R$ and $|\epsilon| \leq \epsilon_0$. We show $q(s,\tau,\eta',y,\epsilon) \neq 0$ if $\operatorname{Im}\tau \leq 0$, $\operatorname{Im}s < 0$ and $|y| \leq R$, $|\epsilon| \leq \epsilon_0$. Suppose otherwise so that there exists $(\hat{s},\hat{\tau},\hat{\eta}',\hat{y},\hat{\epsilon})$ with $\operatorname{Im}\hat{\tau} \leq 0$, $\operatorname{Im}\hat{s} < 0$, $|\hat{y}| \leq R$, $|\hat{\epsilon}| \leq \epsilon_0$ such that $q(\hat{s},\hat{\tau},\hat{\eta}',\hat{y},\hat{\epsilon}) = 0$. From the continuity of the root s, moving $\hat{\tau}$ little bit if necessary we may assume that $\operatorname{Im}\hat{\tau} < 0$. Consider

$$F(\theta) = \min \operatorname{Im}s(\theta)$$

where the minimum is taken over all roots $s(\theta)$ of $q(s,\hat{\tau},\theta\hat{\eta}',\theta\hat{y},\theta\hat{\epsilon}) = 0$. Since $F(1) < 0$ and $F(0) > 0$ because $q(s,\hat{\tau},0,0,0) = (s + \hat{\tau})^m$ and $F(\theta)$ is continuous in θ there exist $\hat{\theta}$ and $s(\hat{\theta})$ such that $\operatorname{Im}s(\hat{\theta}) = 0$ which contradicts with the assumption that p is a hyperbolic polynomial in the direction ξ_0. A repetition of the same arguments shows that $q(s,\tau,\eta',y,\epsilon) \neq 0$ if $\operatorname{Im}\tau \geq 0$, $\operatorname{Im}s > 0$ and $|y| \leq R$, $|\epsilon| \leq \epsilon_0$. We now take $\tau = 0$ then for $|y| \leq R$, $|\epsilon| \leq \epsilon_0$ one has

$$\tilde{p}(y,s,\eta') = q(s,0,\eta',y,\epsilon) = 0 \implies \operatorname{Im}s = 0.$$

Thus we conclude that $\tilde{p}(y,\eta) = 0$ has only real roots η_0 for $|\epsilon| \leq \epsilon_0$ and $|y| \leq R$ which proves the assertion. $\square$

Lemma 5.19 *Assume that the assumption in Theorems 5.1 or 5.2 or 5.3 is satisfied. Then one can find a neighborhood U of the origin, positive numbers $\bar{\epsilon} > 0$ and $\epsilon > 0$ such that for any $f(x) \in C_0^\infty(U)$ with $\operatorname{supp}f \subset \{x \mid x_0 \leq \bar{\epsilon} - \epsilon|x'|^2\}$ there exists $v(x) \in C^m(U)$ with $\operatorname{supp}v \subset \{x \mid x_0 \leq \bar{\epsilon} - \epsilon|x'|^2\}$ which satisfies $P^*v = f$ in U.*

Proof Take $\bar{\epsilon} > 0$ and $\epsilon > 0$ enough small and a new system of local coordinates y in (5.53). Denote $g(y) = f(x)$ then $g(y) \in C_0^\infty(\tilde{U})$, $\tilde{U} = \{y \mid (y_0 - \epsilon|y'|^2, y') \in U\}$ vanishes in $y_0 \geq \bar{\epsilon}$. Note that the assumption in Theorems 5.1 or 5.2 or 5.3 is satisfied by P^* in view of Lemma 5.3 and is independent of the choice of local coordinates system, as remarked above. The principal symbol of P^* in the coordinates y is given by (5.54) which is a hyperbolic polynomial in the direction η_0 by Lemma 5.18 provided that $\epsilon > 0$ is small. Therefore, choosing U and hence $\tilde{U}$ small, one can apply the existence results to the equation $P^*w = g$ which yields a solution $w(y) \in C^m(\tilde{U})$ vanishing in $y_0 \geq \bar{\epsilon}$. Denoting $v(x) = w(y)$ it is clear that the support of v is contained in the set $\{x \mid x_0 \leq \bar{\epsilon} - \epsilon|x'|^2\}$. $\square$

We now prove the uniqueness part. Assume that C^m function u vanishing in $x_0 < 0$ satisfies $Pu = 0$ in a neighborhood of the origin. Take v in Lemma 5.19 and consider

$$0 = \int_0^{\bar{\epsilon}} \int_{\mathbb{R}^n} Pu \cdot v \, dx_0 dx' = \int_0^{\bar{\epsilon}} \int_{\mathbb{R}^n} u \cdot P^*v \, dx_0 dx' = \int_0^{\bar{\epsilon}} \int_{\mathbb{R}^n} u \cdot f \, dx_0 dx'.$$

Since $f \in C_0^\infty$ with supp$f \subset \{x \mid x_0 \leq \bar{\epsilon} - \epsilon|x'|^2\}$ is arbitrary we conclude that $u = 0$ in the set $\{x \mid 0 \leq x_0 \leq \bar{\epsilon} - \epsilon|x'|^2\}$.

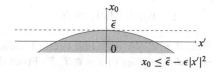

5.8 Melin-Hörmander Inequality

To make more precise study about the sufficiency of the (non-strict) IPH condition for the C^∞ well-posedness we need to improve the Melin's inequality. To do so we first study the lower bound of Q.

We now change notation slightly and study the Weyl quantized pseudodifferential operator A with classical real symbol $A(x, \xi) \in S_{phg}^m(\mathbb{R}^n \times \mathbb{R}^n)$;

$$A(x, \xi) \sim A_2(x, \xi) + A_1(x, \xi) + \cdots$$

where we assume $a(x, \xi) = A_2(x, \xi) \geq 0$. We also assume that the doubly characteristic set Σ of a is a smooth manifold verifying

$$\dim T_\rho \Sigma = \dim \operatorname{Ker} F_a(\rho), \quad \rho \in \Sigma, \tag{5.55}$$

$$\operatorname{rank}\left(\sum_{j=1}^n d\xi_j \wedge dx_j|_\Sigma\right) = \text{const.} \tag{5.56}$$

In what follows we denote by Q_ρ the polar form of the Hesse matrix of a at $\rho \in \Sigma$. Recall that all eigenvalues of $F_a(\rho)$ are pure imaginary for $\rho \in \Sigma$. We denote by V_λ the space of generalized eigenvectors of F_a belonging to the eigenvalue λ. We set

$$V_\rho^+ = \bigoplus_{\mu > 0} V_{i\mu}.$$

Let V_ρ^{0r} denote the real subspace of V_ρ^0, the generalized eigenspace associated to the eigenvalue 0, and N_ρ the kernel of $F_a(\rho)$.

Lemma 5.20 *If $0 \neq v \in V_\rho^+$ then $Q_\rho(v, \bar{v}) > 0$.*

Proof In the proof we drop the suffix ρ. Let $F_a v = i\mu v$. Since $F_a \bar{v} = -i\mu \bar{v}$ then

$$0 \leq Q(v + \bar{v}, v + \bar{v}) = 2Q(v, \bar{v}).$$

This shows that $Q(\operatorname{Re} v, \operatorname{Re} v) = 0$ if $Q(v, \bar{v}) = 0$ and hence $F_a \operatorname{Re} v = 0$. From $F_a \operatorname{Re} v = -\mu \operatorname{Im} v$ it follows that $\operatorname{Im} v = 0$ because $\mu \neq 0$. Repeating the same argument we get $\operatorname{Re} v = 0$. This is a contradiction. $\qquad\square$

By assumption the dimension of V_ρ^+ is constant when $\rho \in \Sigma$ so that one can choose a basis $v_1(\rho), \ldots, v_k(\rho)$ for V_ρ^+ which is smooth in $\rho \in \Sigma$ and verifies

$$Q_\rho(v_i(\rho), \bar{v}_j(\rho)) = 2\delta_{ij}$$

thanks to Lemma 5.20. Note that V_ρ^{0r}/N_ρ is a real vector bundle over Σ. We remark that $Q_\rho(v, v) = 0$ for real $v \neq 0$ implies that $v \in T_\rho\Sigma$. From this one can choose a basis $v_{k+1}(\rho), \ldots, v_{k+\ell}(\rho)$ for V_ρ^{0r}/N_ρ such that

$$Q_\rho(v_i(\rho), v_j(\rho)) = \delta_{ij}, \quad k+1 \leq i, j \leq k+\ell.$$

We set $L_j(\rho; v) = Q_\rho(v_j(\rho), v)$ for $1 \leq j \leq k+\ell$ which is smooth in $\rho \in \Sigma$. We examine that for real v we have

$$\sum_{j=1}^{k+\ell} |L_j(\rho; v)|^2 = Q_\rho(v, v).$$

To see this we note that $Q_\rho(v_i(\rho), v_j(\rho)) = 0$, $1 \leq i, j \leq k$ and $Q_\rho(v_i(\rho), v_j(\rho)) = 0$ for $1 \leq i \leq k$, $k+1 \leq j \leq k+\ell$ because $Q_\rho(V_\lambda, V_\mu) = 0$ if $\lambda + \mu \neq 0$. Writing

$$v = \sum_{j=1}^{k} \alpha_j v_j(\rho) + \sum_{j=1}^{k} \bar{\alpha}_j \bar{v}_j(\rho) + \sum_{j=k+1}^{k+\ell} \gamma_j v_j(\rho)$$

we see that $\sum_{j=1}^{k+\ell} |L_j(\rho; v)|^2 = 2\sum_{j=1}^{k} |\alpha_j|^2 + \sum_{j=k+1}^{k+\ell} \gamma_j^2$. On the other hand we see easily that

$$Q_\rho(v, v) = 2\sum_{j=1}^{k} |\alpha_j|^2 + \sum_{j=k+1}^{k+\ell} \gamma_j^2$$

and hence the assertion. Denote $\Lambda(\rho; v)$ by

$$(\mathsf{Re}\, L_1(v), \mathsf{Im}\, L_1(v), \ldots, \mathsf{Re}\, L_k(v), \mathsf{Im}\, L_k(v), L_{k+1}(v), \ldots, L_{k+\ell}(v))$$

where $L_j(v) = L_j(\rho; v)$ so that we have

$$Q_\rho(v, v) = \sum_{j=1}^{2k+\ell} \Lambda_j(\rho; v)^2.$$

Since one can write $a(\rho) = \sum_{j=1}^{2k+\ell} b_j(\rho)^2$ we have

$$Q_\rho(v) = \sum_{j=1}^{2k+\ell} db_j(\rho; v)^2 = \sum_{j=1}^{2k+\ell} \Lambda_j(\rho; v)^2.$$

Since $db_j(\rho; \cdot)$ are linearly independent one has $\Lambda_j(\rho; \cdot) = \sum_{j=1}^{2k+\ell} O_{jk}(\rho) db_k(\rho; \cdot)$ where $O(\rho) = (O_{jk}(\rho))$ is a non-singular matrix which is smooth in $\rho \in \Sigma$. Since $\mathbb{R}^{2k+\ell} \ni v \mapsto (db_1(\rho; v), \ldots, db_{2k+\ell}(\rho; v))$ is surjective we conclude that $O(\rho)$ is orthogonal. Let us define

$$c_j(\rho) = \sum_{i=1}^{2k+\ell} O_{ji}(\rho) b_i(\rho)$$

and hence $dc_j(\rho; v) = \Lambda_j(\rho; v)$ for $\rho \in \Sigma$ and $a(\rho) = \sum_{j=1}^{2k+\ell} c_j(\rho)^2$. Let $F_a(\rho) v_j(\rho) = i\mu v_j(\rho)$ then

$$\sigma(L_j(\rho; \cdot), \bar{L}_j(\rho; \cdot)) = -\mu^2 \sigma(\bar{v}_j(\rho), v_j(\rho)),$$

$$2 = Q_\rho(v_j(\rho), \bar{v}_j(\rho)) = i\mu\sigma(\bar{v}_j(\rho), v_j(\rho))$$

and hence

$$\sum_{j=1}^{k} \sigma(\operatorname{Im} L_j(\rho; \cdot), \operatorname{Re} L_j(\rho; \cdot)) = \sum_{j=1}^{k} \{\operatorname{Im} L_j(\rho; \cdot), \operatorname{Re} L_j(\rho; \cdot)\} = 2 \operatorname{Tr}^+ F_a(\rho)$$

for $\rho \in \Sigma$. We denote

$$\begin{cases} X_j(x, \xi) = c_{2j-1}(x, \xi) + ic_{2j}(x, \xi), & j = 1, \ldots, k, \\ X_j(x, \xi) = c_{k+j}(x, \xi), & j = k+1, \ldots, k+\ell. \end{cases}$$

Noting that

$$\bar{X}_j \# X_j = |X_j|^2 + \frac{1}{2i}\{\bar{X}_j, X_j\} + R_1, \quad R_1 \in S^0$$

and $A = \operatorname{Op}(a + A_1) + R_2, R_2 \in \operatorname{Op}S^0$ we set

$$B = a + A_1 - \sum_{j=1}^{k+\ell} \bar{X}_j \# X_j = A_1 + \sum_{j=1}^{k} \frac{i}{2}\{\bar{X}_j, X_j\} + R_3$$

where $R_3 \in S^0$. We assume that

$$A_1 + \text{Tr}^+ F_a(\rho) \geq 0, \quad \rho \in \Sigma. \tag{5.57}$$

Denoting the principal symbol of B by b_1 we see from (5.57) that $b_1(x, \xi) \geq 0$ on Σ. Let q be an extension of b_1 outside Σ such that $q(x, \xi) \geq 0$. Then one can write

$$b_1(x, \xi) - q(x, \xi) = \sum_{j=1}^{k+\ell} (\bar{r}_j X_j + r_j \bar{X}_j) = \sum_{j=1}^{k+\ell} (\bar{r}_j \# X_j + \bar{X}_j \# r_j) + R_0$$

with $R_0 \in S^0$ because $b_1(x, \xi) - q(x, \xi)$ is real. Then $a(x, \xi) + A_1(x, \xi)$ is written

$$B(x, \xi) + \sum_{j=1}^{k+\ell} \bar{X}_j \# X_j = q(x, \xi) + \sum_{j=1}^{k+\ell} \overline{(X_j + r_j)} \# (X_j + r_j) + R_0'$$

where $R_0' \in S^0$. Then we have

$$(Au, u) \geq (qu, u) + \sum_{j=1}^{k+\ell} \|(X_j + r_j)u\|^2 - C\|u\|^2 \geq -C\|u\|^2.$$

We summarize what we have proved in

Theorem 5.4 ([32]) *Let A be a pseudodifferential operator with classical symbol $A_2 + A_1 + \cdots$. Assume that $a = A_2 \geq 0$ and A_1 verifies the assumptions (5.55)–(5.57) (in particular A_1 is assumed to be real). Then we have*

$$(Au, u) \geq -C\|u\|^2.$$

In [89], we find more detailed discussions on this inequality, called the Melin-Hörmander inequality.

Let us consider

$$P = -D_0^2 + A$$

where A is a pseudodifferential operator with classical symbol $A_2 + A_1 + \cdots$ which is *real* and satisfies all conditions in Theorem 5.4. From Proposition 5.2 it follows that

$$2\text{Im}(P_\theta u, \Lambda_\theta u) \geq \frac{d}{dx_0}(\|\Lambda_\theta u\|^2 + (Au, u) + \theta^2 \|u\|^2)$$

$$+ \theta \|\Lambda_\theta u\|^2 + 2\theta(Au, u) + \text{Im}([D_0, A]u, u) + \theta^3 \|u\|^2$$

where $\Lambda_\theta = D_0 - i\theta$. Thus, for example, if the inequality

$$2\theta(Au, u) + \mathsf{Im}([D_0, A]u, u) \geq -C\|u\|^2. \tag{5.58}$$

holds for large θ with some $C > 0$ we get an energy estimate. Thanks to Theorem 5.4 one has $(Au, u) \geq -C\|u\|^2$ but this will not be sufficient to obtain (5.58). This is closely related to the fact that the IPH condition is not sufficient in general for the Cauchy problem to be C^∞ well-posed (this point will be discussed in Sect. 8.6), while we find some positive results on the sufficiency for C^∞ well-posedness in [89] when $\{\rho \in \Sigma \mid A_1 + \mathrm{Tr}^+ F_a(\rho) = 0\}$ is not empty (see also Sect. 8.6).

Chapter 6
Tangent Bicharacteristics and Ill-Posedness

Abstract In this chapter we provide a homogeneous second order differential operator P of spectral type 2 on Σ with polynomial coefficients with a bicharacteristic tangent to the double characteristic manifold and satisfies the Levi condition for which the Cauchy problem is ill-posed in the Gevrey class of order $s > 5$, in particular the Cauchy problem is C^∞ ill-posed. We also discuss some open questions on the Cauchy problem for P of spectral type 2 with tangent bicharacteristics and no transition. In the last section we make some remarks on the Cauchy problem for P with transition of spectral type under the assumption of no tangent bicharacteristics.

6.1 Non Solvability in C^∞ and in the Gevrey Classes

In this chapter we study the following model operator

$$P_{mod} = -D_0^2 + 2x_1 D_0 D_n + D_1^2 + x_1^3 D_n^2. \tag{6.1}$$

It is worthwhile to note that if we make the change of coordinates

$$y_j = x_j \ (0 \le j \le n-1), \quad y_n = x_n + x_0 x_1$$

which leaves the initial plane $x_0 = $ const., invariant the operator P_{mod} is written in these coordinates as

$$P_{mod} = -D_0^2 + (D_1 + x_0 D_n)^2 + (x_1\sqrt{1 + x_1}D_n)^2 = -D_0^2 + A^2 + B^2$$

of course, $|x_1|$ must be small. Here we have $A^* = A$ and $B^* = B$ so that P_{mod} is "in divergence form" while $[D_0, A] \ne 0$ and $[A, B] \ne 0$.

Denote by $p(x, \xi)$ the symbol of P_{mod} then $p(x, \xi)$ is nothing but (3.9) with $q = 1$ and $\epsilon = 1$. It is clear that the double characteristic manifold near the double characteristic point $\bar{\rho} = (0, (0, \ldots, 0, 1)) \in \mathbb{R}^{2(n+1)}$ is given by

$$\Sigma = \{(x, \xi) \in \mathbb{R}^{2(n+1)} \mid \xi_0 = 0, x_1 = 0, \xi_1 = 0\} \tag{6.2}$$

© Springer International Publishing AG 2017

T. Nishitani, *Cauchy Problem for Differential Operators with Double Characteristics*, Lecture Notes in Mathematics 2202,
DOI 10.1007/978-3-319-67612-8_6

and the localization of p at $\rho \in \Sigma$ is $p_\rho(x, \xi) = -\xi_0^2 + 2x_1\xi_0 + \xi_1^2$. This is just (3) in Lemma 1.7 with $k = l = 1$ where ξ_1 and x_1 is exchanged. Since $(x_1, \xi_1) \mapsto (\xi_1, -x_1)$ is a symplectic change of the system of coordinates then we see

$$\operatorname{Ker} F_p^2(\rho) \cap \operatorname{Im} F_p^2(\rho) \neq \{0\}, \quad \rho \in \Sigma.$$

Therefore p verifies (1.5) and is of spectral type 2 on Σ. As was seen in Lemma 3.2 there exists a bicharacteristic of p which lands tangentially on Σ. Here we give the explicit form of such a bicharacteristic. Denote the following curve by γ

$$\begin{cases} x_1 = -x_0^2/4, \quad x_n = x_0^5/8, \quad \xi_0 = 0, \quad \xi_1 = x_0^3/8, \quad \xi_n = 1, \\ x_j = \text{constant}, \quad \xi_j = \text{constant}, \quad 2 \le j \le n-1 \end{cases} \tag{6.3}$$

which is parametrized by x_0. Since it is easy to check that

$$\frac{dx_1}{dx_0} = \frac{\partial p}{\partial \xi_1}(\gamma) \Big/ \frac{\partial p}{\partial \xi_0}(\gamma), \quad \frac{dx_n}{dx_0} = \frac{\partial p}{\partial \xi_n}(\gamma) \Big/ \frac{\partial p}{\partial \xi_0}(\gamma), \quad \frac{d\xi_1}{dx_0} = -\frac{\partial p}{\partial x_1}(\gamma) \Big/ \frac{\partial p}{\partial \xi_0}(\gamma)$$

so that γ is an integral curve of H_p. Since Σ is given by (6.2) then it is clear that $\gamma(x_0)$ lands on Σ tangentially as $\pm x_0 \downarrow 0$.

We are now concerned with the Cauchy problem for P_{mod}. We start with the definition of the Gevrey classes [28, 30, 33].

Definition 6.1 We say $f(x) \in \gamma^{(s)}(\mathbb{R}^n)$, the Gevrey class of order $s \,(\ge 1)$ if for any compact set $K \subset \mathbb{R}^n$ there exist $C > 0, h > 0$ such that

$$|\partial_x^\alpha f(x)| \le Ch^{-|\alpha|}(\alpha!)^s, \quad x \in K, \quad \forall \alpha \in \mathbb{N}^n.$$

We also denote $\gamma_0^{(s)}(\mathbb{R}^n) = C_0^\infty(\mathbb{R}^n) \cap \gamma^{(s)}(\mathbb{R}^n)$.

Definition 6.2 We say that the Cauchy problem for second order differential operator P is locally solvable in $\gamma^{(s)}$ at the origin if for any $\Phi = (u_0, u_1) \in (\gamma^{(s)}(\mathbb{R}^n))^2$, there exists a neighborhood U_Φ of the origin such that there exists $u(x) \in C^\infty(U_\Phi)$ satisfying

$$\begin{cases} Pu = 0 \quad \text{in} \quad U_\Phi, \\ D_0^j u(0, x') = u_j(x'), \quad j = 0, 1, \quad (0, x') \in U_\Phi \cap \{x_0 = 0\}. \end{cases}$$

We prove the next result following [6] and [82].

Theorem 6.1 *If $s > 5$ then the Cauchy problem for P_{mod} is not locally solvable in $\gamma^{(s)}$. In particular the Cauchy problem for P_{mod} is not C^∞ solvable.*

The idea to prove Theorem 6.1 is to find a family of exact solutions U_ρ to $P_{mod}U_\rho = 0$ such that $U_\rho(0, x')$ behaves like $e^{\zeta(\rho)}e^{i\rho^5 x_n}V(\tilde{x}, \rho)$, $\tilde{x} = (x_1, \ldots, x_{n-1})$, where $\zeta(\rho)$ verifies $\text{Re}\,\zeta(\rho) \geq c\rho$ and $V(\tilde{x}, \rho) \in \mathscr{S}(\mathbb{R}^{n-1})$ uniformly in ρ. Then applying some duality arguments to $0 = \int_0^T (P_{mod}U_\rho, u)dx_0$ where $u(x)$ is assumed to be a solution to $P_{mod}u = 0$, we can estimate the decay rate in ρ of the integral

$$\int e^{i\rho^5 x_n}u(0, x')dx_n$$

which shows that the Cauchy data $u(0, x')$ must belong to $\gamma^{(5)}$ with respect to x_n.

6.2 Construction of Solutions, Zeros of Stokes Multipliers

We look for a solution to $P_{mod}U = 0$ of the form

$$U(x) = \exp(i\rho^5 x_n + \frac{i}{2}\zeta\rho x_0)w(x_1\rho^2), \quad \zeta \in \mathbb{C}, \ \rho > 0.$$

It is clear that if w verifies

$$w''(x) = (x^3 + \zeta x - \zeta^2\rho^{-2}/4)w(x) \tag{6.4}$$

then $P_{mod}U = 0$. Taking this into account we study the following ordinary differential equation

$$w''(x) = (x^3 + \zeta x + \epsilon)w(x) \tag{6.5}$$

where ζ, ϵ are complex numbers and ϵ will be eventually small enough. We briefly recall, for this special situation, the general theory of subdominant solutions of Eq. (6.5), according to the exposition for instance in the book of Sibuya [93]. Recall that [93, Theorem 6.1] states that the differential equation (6.5) has a solution $w(x; \zeta, \epsilon) = \mathscr{Y}(x; \zeta, \epsilon)$ such that

(i) $\mathscr{Y}(x; \zeta, \epsilon)$ is an entire function of (x, ζ, ϵ),
(ii) $\mathscr{Y}(x; \zeta, \epsilon)$ admits an asymptotic representation

$$\mathscr{Y}(x; \zeta, \epsilon) \sim x^{-3/4}\left(1 + \sum_{N=1}^{\infty} B_N x^{-N/2}\right)\exp\{-E(x; \zeta)\}$$

uniformly on each compact set in the (ζ, ϵ) space as x goes to infinity in any closed subsector of the open sector $|\arg x| < 3\pi/5$ where B_N are polynomials in (ζ, ϵ) and

$$E(x; \zeta) = \frac{2}{5}x^{5/2} + \zeta x^{1/2}.$$

We note that with $\omega = \exp{(i\frac{2\pi}{5})}$ and

$$\mathcal{Y}_k(x; \zeta, \epsilon) = \mathcal{Y}(\omega^{-k}x; \omega^{-2k}\zeta, \omega^{-3k}\epsilon) \tag{6.6}$$

where $k = 0, 1, 2, 3, 4$ then all the five functions $\mathcal{Y}_k(x; \zeta, \epsilon)$ solve (6.5). In particular $\mathcal{Y}_0(x; \zeta, \epsilon) = \mathcal{Y}(x; \zeta, \epsilon)$. Denote

$$Y = x^{-3/4}\left(1 + \sum_{N=1}^{\infty} B_N x^{-N/2}\right) \exp{\{-E(x; \zeta)\}}$$

then we have immediately

(i) $\mathcal{Y}_k(x; \zeta, \epsilon)$ is an entire function of (x, ζ, ϵ),
(ii) $\mathcal{Y}(x; \zeta, \epsilon) \sim Y(\omega^{-k}x; \omega^{-2k}\zeta, \omega^{-3k}\epsilon)$ uniformly on each compact set in the (ζ, ϵ) space as x goes to infinity in any closed subsector of the open sector

$$|\arg x - 2k\pi/5| < 3\pi/5.$$

Let S_k denote the open sector defined by $|\arg x - 2k\pi/5| < \pi/5$. We say that a solution of (6.5) is subdominant in the sector S_k if it tends to 0 as x tends to infinity along any direction in the sector S_k. Analogously a solution is called dominant in the sector S_k if this solution tends to ∞ as x tends to infinity along any direction in the sector S_k. Since

$$\text{Re}\, x^{5/2} > 0 \quad \text{for } x \in S_0 \tag{6.7}$$

and $\text{Re}\, x^{5/2} < 0$ for $x \in S_{-1} = S_4$ and $x \in S_1$ the solution $\mathcal{Y}_0(x; \zeta, \epsilon)$ is subdominant in S_0 and dominant in S_4 and S_1. Similarly $\mathcal{Y}_k(x; \zeta, \epsilon)$ is subdominant in S_k and dominant in S_{k-1} and S_{k+1}. From (6.7) and (6.6) we conclude that $\mathcal{Y}_{k+1}$ and $\mathcal{Y}_{k+2}$ are linearly independent. Therefore $\mathcal{Y}_k$ is a linear combination of those two $\mathcal{Y}_{k+1}$ and $\mathcal{Y}_{k+2}$

$$\mathcal{Y}_k(x; \zeta, \epsilon) = C_k(\zeta, \epsilon)\mathcal{Y}_{k+1}(x; \zeta, \epsilon) + \tilde{C}_k(\zeta, \epsilon)\mathcal{Y}_{k+2}(x; \zeta, \epsilon)$$

where $C_k, \tilde{C}_k$ are called the Stokes multipliers for $\mathcal{Y}_k(x; \zeta, \epsilon)$.

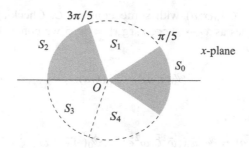

The next proposition is the key to the proof of Theorem 6.1.

Proposition 6.1 $C_0(\zeta, 0)$ *has a zero with negative imaginary part.*
Here we first summarize in the following statement some of known and useful facts about the Stokes multipliers for our particular equation (6.5). Proofs can be found in [93, Chapter 5].

Proposition 6.2 *The following properties hold.*

 (i) $C_0(0, 0) = 1 + \omega$,
 (ii) $\tilde{C}_k(\zeta, \epsilon) = -\omega$, *for all* k, ϵ *and* ζ,
 (iii) $C_k(\zeta, \epsilon) = C_0(\omega^{-2k}\zeta, \omega^{-3k}\epsilon)$, *for all* k, ϵ, ζ *and* $C_0(\zeta, \epsilon)$ *is an entire function of* (ζ, ϵ),
 (iv) $\partial_\zeta C_0(\zeta, \epsilon)\big|_{(\zeta,\epsilon)=(0,0)} \neq 0$.

We now prove key lemmas to the proof of Proposition 6.1.

Lemma 6.1 *Denote* $c_k(\zeta) = C_k(\zeta, 0)$. *Then we have*

$$c_k(\zeta) + \omega^2 c_{k+2}(\zeta)c_{k+3}(\zeta) - \omega^3 = 0 \quad \mod 5.$$

Or otherwise stated with $c(\zeta) = c_0(\zeta) = C_0(\zeta, 0)$ *one has*

$$c(\zeta) + \omega^2 c(\omega\zeta)c(\omega^4\zeta) - \omega^3 = 0, \quad \forall \zeta \in \mathbb{C}.$$

Proof For the proof, see [93, Section 5]. □

Lemma 6.2 ([95]) *We have*

$$\overline{C_0(\zeta, \epsilon)} = \bar{\omega} C_0(\bar{\omega}\bar{\zeta}, \omega\bar{\epsilon}).$$

In particular we have $\overline{c(\zeta)} = \bar{\omega} c(\bar{\omega}\bar{\zeta})$.

Proof Write $a = (\zeta, \epsilon)$ and $\bar{a} = (\bar{\zeta}, \bar{\epsilon})$. Since $w(x) = \overline{\mathscr{Y}_0(\bar{x}; \bar{a})}$ verifies the equation

$$w''(x) = (x^3 + \zeta x + \epsilon)w(x)$$

and hence $w(x) = C\mathscr{Y}_0(x; a)$ with some constant C. Checking the asymptotic behavior of both sides as $x \to +\infty$, $|\arg x| < \pi/5$ we conclude $C = 1$ so that $w(x) = \mathscr{Y}_0(x; a)$ that is

$$\mathscr{Y}_0(\bar{x}; \bar{a}) = \overline{\mathscr{Y}_0(x; a)}.$$

From this we see

$$\overline{\mathscr{Y}_4(x; a)} = \mathscr{Y}_0(\bar{\omega}\bar{x}; \bar{\omega}^2\bar{\zeta}, \omega^2\bar{\epsilon}) = \mathscr{Y}_0(\omega^{-1}\bar{x}; \omega^{-2}\bar{\zeta}, \omega^{-3}\bar{\epsilon})$$

$$= \mathscr{Y}_1(\bar{x}; \bar{\zeta}, \bar{\epsilon}) = \mathscr{Y}_1(\bar{x}; \bar{a}).$$

Similarly $\overline{\mathscr{Y}_1(x; a)} = \mathscr{Y}_4(\bar{x}; \bar{a})$. Thus from $\mathscr{Y}_4(x; a) = C_4(a)\mathscr{Y}_0(x; a) - \omega\mathscr{Y}_1(x; a)$ it follows that

$$\mathscr{Y}_1(\bar{x}; \bar{a}) = \overline{C_4(a)}\mathscr{Y}_0(\bar{x}; \bar{a}) - \bar{\omega}\mathscr{Y}_4(\bar{x}; \bar{a}),$$

$$\mathscr{Y}_4(\bar{x}; \bar{a}) = C_4(\bar{a})\mathscr{Y}_0(\bar{x}; \bar{a}) - \omega\mathscr{Y}_1(\bar{x}; \bar{a}).$$

Multiplying the first identity by ω we get

$$\mathscr{Y}_4(\bar{x}; \bar{a}) = \omega\,\overline{C_4(a)}\mathscr{Y}_0(\bar{x}; \bar{a}) - \omega\mathscr{Y}_1(\bar{x}; \bar{a})$$

which proves $C_4(\bar{\zeta}, \bar{\epsilon}) = \omega\,\overline{C_4(\zeta, \epsilon)}$ and hence the assertion.   □

Lemma 6.3 *The Stokes multiplier $C_0(\zeta, 0)$ has at least one zero $\zeta_0(\neq 0)$.*

Proof Suppose that $c(\zeta) \neq 0$ for all $\zeta \in \mathbb{C}$. Then from Lemma 6.1 it follows that $c(\zeta) \neq \omega^3$ for all $\zeta \in \mathbb{C}$. Since $c(\zeta)$ is an entire function then Picard's Little Theorem implies that $c(\zeta)$ would be constant because $c(\zeta)$ avoids two distinct values 0 and ω^3. But this contradicts (vi) of Proposition 6.2. Since $C_0(0, 0) = 1 + \omega$ from Proposition 6.2 we see that $\zeta_0 \neq 0$.   □

Lemma 6.4 *For real ζ and ϵ we have $C_0(\zeta, \epsilon) \neq 0$. In particular $c(\zeta) \neq 0$ for real ζ.*

Proof Suppose that $C_0(\zeta, \epsilon) = 0$ for some real ζ and ϵ. From Lemma 6.2 it follows that $C_0(\bar{\omega}\zeta, \omega\bar{\epsilon}) = C_0(\bar{\omega}\zeta, \omega\epsilon) = 0$ which contradicts Lemma 6.1.   □

Lemma 6.5 *The closed sector $3\pi/5 \leq \arg\zeta \leq \pi$ is a zero free region of $c(\zeta)$.*

Proof Recall that $\mathscr{Y}_0(x; \zeta, 0)$ is subdominant in the sector $|\arg x| < \pi/5$ and verifies $\mathscr{Y}_0''(x; \zeta, 0) = (x^3 + \zeta x)\mathscr{Y}_0(x; \zeta, 0)$. We put

$$u(x) = \mathscr{Y}_0(\alpha(x + 1); -3\alpha^2, 0)$$

where $-\pi/5 < \arg \alpha < 0$ then we have

$$u''(x) = (\alpha^5 x^3 + 3\alpha^5 x^2 - 2\alpha^5)u(x) = \alpha^5(x^3 + 3x^2 - 2)u(x). \tag{6.8}$$

Note that

$$\mathscr{Y}_0(\alpha(x+1); -3\alpha^2, 0) = c(-3\alpha^2)\mathscr{Y}_1(\alpha(x+1); -3\alpha^2, 0)$$
$$-\omega\mathscr{Y}_2(\alpha(x+1); -3\alpha^2, 0).$$

Suppose that $c(-3\alpha^2) = 0$ so that

$$\mathscr{Y}_0(\alpha(x+1); -3\alpha^2, 0) = -\omega\mathscr{Y}_2(\alpha(x+1); -3\alpha^2, 0)$$
$$= -\omega\mathscr{Y}_0(\omega^{-2}\alpha(x+1); -3\omega^{-4}\alpha^2, 0).$$

Since $\mathsf{Re}\,(\omega^{-2}\alpha x)^{5/2} = \mathsf{Re}\,(e^{i\pi/5}|x|\alpha)^{5/2} > 0$ for $x < 0$ it is clear from the asymptotic behavior that $\mathscr{Y}_0(\alpha(x+1); -3\alpha^2, 0)$ is exponentially decaying in $\mathbb{R}$ as $|x| \to \infty$ and in particular $u(x) \in \mathscr{S}(\mathbb{R})$. We multiply by $\bar{u}(x)$ in (6.8) and integrate by parts which yields

$$-\int_{\mathbb{R}} |u'(x)|^2 dx = \alpha^5 \int_{\mathbb{R}} (x^3 + 3x^2 - 2)|u(x)|^2 dx. \tag{6.9}$$

Since $\mathsf{Im}\,\alpha^5 \neq 0$, taking the imaginary part of (6.9) we get

$$\int_{\mathbb{R}} (x^3 + 3x^2 - 2)|u(x)|^2 dx = 0$$

hence $u'(x) = 0$ so that $u(x) = 0$ for $u(x) \in \mathscr{S}(\mathbb{R})$. This is a contradiction. So we conclude that

$$c(-3\alpha^2) \neq 0 \quad \text{if} \quad -\pi/5 < \arg \alpha < 0$$

which proves that $c(\zeta) \neq 0$ for $3\pi/5 < \arg \zeta < \pi$. From Lemma 6.4 we see $c(\zeta) \neq 0$ if $\arg \zeta = \pi$. We finally check that $c(\zeta) \neq 0$ if $\arg \zeta = 3\pi/5$. Indeed if $c(\zeta) = 0$ with $\arg \zeta = 3\pi/5$ then $c(\bar{\omega}\zeta) = 0$ by Lemma 6.2 but $\bar{\omega}\zeta \in \mathbb{R}$ contradicts Lemma 6.4 again and hence the assertion. $\qquad\square$

Proof of Proposition 6.1 From Lemma 6.3 there exists $\zeta \neq 0$ with $c(\zeta) = 0$. From Lemma 6.5 we see $-\pi < \arg \zeta < 3\pi/5$. If $-\pi < \arg \zeta < 0$ then this ζ is a desired one. If $0 \leq \arg \zeta < 3\pi/5$ then it follows that $-\pi < \arg \bar{\omega}\zeta < -2\pi/5$ which proves the assertion because $c(\bar{\omega}\zeta) = 0$ by Lemma 6.2.

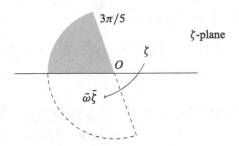

This ends the proof of Proposition 6.1.   □

We now consider the equation

$$C_0(\zeta, -\zeta^2 s/4) = 0 \tag{6.10}$$

with small s. Let ζ_0 be a zero of $c(\zeta) = C_0(\zeta, 0)$ with negative imaginary part and μ be the multiplicity of the root ζ_0. Since $C_0(\zeta, 0)$ is holomorphic by the Weierstrass preparation theorem we can write

$$C_0(\zeta, -\zeta^2 s/4) = \gamma(\zeta, s)\Big\{(\zeta - \zeta_0)^\mu + \sum_{j=1}^{\mu} a_j(s)(\zeta - \zeta_0)^{\mu-j}\Big\}$$

where $\gamma(\zeta_0, 0) \neq 0$, $a_j(0) = 0$ and $a_j(s)$ is holomorphic at $s = 0$. Then each root $\zeta(s)$ of $C_0(\zeta, -\zeta^2 s/4) = 0$ admits the Puiseux expansion

$$\zeta(s) = \zeta_0 + \sum_{j=0}^{\infty} \zeta_j(s^{1/p})^j = \tilde{\zeta}(s^{1/p}) \tag{6.11}$$

with some positive integer p where $\tilde{\zeta}(s)$ is holomorphic at $s = 0$.

Returning to Eq. (6.4) and replacing ρ^{-2} by s^p we consider

$$w''(x) = \big(x^3 + \zeta x - \zeta^2 s^p/4\big)w(x). \tag{6.12}$$

Since the equation $C_0(\zeta, -\zeta^2 s^p/4) = 0$ has a solution $\zeta(s^p) = \tilde{\zeta}(s)$ where

$$\tilde{\zeta}(0) = \zeta_0, \quad \mathsf{Im}\,\zeta_0 < 0 \tag{6.13}$$

then with $\eta(s) = -\tilde{\zeta}(s)^2 s^p/4$ we have for $|s| \ll 1$

$$\mathscr{Y}_0(x; \tilde{\zeta}(s), \eta(s)) = -\omega \mathscr{Y}_2(x; \tilde{\zeta}(s), \eta(s)), \quad \forall x \in \mathbb{C}. \tag{6.14}$$

We now examine the behavior of $\mathscr{Y}_0(x; \tilde{\zeta}(s), \eta(s))$ as $\mathbb{R} \ni x \to \pm\infty$. Recall

$$\mathscr{Y}_0(x; \tilde{\zeta}, \eta) = x^{-3/4}(1 + R(x, \tilde{\zeta}, \eta))e^{-(\frac{2}{5}x^{5/2} + \tilde{\zeta}x^{1/2})} \quad \text{in } |\arg x| < 3\pi/5$$

and hence $\mathcal{Y}_0(x; \tilde{\zeta}, \eta)$ decays as $\exp(-2x^{5/2}/5)$ when $\mathbb{R} \ni x \to +\infty$. On the other hand from (6.14) we have

$$\mathcal{Y}_0(x; \tilde{\zeta}(s), \eta(s)) = -\omega \mathcal{Y}_0(\omega^{-2}x; \omega^{-4}\tilde{\zeta}(s), \omega^{-6}\eta(s))$$

and for negative $x < 0$ since $\omega^{-2}x = e^{\pi i/5}|x|$ and

$$(\omega^{-2}x)^{5/2} = i|x|^{5/2}, \quad \omega^{-4}\tilde{\zeta}(\omega^{-2}x)^{1/2} = i\tilde{\zeta}|x|^{1/2}$$

it follows that $\mathcal{Y}_0(x; \tilde{\zeta}, \eta)$ decays as $\exp(\mathrm{Im}\,\tilde{\zeta}|x|^{1/2})$ when $\mathbb{R} \ni x \to -\infty$ because of (6.13). This is one of the main reasons that we need to find a zero with *negative* imaginary part (non-real root is insufficient). We conclude that $\mathcal{Y}_0(x; \tilde{\zeta}(s), \eta(s)) \in \mathscr{S}(\mathbb{R})$ and in particular is bounded uniformly in $x \in \mathbb{R}$ and $|s| \ll 1$:

$$|\mathcal{Y}_0(x; \tilde{\zeta}(s), \eta(s))| \leq B, \quad x \in \mathbb{R}, \quad |s| \ll 1.$$

6.3   Proof of Non Solvability of the Cauchy Problem

Take $T > 0$ small and recalling $s = \rho^{-2/p}$ we set for $\rho > 0$

$$U_\rho(x) = \exp\left(-i\rho^5 x_n + \frac{i}{2}\tilde{\zeta}(\rho^{-2/p})\rho(T - x_0)\right) \tag{6.15}$$
$$\times \mathcal{Y}(x_1\rho^2; \tilde{\zeta}(\rho^{-2/p}), \eta(\rho^{-2/p})).$$

It is clear that $P_{mod}U_\rho = 0$. We now consider the following Cauchy problem

$$\begin{cases} P_{mod}u = 0, \\ u(0, x') = 0, \\ D_0u(0, x') = \phi(x_1)\psi(x'')\theta(x_n) \end{cases} \tag{6.16}$$

where $x'' = (x_2, \dots, x_{n-1})$ and $\phi \in C_0^\infty(\mathbb{R})$, $\psi \in C_0^\infty(\mathbb{R}^{n-2})$ and $\theta \in C_0^\infty(\mathbb{R})$ are real valued.

Before going into details of the proof of non solvability of (6.16) we remark that we can assume that solutions u to (6.16) have compact supports with respect to x'. To examine this we recall the Holmgren uniqueness theorem (see, for example [66, Theorem 4.2]). For $\delta > 0$ we denote

$$D_\delta = \{x \in \mathbb{R}^{n+1} \mid |x'|^2 + |x_0| < \delta\}$$

then we have

Proposition 6.3 (Holmgren) *There exists $\delta_0 > 0$ such that if $u(x) \in C^2(D_\delta)$ with $0 < \delta \le \delta_0$ verifies*

$$\begin{cases} P_{mod}u = 0 & in\ D_\delta, \\ D_0^j u(0, x') = 0, & j = 0, 1, \quad x' \in D_\delta \cap \{x_0 = 0\} \end{cases}$$

then $u(x)$ vanishes identically in D_δ.

To state the non-solvability assertion we denote

$$a_j = \int \mathscr{Y}(x_1; \zeta_0, 0) x_1^j dx_1, \quad j = 0, 1, 2$$

and note that at least one of a_j ($j = 0, 1, 2$) is different from zero which is checked later. Denote $k = \min\{j \mid a_j \ne 0\}$ so that $a_k \ne 0$ and $a_j = 0$ for $j = 0, \dots, k - 1$.

Proposition 6.4 *Assume that $\theta \in C_0^\infty(\mathbb{R})$ is an even function such that $\theta \notin \gamma_0^{(5)}(\mathbb{R})$. We assume also $\int \psi(x'')dx'' \ne 0$ and $\phi^{(j)}(0) = 0$ for $0 \le j \le k-1$ and $\phi^{(k)}(0) \ne 0$. Let Ω be a neighborhood of the origin of $\mathbb{R}^{n+1}$ such that $\mathrm{supp}\,\phi\psi\theta \subset \Omega \cap \{x_0 = 0\}$. Then the Cauchy problem (6.16) has no $C^2(\Omega)$ solution.*

Proof Suppose that (6.16) has a $C^2(\Omega)$ solution. Applying Proposition 6.3 we conclude that we can assume $u(x) = 0$ if $|x_0| \le T$ and $|x'| \ge r$ with some small $T > 0$ and $r > 0$.

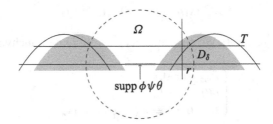

We note that

$$\int_0^T (P_{mod}U_\rho, u)dx_0 = \int_0^T (U_\rho, P_{mod}u)dx_0 + i(D_0 U_\rho(T), u(T))$$

$$+ i(U_\rho(T), D_0 u(T)) - i(U_\rho(0), D_0 u(0)) - i(2x_1 D_n U_\rho(T), u(T))$$

because $u(0) = 0$. From this we have

$$\begin{aligned} &(D_0 U_\rho(T), u(T)) + (U_\rho(T), D_0 u(T)) \\ &- (2x_1 D_n U_\rho(T), u(T)) = (U_\rho(0), D_0 u(0)). \end{aligned} \tag{6.17}$$

Recalling that $\mathscr{Y}(\rho^2 x_1; \tilde{\xi}, \eta)$ is bounded uniformly in ρ and x_1 we see that the left-hand side on (6.17) is $O(\rho^5)$. On the other hand the right-hand side is

$$\int_{\mathbb{R}^n} e^{-i\rho^5 x_n + i\tilde{\xi}\rho T/2} \mathscr{Y}(\rho^2 x_1; \tilde{\xi}, \eta)\phi(x_1)\psi(x'')\theta(x_n)dx'$$

$$= e^{i\tilde{\xi}\rho T/2}\hat{\theta}(\rho^5)\left(\int_{\mathbb{R}^{n-2}} \psi(x'')dx''\right)\rho^{-2}\int \mathscr{Y}(x_1; \tilde{\xi}, \eta)\phi(\rho^{-2}x_1)dx_1 \tag{6.18}$$

where $\hat{\theta}$ is the Fourier transform of θ. We note that for large ρ one has

$$\left|e^{i\tilde{\xi}\rho T/2}\right| \geq e^{c\rho T}$$

with some $c > 0$ because $\tilde{\xi}(\rho^{-2/p}) \to \zeta_0$ as $\rho \to \infty$ and $\mathrm{Im}\,\zeta_0 < 0$. Thus we conclude that

$$\rho^{-7}e^{c\rho T}|\hat{\theta}(\rho^5)|\left|\int \mathscr{Y}(x_1; \tilde{\xi}, \eta)\phi(\rho^{-2}x_1)dx_1\right| = O(1). \tag{6.19}$$

Write

$$\int \mathscr{Y}(x_1; \tilde{\xi}, \eta)\phi(\rho^{-2}x_1)dx_1 = \sum_{j=0}^{2} \frac{\rho^{-2j}}{j!}\phi^{(j)}(0)\int \mathscr{Y}(x_1; \tilde{\xi}, \eta)x_1^j dx_1 + O(\rho^{-6})$$

then noting

$$\int \mathscr{Y}(x_1; \tilde{\xi}, \eta)x_1^j dx_1 \to \int \mathscr{Y}(x_1; \zeta_0, 0)x_1^j dx_1, \quad \rho \to \infty$$

the right-hand side is $\rho^{-2k}(a_k\phi^{(k)}(0)/k! + O(\rho^{-2}))$ as $\rho \to \infty$. Therefore (6.19) implies that $|\hat{\theta}(\rho^5)| \leq C\rho^{7+2k}e^{-c\rho T}$ with some $C > 0$ for large $\rho > 0$, that is

$$|\hat{\theta}(\rho)| \leq C\rho^{(7+2k)/5}e^{-c\rho^{1/5}} \leq C'e^{-c'\rho^{1/5}}$$

with some $c' > 0$. Since θ is even we conclude that $\theta \in \gamma_0^{(5)}(\mathbb{R})$ which is a contradiction. □

Lemma 6.6 *For one of $j = 0, 1, 2$ we have*

$$a_j = \int \mathscr{Y}(x_1; \zeta_0, 0)x_1^j dx_1 \neq 0.$$

Proof We denote by $w(\xi)$ the Fourier transform of $\mathscr{Y}(x; \zeta_0, 0)$;

$$w(\xi) = \int e^{-ix\xi} \mathscr{Y}(x; \zeta_0, 0)dx.$$

Since $\mathscr{Y}(x; \zeta_0, 0)$ verifies $\mathscr{Y}'' = (x^3 + \zeta_0 x)\mathscr{Y}$ and $\mathscr{Y}(x; \zeta_0, 0) \in \mathscr{S}(\mathbb{R})$ then $w(\xi)$ satisfies

$$w'''(\xi) - \zeta_0 w'(\xi) + i\xi^2 w(\xi) = 0. \tag{6.20}$$

Noting that

$$w^{(j)}(0) = \int \mathscr{Y}(x; \zeta_0, 0)x^j dx$$

the proof follows from the uniqueness of solution to the initial value problem for the ordinary differential equation (6.20).   □

To prove Theorem 6.1 we prepare a proposition.

Definition 6.3 Let $h > 0$ be fixed and L be a compact set in $\mathbb{R}^n$. We say $f(x) \in \gamma_0^{(s),h}(L)$ if $f(x) \in C_0^\infty(L)$ and

$$\sum_\alpha \sup_x \frac{h^{|\alpha|}|\partial_x^\alpha f(x)|}{(\alpha!)^s}$$

is finite.

Note that $\gamma_0^{(s),h}(L) \subset \gamma^{(s)}(\mathbb{R}^n)$ and $\gamma_0^{(s),h}(L)$ is a Banach space with the norm

$$\sum_\alpha \sup_x \frac{h^{|\alpha|}|\partial_x^\alpha f(x)|}{(\alpha!)^s}.$$

Then following [54] (see also [66, Proposition 4.1]) we have

Proposition 6.5 *Let L be a compact set of $\mathbb{R}^n$ and $h > 0$ be fixed. Assume that the Cauchy problem for P_{mod} is locally solvable in $\gamma^{(s)}$ at the origin. Then there is $\delta > 0$ such that for any $(u_0(x'), u_1(x')) \in (\gamma^{(s),h}(L))^2$ there exists a unique $u(x) \in C^2(D_\delta)$ verifying $P_{mod}u = 0$ in D_δ and $D_0^j u(0, x') = u_j(x')$ on $D_\delta \cap \{x_0 = 0\}$.*

Proof of Theorem 6.1 Suppose that the Cauchy problem for P_{mod} is locally solvable in $\gamma^{(s)}$ at the origin with some $s > 5$. Take s' so that $s > s' > 5$. We fix a compact neighborhood L of the origin of $\mathbb{R}^n$ and a positive $h > 0$. Then from Proposition 6.5 there exists D_δ such that the Cauchy problem for P_{mod} has a $C^2(D_\delta)$ solution for any Cauchy data in $(\gamma^{(s),h}(L))^2$. We now choose $\phi \in \gamma_0^{(s')}(\mathbb{R})$, $\psi \in \gamma_0^{(s')}(\mathbb{R}^{n-1})$ and $\theta \in \gamma_0^{(s')}(\mathbb{R})$ so that $\text{supp}\,\phi\psi\theta \subset L \cap (D_\delta \cap \{x_0 = 0\})$ which satisfy the conditions in Proposition 6.4. Since it is clear that $\phi\psi\theta \in \gamma_0^{(s),h}(L)$ because $s > s'$ one can apply Proposition 6.5 to conclude that there is a $C^2(D_\delta)$ solution to (6.16), while this contradicts with Proposition 6.4.   □

6.4 Open Questions and Remarks

As far as the Cauchy problem is concerned, the main question remaining to be unclear is *what will happen to the Cauchy problem for P of spectral type 2 when* $\mathrm{Tr}^+ F_p > 0$ *if there is a tangent bicharacteristic*. A model operator satisfying these conditions is

$$P = -D_0^2 + 2x_1 D_0 D_n + D_1^2 + x_1^3 D_n^2 + a(x_3^2 D_n^2 + D_3^2)$$

where $a > 0$ is a positive constant and hence $\mathrm{Tr}^+ F_p = a$, which coincides with P in (6.1) when $a = 0$. The doubly characteristic manifold is $\Sigma = \{\xi_0 = \xi_1 = \xi_3 = 0, x_1 = x_3 = 0\}$. Since $P_{sub} = 0$ the strict IPH condition is clearly verified. In the case $a = 0$ it seems that the non solvability in $\gamma^{(s)}$ for $s > 5$ of the Cauchy problem is caused by the existence of the singular orbit (6.3). If we define $(x_1, x_2, \xi_0, \xi_1, \xi_2)$ by (6.3) and (x_3, ξ_3) by $x_3 = 0, \xi_3 = 0$ then this curve is still a bicharacteristic of P even if $a > 0$. That is, there exists a "singular" orbit (in the view point of "classical mechanics") for P even if $a > 0$. From this point of view it is expected that the Cauchy problem for P with $a > 0$ is still $\gamma^{(s)}$ ill-posed for some $s > 5$. On the other hand, in the view point of "quantum mechanics" it is forbidden to choose $x_3 = 0$, $\xi_3 = 0$ at the same time by Heisenberg's uncertainty principle. Up to now it is only known that the Cauchy problem for P with $a > 0$ is $\gamma^{(6)}$ well-posed (see [5, 80]).

To make more detailed study when p is of spectral type 2 on Σ, including both the case with and without tangent bicharacteristics, one can introduce a well-posedness notion given below which is finer than the C^∞ strong hyperbolicity (Definition 1.3). In what follows we study differential operators of second order

$$P = -D_0^2 + A_1(x, D')D_0 + A_2(x, D') \tag{6.21}$$

where we assume that the coefficients are real analytic or in the Gevrey class of order $s > 1$, assumed to be enough close to 1.

Definition 6.4 Let $s > 1$. Then P is said to be Gevrey s strongly hyperbolic at the origin if the Cauchy problem for $P + Q$ with any differential operator of order at most 1 is locally $\gamma^{(\kappa)}$ solvable at the origin for every κ less than s. We assume (1.5) and that

$$\text{codimension } \Sigma = 3. \tag{6.22}$$

Proposition 6.6 ([8]) *Assume* (1.5), (6.22) *and that p is of spectral type 2 on* Σ. *Then P is Gevrey* 3 *strongly hyperbolic at the origin.*
The Gevrey order 3 is optimal in the following sense. We again consider $P = P_{mod}$ given by (6.1)

$$P = -D_0^2 + 2x_1 D_0 D_n + D_1^2 + x_1^3 D_n^2$$

near the origin of $\mathbb{R}^n$ which verifies all the assumptions in Proposition 6.6.

Proposition 6.7 ([8]) *The Cauchy problem for $P + AD_n$, $A \in \mathbb{C} \setminus \mathbb{R}_+$ is not locally $\gamma^{(s)}$ solvable at the origin if $s > 3$, where $\mathbb{R}_+$ is the set of all nonnegative real numbers.*

Proposition 6.8 ([7]) *Assume (1.5), (6.22) and that p is of spectral type 2 on Σ and there is no tangent bicharacteristic. Then P is Gevrey 4 strongly hyperbolic at the origin.*

Consider the following model operator near the origin of $\mathbb{R}^n$ for which all the assumptions in Proposition 6.8 are fulfilled;

$$P = -D_0^2 + 2x_1 D_0 D_n + D_1^2. \tag{6.23}$$

The Gevrey order 4 in Proposition 6.8 is optimal in the following sense of which proof is sketched in the end of Sect. 8.5.

Proposition 6.9 ([35, 81]) *The Cauchy problem for $P + AD_n$, $A \in \mathbb{C} \setminus \mathbb{R}_+$ is not locally $\gamma^{(s)}$ solvable at the origin if $s > 4$.*

To our knowledge, the positive trace $\mathrm{Tr}^+ F_p$, concerned with well-posedness of the Cauchy problem, appears in the sum with the subprincipal symbol P_{sub}; $\mathrm{Tr}^+ F_p + P_{sub}$, that is the effect of $\mathrm{Tr}^+ F_p$ could be absorbed by suitably chosen lower order terms. Therefore *it would be quite reasonable to expect that Propositions 6.6 and 6.8 can be generalized to higher codimensional case with $\mathrm{Tr}^+ F_p \neq 0$, although $\mathrm{Tr}^+ F_p = 0$ in the present case.*

   The following result is a particular case of the general result proved in [12] (see also [36]).

Proposition 6.10 *P is Gevrey 2 strongly hyperbolic at the origin.*

When p is of spectral type 2 on Σ and (1.5) and (6.22) are verified then model operator verifying these conditions is either $P = -D_0^2 + D_1^2 + D_2^2$ in $\mathbb{R}^{1+n}$ with $n \geq 3$ or

$$P = -D_0^2 + a(x_1^2 D_n^2 + D_1^2) \tag{6.24}$$

near the origin of $\mathbb{R}^{1+n}$ with $n \geq 2$ where $a > 0$. For $P = -D_0^2 + D_1^2 + D_2^2$ it is well known that the Cauchy problem for $P + AD_n$ with $A \in \mathbb{C} \setminus \mathbb{R}_+$ is not locally $\gamma^{(s)}$ solvable at the origin if $s > 2$. As for P in (6.24) one has

Lemma 6.7 *The Cauchy problem for $P + AD_n$ with $A + a \in \mathbb{C} \setminus \mathbb{R}_+$ is not locally $\gamma^{(s)}$ solvable at the origin if $s > 2$.*

Proof Consider the following Cauchy problem

$$\begin{cases} (P + AD_n)v = 0, \\ v(0, x') = 0, \quad D_0 v(0, x') = \phi(x_1)\psi(x'')\theta(x_n) \end{cases}$$

where $x'' = (x_2, \ldots, x_{n-1})$ and $\phi(x_1) \in C_0^\infty(\mathbb{R})$, $\psi(x'') \in C_0^\infty(\mathbb{R}^{n-1})$, $\theta(x_n) \in C_0^\infty(\mathbb{R})$ are real valued. Assume that there is a neighborhood Ω of the origin verifying $\operatorname{supp} \phi \psi \theta \subset \Omega \cap \{x_0 = 0\}$ such that the Cauchy problem has $C^2(\Omega)$ solution v. Repeating the same arguments using Proposition 6.3 one can assume $v(x) = 0$ for $|x'| \geq r$ if $|x_0| \leq T$ with some small $T > 0$ and $r > 0$. Since $\bar{A} + a \in \mathbb{C} \setminus \mathbb{R}_+$ there exists ζ such that $\zeta^2 = \bar{A} + a$ with $\operatorname{Im} \zeta \neq 0$. Since the arguments is completely parallel we may assume $\operatorname{Im} \zeta < 0$ without restrictions. Denote with $\lambda > 0$

$$U_\lambda(x) = e^{i\lambda^2 x_n + i\zeta\lambda(T - x_0)} e^{-\lambda^2 x_1^2/2}$$

then it is easy to check that $(P + \bar{A}D_n)U_\lambda = 0$. From

$$0 = \int_0^T ((P + \bar{A}D_n)U_\lambda, v)dx_0 = \int_0^T (U_\lambda, (P + AD_n)v)dx_0$$

$$+ i \sum_{j=0}^{1} (D_0^{1-j}U_\lambda(T), D_0^j v(T)) - i(U_\lambda(0), D_0 v(0))$$

it follows that

$$\sum_{j=0}^{1} (D_0^{1-j}U_\lambda(T), D_0^j v(T)) = (U_\lambda(0), D_0 v(0)). \tag{6.25}$$

Note that the left-hand side on (6.25) is $O(\lambda)$ as $\lambda \to \infty$ while the right-hand side is

$$\lambda^{-1} e^{i\zeta\lambda T} \hat{\theta}(\lambda^2) \left(\int \psi(x'')dx'' \right) \int e^{-x_1^2/2} \phi(\lambda^{-1}x_1)dx_1 \tag{6.26}$$

where $\hat{\theta}$ denotes the Fourier transform of θ. We choose ϕ, ψ and θ so that $\phi(0) \neq 0$, $\int \psi(x'')dx'' \neq 0$ and θ is even. Then from (6.26) we conclude that there is $C > 0$ such that for large λ one has

$$|\hat{\theta}(\lambda^2)| \leq C\lambda^2 e^{(\operatorname{Im} \zeta)\lambda T}.$$

Since θ is even this implies that $|\hat{\theta}(\xi_n)| \leq C' e^{-c|\xi_n|^{1/2}}$ with some $c > 0$ and hence $\theta \in \gamma_0^{(2)}(\mathbb{R})$. The rest of the proof is just a repetition of that of Theorem 6.1. $\square$
Therefore, assuming (1.5), (6.22) and no spectral transition, the threshold of Gevrey s strong hyperbolicity occurs only at $s = 2, 3, 4$ and that these thresholds completely determine the structure of the Hamilton map F_p and the geometry of bicharacteristics near Σ. *It is then quite natural to ask whether the same result holds without the restriction* (6.22).

In this monograph we always assume that there is no transition of spectral type of F_p. Study of the case when spectral transition occurs is widely open and only some special cases are considered. See [1, 4, 10, 22, 83, 86, 90]. Here we explain a typical situation when the spectral transition occurs, taking P of (6.21) and considering a simple but very suggestive case. We assume (6.22) and the principal symbol $p(x, \xi)$ vanishes exactly of order 2 on Σ and

$$\begin{cases} \text{rank } (\sum_{j=0}^{n} d\xi_j \wedge dx_j|_{\Sigma}) = \text{constant on } \Sigma, \\ \text{the spectral type of } F_p \text{ changes simply} \\ \text{across a submanifold } S \text{ of codimension 1 of} \Sigma \end{cases} \qquad (6.27)$$

where the meaning of "changes simply" will be made clear just below. By conjugation with a Fourier integral operator one can assume $A_1 = 0$ then, near any point $\rho \in \Sigma$, and hence one has

$$p(x, \xi) = -\xi_0^2 + \phi_1(x, \xi')^2 + \phi_2(x, \xi')^2$$

where the differentials $d\phi_1$ and $d\phi_2$ are linearly independent at ρ and $\Sigma = \{\xi_0 = 0, \phi_1 = 0, \phi_2 = 0\}$.

Lemma 6.8 *Assume* (6.22) *and* (6.27). *Then we have either* $\{\xi_0, \phi_1\} \neq 0$ *or* $\{\xi_0, \phi_2\} \neq 0$ *on S.*

Proof Suppose that $\{\xi_0, \phi_j\} = 0$ at $\rho \in S$ for $j = 1, 2$. Then in virtue of Lemma 1.7 one can choose a symplectic basis such that

$$p_\rho = -\xi_0^2 + \xi_1^2 + \xi_2^2, \qquad (6.28)$$

$$p_\rho = -\xi_0^2 + \mu(x_1^2 + \xi_1^2), \quad \mu > 0 \qquad (6.29)$$

according to $\{\phi_1, \phi_2\}(\rho) = 0$ or $\{\phi_1, \phi_2\}(\rho) \neq 0$. If the first case occurs then from (6.27) we have rank $(d\xi \wedge dx|_{\Sigma}) = 0$ and hence p_ρ takes the form (6.28) everywhere Σ in a suitable symplectic basis. In the second case we note that $\pm i\mu$ are the eigenvalues of F_p. From the continuity of $F_p(\rho)$ with respect to $\rho \in \Sigma$ it follows that F_p has still non-zero pure imaginary eigenvalues near ρ on Σ and hence p_ρ takes the form (6.29) in a suitable symplectic basis. Therefore, in both cases, the spectral structure of F_p does not change near ρ. This proves the assertion. $\qquad \square$

We now assume that $\{\xi_0, \phi_j\}(\rho) \neq 0$ with some j. Considering

$$\tilde{\phi}_i = \sum_{j=1}^{2} O_{ij} \phi_j$$

with a smooth orthogonal matrix (O_{ij}) we may assume without restrictions that

$$\{\xi_0, \phi_2\} > 0, \quad \{\xi_0, \phi_1\} = O(|\phi|) \qquad (6.30)$$

near ρ where $f = O(|\phi|)$, $\phi = (\phi_1, \phi_2)$ means that f is a linear combination of ϕ_1 and ϕ_2 near the reference point. We look for non-zero eigenvalues μ of F_p; $F_pX = \mu X$. Since $\mu \neq 0$ it suffices to consider F_p on the image of F_p, that is on the vector space spanned by $H_{\xi_0}, H_{\phi_1}, H_{\phi_2}$. Let $X = \alpha H_{\xi_0} + \beta H_{\phi_1} + \gamma H_{\phi_2}$ and consider $F_pX = \mu X$. Since

$$F_pX = -\gamma\{\phi_2, \xi_0\}H_{\xi_0} + \gamma\{\phi_2, \phi_1\}H_{\phi_1} + (\alpha\{\xi_0, \phi_2\} + \beta\{\phi_1, \phi_2\})H_{\phi_2}$$

we have

$$-\gamma\{\phi_2, \xi_0\} = \mu\alpha, \quad \gamma\{\phi_2, \phi_1\} = \mu\beta, \quad \alpha\{\xi_0, \phi_2\} + \beta\{\phi_1, \phi_2\} = \mu\gamma$$

so that

$$\begin{pmatrix} 0 & 0 & \{\xi_0, \phi_2\} \\ 0 & 0 & -\{\phi_1, \phi_2\} \\ \{\xi_0, \phi_2\} & \{\phi_1, \phi_2\} & 0 \end{pmatrix} \begin{pmatrix} \alpha \\ \beta \\ \gamma \end{pmatrix} = \mu \begin{pmatrix} \alpha \\ \beta \\ \gamma \end{pmatrix}.$$

Thus the characteristic equation is

$$\mu\big(\mu^2 - (\{\xi_0, \phi_2\}^2 - \{\phi_1, \phi_2\}^2)\big) = 0. \tag{6.31}$$

Lemma 6.9 *If the spectral type of F_p changes across S then $\{\xi_0, \phi_2\}^2 - \{\phi_1, \phi_2\}^2$ vanishes on S and one and only one of the following cases occurs;*

(i) $\{\xi_0, \phi_2\}^2 - \{\phi_1, \phi_2\}^2 < 0$ *in* $\Sigma \setminus S$ *so that p is non-effectively hyperbolic in Σ and of spectral type 1 on $\Sigma \setminus S$ and of spectral type 2 on S,*

(ii) $\{\xi_0, \phi_2\}^2 - \{\phi_1, \phi_2\}^2 > 0$ *in* $\Sigma \setminus S$ *so that p is effectively hyperbolic in $\Sigma \setminus S$ and non-effectively hyperbolic of spectral type 2 on S,*

(iii) $\{\xi_0, \phi_2\}^2 - \{\phi_1, \phi_2\}^2$ *changes the sign across S, that is p is non-effectively hyperbolic of spectral type 1 in the one side of $\Sigma \setminus S$ and of spectral type 2 on S and effectively hyperbolic in the other side of $\Sigma \setminus S$.*

Proof Note that if $\{\xi_0, \phi_2\}^2 - \{\phi_1, \phi_2\}^2 \neq 0$ at $\rho \in S$ then $F_p(\rho)$ has non-zero real eigenvalues or non-zero pure imaginary eigenvalues according to $\{\xi_0, \phi_2\}^2 - \{\phi_1, \phi_2\}^2 > 0$ or $\{\xi_0, \phi_2\}^2 - \{\phi_1, \phi_2\}^2 < 0$. By the continuity of F_p with respect to ρ, $F_p(\rho)$ has still non-zero real eigenvalues or pure imaginary eigenvalues near ρ on Σ and then the spectral structure of F_p does not change near ρ on Σ. This proves the first assertion. Since

$$F_p^2 H_{\phi_2} = (\{\xi_0, \phi_2\}^2 - \{\phi_1, \phi_2\}^2)H_{\phi_2}$$

it is clear that $0 \neq H_{\phi_2} \in \mathrm{Ker}\, F_p^2 \cap \mathrm{Im}\, F_p^2$ if $\{\xi_0, \phi_2\}^2 - \{\phi_1, \phi_2\}^2 = 0$ and this proves the assertion. $\qquad\square$

By "changes simply" in (6.27) we mean that one can write

$$\{\xi_0, \phi_2\}^2 - \{\phi_1, \phi_2\}^2 = -\theta^2, \quad \theta^2, \quad \theta \tag{6.32}$$

near $\rho \in S$ according to the case (i), (ii) and (iii) respectively where S is defined by $\{\theta = 0\} \cap \Sigma$ and $d\theta \neq 0$ on S. Note that, denoting $\kappa(\rho)^2 = |\{\xi_0, \phi_2\}^2 - \{\phi_1, \phi_2\}^2|$, in the case (i) F_p has pure imaginary eigenvalues $\pm i\kappa(\rho)$ on Σ which are 0 on S and in the case (ii) F_p has real eigenvalues $\pm\kappa(\rho)$ on Σ which are 0 on S. In the case (iii) F_p has non-zero real eigenvalues $\pm\kappa(\rho)$ on Σ^+ and non-zero pure imaginary eigenvalues $\pm i\kappa(\rho)$ on Σ^- where

$$\Sigma^\pm = \{(x, \xi) \in \Sigma \mid \pm(\{\xi_0, \phi_2\}^2 - \{\phi_1, \phi_2\}^2) > 0\}.$$

Consider the case (i). From (6.32) we can write

$$\{\xi_0, \phi_2\}^2 - \{\phi_1, \phi_2\}^2 = -\theta^2 + c_1\phi_1 + c_2\phi_2$$

in a neighborhood of ρ. Since

$$\{\xi_0, \phi_2\}^2 - \{\phi_1, \phi_2\}^2 = \{\xi_0 + \phi_1, \phi_2\}\{\xi_0 - \phi_1, \phi_2\}$$

and hence we have either $\{\xi_0 + \phi_1, \phi_2\} = 0$ or $\{\xi_0 - \phi_1, \phi_2\} = 0$ on S. Since the arguments are completely parallel we may assume without restrictions that

$$\{\xi_0 - \phi_1, \phi_2\} = 0 \quad \text{on} \quad S$$

and hence one has $\{\xi_0, \phi_2\} = \{\phi_1, \phi_2\} > 0$ on S by (6.30). Thus we can write

$$\{\xi_0 - \phi_1, \phi_2\} = -\tilde{\theta}^2 + c_1'\phi_1 + c_2'\phi_2 \tag{6.33}$$

near ρ where $\tilde{\theta} = c\theta$ with a non-zero factor c. For the case (ii), repeating the same arguments as above one obtains

$$\{\xi_0 - \phi_1, \phi_2\} = \tilde{\theta}^2 + c_1'\phi_1 + c_2'\phi_2. \tag{6.34}$$

Similarly in the case (iii) we can write near ρ

$$\{\xi_0 - \phi_1, \phi_2\} = \theta + c_1'\phi_1 + c_2'\phi_2. \tag{6.35}$$

Writing $p = -(\xi_0 - \phi_1)(\xi_0 + \phi_1) + \phi_2^2$ the relations (6.33)–(6.35) show how commutes p against $\xi_0 - \phi_1$.

In [62] R. Melrose has conjectured that the condition

$$\mathrm{dist}_{\mathbb{C}}\left(P_{sub}(\rho), [-\mathrm{Tr}^+ F_p(\rho), \mathrm{Tr}^+ F_p(\rho)]\right) \leq Ce(\rho) \tag{6.36}$$

where $\pm e(\rho)$, $e(\rho) > 0$ is a pair of non-zero real eigenvalues of $F_p(\rho)$, is necessary for the Cauchy problem to be C^∞ well-posed, but little is known about necessary conditions for the well-posedness when the spectral type of F_p changes. In the case (i) the condition (6.36) with strict inequality implies

$$\operatorname{Im} P_{sub}(\rho) = 0 \quad \text{and} \quad |\operatorname{Re} P_{sub}(\rho)| < \kappa(\rho) \quad \text{on} \quad \Sigma$$

and in the case (ii) the condition (6.36) with strict inequality reads

$$||\operatorname{Im} P_{sub}(\rho)| \leq C\kappa(\rho) \text{ on } \Sigma \quad \text{and} \quad \operatorname{Re} P_{sub}(\rho) > 0 \quad \text{on} \quad S$$

with some $C > 0$. The sufficiency of these conditions for the C^∞ well-posedness of the Cauchy problem was proved in [10] and [83, 86], where the assumption of *non existence of tangent bicharacteristics* is crucial. As for the case (iii) this condition implies that there is $C > 0$ such that

$$\begin{cases} ||\operatorname{Im} P_{sub}(\rho)| \leq C\kappa(\rho) & \text{and} \quad \operatorname{Re} P_{sub}(\rho) = 0 \quad \text{on} \quad \Sigma^+, \\ \operatorname{Im} P_{sub}(\rho) = 0 & \text{and} \quad |\operatorname{Re} P_{sub}(\rho)| \leq \kappa(\rho) \quad \text{on} \quad \Sigma^-. \end{cases} \tag{6.37}$$

It is easy to see

$$F_p(\rho)X^\pm = \pm e(\rho)X^\pm, \quad \rho \in \Sigma^+$$

with $X^\pm = \{\xi_0, \phi_2\}H_{\xi_0} - \{\phi_1, \phi_2\}H_{\phi_1} \pm \kappa(\rho)H_{\phi_2}$ that converges to $\{\phi_1, \phi_2\}H_{\xi_0-\phi_1} \in T\Sigma$ when ρ approaches to S. Thanks to Proposition 3.1 there exist exactly two bicharacteristics passing $\rho \in \Sigma^+$ transversally to Σ^+ with tangents $X^\pm$. On the other hand there is no bicharacteristic reaching Σ^- thanks to Proposition 3.3. Therefore in the case (iii) the geometry of bicharacteristic near Σ changes drastically from Σ^- to Σ^+. This suggests that the case (iii) will be much more complicated compared to the case (i) or (ii). We can find in [1, 22] an example of the case (iii) having no tangent bicharacteristics, for which the Cauchy problem is C^∞ well-posed, although the condition (6.37) was not discussed there.

Chapter 7
Cauchy Problem in the Gevrey Classes

Abstract In Chap. 6 we showed that there exists a second order differential operator of spectral type 2 on Σ with bicharacteristics tangent to the double characteristic manifold for which the Cauchy problem is ill-posed in the Gevrey class of order s for any $s > 5$ even though the Levi condition is satisfied. The best we can expect is the well-posedness in the Gevrey class of order 5 under the Levi condition. This is indeed the case. We prove that for general second order differential operator of spectral type 2 on Σ which may have tangent bicharacteristics, the Cauchy problem is well-posed in the Gevrey class of order 5 under the Levi condition.

7.1 Pseudodifferential Operators, Revisited

In this chapter we use more general classes of pseudodifferential operators than we introduced in Sect. 1.3. We make a brief summary of calculus of pseudodifferential operators which we use in this chapter. For details we refer to [2, 34, 57].

Definition 7.1 Let $\Phi(x', \xi')$, $\phi(x', \xi')$ and $m(x', \xi')$ be positive functions on $\mathbb{R}^{2n}$. Then $S_{\Phi,\phi}(m)$ is defined as the set of all $a(x', \xi') \in C^{\infty}(\mathbb{R}^{2n})$ such that for all multi-indices α, $\beta \in \mathbb{N}^n$ there exists $C_{\alpha\beta}$ such that

$$|\partial_{x'}^{\alpha} \partial_{\xi'}^{\beta} a(x', \xi')| \leq C_{\alpha\beta} m(x', \xi') \phi(x', \xi')^{-|\alpha|} \Phi(x', \xi')^{-|\beta|} \tag{7.1}$$

holds in $\mathbb{R}^{2n}$. If ϕ, Φ and m depend on several parameters then $S_{\Phi,\phi}(m)$ is defined as the set of all $a(x', \xi')$ verifying (7.1) uniformly in the parameters.

For calculus of pseudodifferential operators with symbol $S_{\Phi,\phi}(m)$ several conditions are required on Φ, ϕ and m. To formulate those, following to [34, 57] we introduce

© Springer International Publishing AG 2017

T. Nishitani, *Cauchy Problem for Differential Operators with Double Characteristics*, Lecture Notes in Mathematics 2202,
DOI 10.1007/978-3-319-67612-8_7

Definition 7.2 A metric $g_{(x',\xi')}(z',\zeta') = \phi(x',\xi')^{-2}|z'|^2 + \Phi(x',\xi')^{-2}|\zeta'|^2$ on $\mathbb{R}^{2n}$ is called admissible if one can find positive constants $c, C > 0$ and N such that

$$g_{(x',\xi')}(y',\eta') \leq c \Longrightarrow g_{(x'+y',\xi'+\eta')}(z',\zeta') \leq Cg_{(x',\xi')}(z',\zeta'),$$

$$g_{(y',\eta')}(z',\zeta') \leq Cg_{(x',\xi')}(z',\zeta')\big(1 + g^{\sigma}_{(y',\eta')}(x'-y',\xi'-\eta')\big)^N, \qquad (7.2)$$

$$\sup_{(y',\eta')} g_{(x',\xi')}(y',\eta')/g^{\sigma}_{(x',\xi')}(y',\eta') = (\phi(x',\xi')\Phi(x',\xi'))^{-2} \leq 1$$

where $g^{\sigma}_{(x',\xi')}(z',\zeta') = \Phi(x',\xi')^2|z'|^2 + \phi(x',\xi')^2|\zeta'|^2$, the dual metric with respect to the symplectic form σ. If Φ and ϕ depend on several parameters we call g an admissible metric when (7.2) holds uniformly in the parameters.

Definition 7.3 A positive function $m(x',\xi')$ is called g-admissible weight if there are positive constants c, C and N such that

$$g_{(x',\xi')}(y',\eta') \leq c \Longrightarrow m(x',\xi')/C \leq m(x'+y',\xi'+\eta') \leq Cm(x',\xi'),$$

$$m(y',\eta') \leq Cm(x',\xi')\big(1 + g^{\sigma}_{(y',\eta')}(x'-y',\xi'-\eta')\big)^N. \qquad (7.3)$$

If $m(x',\xi')$ depends on several parameters we call m a g-admissible weight when (7.3) holds uniformly in the parameters.

From now on we assume that g is an admissible metric and m is a g-admissible weight. If $g = \langle\xi'\rangle^{2\delta}|dx'|^2 + \langle\xi'\rangle^{-2\rho}|d\xi'|^2$ then $S_{\Phi,\phi}(\langle\xi'\rangle^m) = S^m_{\rho,\delta}$. To $a(x',\xi') \in S_{\Phi,\phi}(m)$ we associate the Weyl quantized pseudodifferential operator $a(x',D')$ by Definition 1.9. Let $a_i(x',\xi') \in S_{\Phi,\phi}(m_i), i = 1, 2$ then there is $b(x',\xi') \in S_{\Phi,\phi}(m_1 m_2)$ such that

$$a_1(x',D')a_2(x',D') = b(x',D').$$

We denote $b(x',\xi')$ by $a_1(x',\xi')\#a_2(x',\xi')$.

Proposition 7.1 *Let $a_i \in S_{\Phi,\phi}(m_i)$. Then we have*

$$a_1\#a_2 - a_2\#a_1 - \{a_1, a_2\}/i \in S_{\Phi,\phi}(m_1 m_2(\phi\Phi)^{-3}),$$

$$a_1\#a_2 + a_2\#a_1 - 2a_1 a_2 \in S_{\Phi,\phi}(m_1 m_2(\phi\Phi)^{-2}),$$

$$a_1\#a_2\#a_1 - a_1^2 a_2 \in S_{\Phi,\phi}(m_1^2 m_2(\phi\Phi)^{-2}).$$

Proposition 7.2 *If $a \in S_{\Phi,\phi}(m)$ then one has*

$$\mathrm{Re}(au, u) = ((\mathrm{Re}\, a)u, u), \quad \mathrm{Im}(au, u) = ((\mathrm{Im}\, a)u, u).$$

In particular if $a(x',\xi')$ is real valued then $(au, u) = (u, au)$.

Theorem 7.1 (L^2-Boundedness) *Let $a \in S_{\Phi,\phi}(1)$. Then there is $C > 0$ such that*

$$\|au\| \leq C\|u\|, \quad u \in \mathscr{S}(\mathbb{R}^n).$$

Theorem 7.2 (Fefferman-Phong Inequality) *Assume that $a(x', \xi')$ is non-negative and $a \in S_{\Phi,\phi}((\phi\Phi)^2)$. Then there is $C > 0$ such that*

$$(au, u) \geq -C\|u\|^2, \quad u \in \mathscr{S}(\mathbb{R}^n).$$

Lemma 7.1 *Assume that g is an admissible metric and $m(x', \xi') \geq c_1 > 0$ is a g-admissible weight. Assume $m(\phi\Phi)^{-1} \leq 1$ then $\tilde{g} = mg$ is an admissible metric and any g-admissible weight $\tilde{m}$ is $\tilde{g}$-admissible.*

Proof In the proof we write $X = (x', \xi')$, $Y = (y', \eta')$. Since $\tilde{g}_X(Y) < c_1 c$ implies $g_X(Y) < c$ and m is g-admissible then it is clear that $\tilde{g}$ verifies the first inequality of (7.2). Since $\tilde{g}^\sigma = m^{-1}g^\sigma$ it is clear that the third inequality of (7.2) holds for $\tilde{g}$. When $g_Y(Y-X) < c$ the second inequality of (7.2) holds for $\tilde{g}$ clearly. If $g_Y(Y-X) \geq c$ then from $g_Y^\sigma(\phi\Phi)^{-2} \geq g_Y$ and $m(\phi\Phi)^{-1} \leq 1$ one has

$$\tilde{g}_Y^\sigma(X - Y)^2 = m(Y)^{-2}g_Y^\sigma(X - Y)^2$$
$$\geq m(Y)^{-2}(\phi(Y)\Phi(Y))^2 g_Y(X - Y)g_Y^\sigma(X - Y) \geq c\,g_Y^\sigma(X - Y). \tag{7.4}$$

This proves that $\tilde{g}$ verifies the second condition of (7.2) since m is g-admissible. We now assume that $\tilde{m}$ is a g-admissible weight. The first inequality of (7.3) for $\tilde{m}$ and $\tilde{g}$ is clear because $c_1^{-1}\tilde{g}_X(Y) \geq g_X(Y)$. If $g_Y(Y - X) < c$ the second inequality of (7.3) for $\tilde{m}$ and $\tilde{g}$ is obvious because $\tilde{m}$ is g-admissible. Assume $g_Y(Y - X) \geq c$. Noting $\tilde{m}(Y) \leq C\tilde{m}(X)(1 + g_Y^\sigma(Y - X))^N$ we conclude the second inequality for $\tilde{m}$ and $\tilde{g}$ thanks to (7.4). $\qquad\square$

7.2 Pseudodifferential Weights and Factorization

We study a second order differential operator P

$$P(x, D) = -D_0^2 + \sum_{|\alpha|\leq 2, \alpha_0 < 2} a_\alpha(x)D^\alpha = P_2 + P_1 + P_0$$
$$= -D_0^2 + A_1(x, D')D_0 + A_2(x, D') \tag{7.5}$$

where we assume that the coefficients are in the Gevrey class of order s, $1 < s \leq 5$. More precisely we assume that for every $\rho \in \Sigma$ one can find a conic neighborhood V of ρ, $r \in \mathbb{N}$ and $\xi_0 = \phi_0$, $\phi_j(x, \xi')$, $j = 1, \ldots, r$ defined in V, homogeneous of degree 1 in ξ' with linearly independent differentials verifying

$$|\partial_x^\beta \partial_{\xi'}^\alpha \phi_j(x, \xi')| \leq C_\alpha A^{|\beta|} |\xi'|^{1-|\alpha|} |\beta|!^s, \quad \forall \alpha, \beta$$

in V with some $C > 0$, $A > 0$ and $V \cap \Sigma$ is given by $\phi_j = 0, j = 0, \ldots, r$.
We also assume that p satisfies (1.5) and the spectral type is 2 on Σ. We do not
assume the non existence of tangent bicharacteristics any more. We fix any $\rho \in \Sigma$.
Thanks to Proposition 2.4, p admits a factorization near ρ verifying the conditions
in Proposition 2.4. We extend these ϕ_j, which are given in Proposition 2.4, to be 0
outside a conic neighborhood of $(0, \rho')$ so that they satisfy

$$|\partial_x^\beta \partial_{\xi'}^\alpha \phi_j(x, \xi')| \leq C_\alpha \langle \xi' \rangle^{1-|\alpha|} A^{|\beta|} |\beta|!^s, \quad \forall \alpha, \beta. \tag{7.6}$$

We now define $f(x, \xi')$ by (5.7) with $\chi(x', \xi')$ such that $\langle \xi' \rangle \chi(x', \xi')$ verifies (7.6).
Then it follows that $f(x, \xi')$ satisfies the estimate (7.6).

Lemma 7.2 *Let $f(x, \xi')$ be as above. Taking $k > 0$ large and $\tau > 0$ small we put*

$$p = -(\xi_0 + \phi_1)(\xi_0 - \phi_1) + \sum_{j=2}^{r+1} \phi_j^2, \quad \phi_{r+1} = kf(x, \xi')$$

*which coincides with the original p in a conic neighborhood of ρ' and with some
$c > 0$ we have*

$$\{\xi_0 - \phi_1, \phi_j\} = \sum_{k=1}^{r+1} C_{jk}\phi_k, \quad \{\phi_2, \phi_1\} + |\phi_{r+1}| \geq c|\xi'|$$

for $|x_0| < \tau$.

Proof It suffices to repeat the same arguments in the proof of Lemma 5.6.   □

Lemma 7.3 *Assume that P satisfies the Levi condition on Σ. Write $\phi_0 = \xi_0$ then
P_{sub} can be written*

$$P_{sub} = \sum_{j=0}^{r+1} C_j\phi_j.$$

Proof It is clear from Proposition 5.7.   □

We now make a dilation of the variable; $(x_0, x') \mapsto (\mu x_0, x')$ with small $\mu > 0$ so
that $p(x, \xi) \mapsto p(x, \xi, \mu)$ where

$$\mu^2 p(x, \xi, \mu) = -(\xi_0 + \phi_1(\mu x_0, x', \mu \xi'))(\xi_0 - \phi_1(\mu x_0, x', \mu \xi'))$$

$$+ \sum_{j=2}^{r+1} \phi_j(\mu x_0, x', \mu \xi')^2. \tag{7.7}$$

We simply write $\phi_j(x, \xi')$ or $\phi_j(x, \xi', \mu)$ for $\phi_j(\mu x_0, x', \mu \xi')$. Define

$$w = \sqrt{\langle \mu \xi' \rangle^{-2} \phi_1^2 + \langle \mu \xi' \rangle^{-2\delta}}, \quad k = \sqrt{1 - cw}. \tag{7.8}$$

with $0 < \delta < 1/2$ and a constant $c > 0$ such that $1 - cw \geq c_1 > 0$. Then one can rewrite p as

$$p = -(\xi_0 + \phi_1 k)(\xi_0 - \phi_1 k) + \sum_{j=2}^{r+1} \phi_j^2 + cw\phi_1^2$$

because $1 - k^2 = cw$. Remark that

$$\xi_0 - \phi_1 k = \xi_0 - \phi_1 - \phi_1(k - 1) = \xi_0 - \phi_1 + \phi_1 \bar{k}$$

where $\bar{k} = 1 - k = cw/(1 + \sqrt{1 - cw})$ and in what follows one can assume that $c = 1$ without restrictions. We denote

$$\langle \xi' \rangle_\mu^2 = \mu^{-2} + |\xi'|^2 = \mu^{-2} \langle \mu \xi' \rangle^2, \quad 0 < \mu < 1.$$

Definition 7.4 Assume $0 \leq \delta < 1/2$. We say $a = a(x, \xi', \mu) \in \tilde{S}_\delta^{(s)}(m)$ where $m = m(x, \xi', \mu)$ if a verifies the following estimates for all α, β with A independent of α, β, μ and C_α independent of β, μ.

$$|\partial_x^\beta \partial_{\xi'}^\alpha a| \leq C_\alpha m A^{|\beta|} |\beta|!^{s/2} (|\beta|^{s/2} + \langle \mu \xi' \rangle^\delta)^{|\beta|} (\langle \mu \xi' \rangle^\delta \langle \xi' \rangle_\mu^{-1})^{|\alpha|}.$$

We assume that $a(x, \xi', \mu)$ is independent of x for $|x| \geq M$ with a large M. Here we note that if

$$|\partial_x^\beta \partial_{\xi'}^\alpha a(x, \xi', \mu)| \leq C_\alpha m(x, \xi', \mu) \langle \xi' \rangle_\mu^{-|\alpha|} A^{|\beta|} |\beta|!^s, \quad \forall \alpha, \beta$$

then it is clear that $a(x, \xi', \mu) \in \tilde{S}_\delta^{(s)}(m)$ for all $0 \leq \delta < 1/2$.

Lemma 7.4 *Assume that $a(x, \xi')$ satisfy $|\partial_x^\beta \partial_{\xi'}^\alpha a(x, \xi')| \leq C_\alpha \langle \xi' \rangle^{k-|\alpha|} A^{|\beta|} |\beta|!^s$ for all α, β. Then we have $a(\mu x_0, x', \mu \xi') \in \tilde{S}_0^{(s)}(\langle \mu \xi' \rangle^k)$.*

Proof Easy.   □

Lemma 7.5 *Assume that $a_i(x, \xi', \mu) \in \tilde{S}_\delta^{(s)}(m_i)$, $i = 1, 2$ where $m_i = m_i(x, \xi', \mu)$ then $a_1(x, \xi', \mu) a_2(x, \xi', \mu) \in \tilde{S}_\delta^{(s)}(m_1 m_2)$.*

Proof It is enough to note that for $A_1 > A_2$ we have

$$\sum_{\beta' \leq \beta} \binom{\beta}{\beta'} A_1^{|\beta'|} |\beta'|!^{s/2} A_2^{|\beta-\beta'|} |\beta - \beta'|!^{s/2} \leq A_1^{1+|\beta|} (A_1 - A_2)^{-1} |\beta|!^{s/2}$$

and $(|\beta'|^{s/2} + \langle \mu\xi' \rangle^\delta)^{|\beta'|} (|\beta - \beta'|^{s/2} + \langle \mu\xi' \rangle^\delta)^{|\beta-\beta'|} \leq (|\beta|^{s/2} + \langle \mu\xi' \rangle^\delta)^{|\beta|}$. $\square$

Lemma 7.6 *Let $s > 2$ and assume $|\partial_x^\beta \partial_\xi^\alpha f(x, \xi', \mu)| \leq C_\alpha \langle \xi' \rangle_\mu^{-|\alpha|} A^{|\beta|} |\beta|!^s$ for all α, β. Then*

$$w(x, \xi', \mu) = \left(f(x, \xi', \mu)^2 + \langle \mu\xi' \rangle^{-2\delta} \right)^{1/2} \in \tilde{S}_\delta^{(s)}(w).$$

To prove this lemma we first show the next

Lemma 7.7 *Let $s > 2$ and assume that with positive constants C_1, C_2*

$$|\partial_x^\alpha f(x)| \leq C_1 C_2^{|\alpha|} |\alpha|!^s, \quad |\alpha| \geq 1, \quad x \in \mathbb{R}^n \tag{7.9}$$

and define $w(x) = \sqrt{f(x)^2 + B^{-2}}$ with a positive constant B. Then one can find positive constant A such that

$$|\partial_x^\alpha w^{\pm 1}(x)| \leq w^{\pm 1}(x) A^{|\alpha|} |\alpha|!^{s/2} (|\alpha|^{s/2} + B)^{|\alpha|}, \quad |\alpha| \geq 1. \tag{7.10}$$

Proof Note that $w(x)\partial_x^e w(x) = f(x)\partial_x^e f(x) = F(x)$ for $|e| = 1$ where we may assume that $|\partial_x^\alpha F(x)| \leq A_1^{|\alpha|+1} |\alpha|!^s$ for $|\alpha| \geq 1$. Write $s = 2 + 2\delta$ with $\delta > 0$ then one can find $N(\delta)$ such that for $n \geq N(\delta)$

$$\sum_{j=1}^n \binom{n+1}{j}^{-\delta} \leq \frac{1}{2}.$$

Indeed it suffices to note that for $n \geq N$ with a fixed N

$$\sum_{j=1}^n \binom{n+1}{j}^{-\delta} \leq 2\sum_{j=1}^N \binom{n+1}{j}^{-\delta} + 2\sum_{j=N+1}^{[(n+1)/2]} \binom{n+1}{j}^{-\delta}$$

$$\leq 2\sum_{j=1}^N (n+1)^{-\delta} + 2\sum_{j=N+1}^\infty 2^{-\delta j} \leq 2N/(n+1)^\delta + 2^{-N\delta+1}/(2^\delta - 1).$$

Choose A such that $A_1/A \leq 1/2$ and that (7.10) holds for $n \leq N(\delta)$. Assume that the inequalities (7.10) for $w(x)$ hold for $|\alpha| \leq n$ where $n \geq N(\delta)$ and study $\partial_x^\alpha w(x)$

with $|\alpha| = n + 1$. Then we have

$$w\partial_x^{\alpha+e}w = - \sum_{|\beta|\neq 0, \beta \leq \alpha} \frac{\alpha!}{\beta!(\alpha - \beta)!} \partial_x^\beta w \partial_x^{\alpha+e-\beta}w + \partial_x^\alpha F \tag{7.11}$$

which gives

$$|w\partial_x^{\alpha+e}w| \leq \sum_{j=1}^n \binom{n}{j} wA^j j! j^{s/2} (j^{s/2} + B)^j wA^{n+1-j}(n + 1 - j)!^{s/2}$$

$$\times ((n + 1 - j)^{s/2} + B)^{n+1-j} + A_1^{n+1} n!^s$$

$$\leq w^2 A^{n+1}\Big(\sum_{j=1}^n \binom{n}{j} j!^{s/2}(n + 1 - j)!^{s/2}(j^{s/2} + B)^j$$

$$\times ((n + 1 - j)^{s/2} + B)^{n+1-j} + w^{-2}(A_1/A)^{n+1} n!^s\Big).$$

Note that $(j^{s/2} + B)^j((n + 1 - j)^{s/2} + B)^{n+1-j} \leq ((n + 1)^{s/2} + B)^{n+1}$ and

$$\frac{n(n + 1)}{2}B^2 n!^s \leq \frac{(n + 1)n}{2}(n + 1)^{(n-1)s/2}B^2(n + 1)!^{s/2}$$

$$\leq ((n + 1)^{s/2} + B)^{n+1}(n + 1)!^{s/2}.$$

Since $w^{-2} \leq B^2$ one has

$$|\partial_x^{\alpha+e}w| \leq wA^{n+1}((n + 1)^{s/2} + B)^{n+1}(n + 1)!^{s/2}$$

$$\times \Big(\sum_{j=1}^n \binom{n}{j}\binom{n + 1}{j}^{-s/2} + \Big(\frac{A_1}{A}\Big)^{n+1}\frac{2}{n(n + 1)}\Big).$$

We now check that $\sum_{j=1}^n \binom{n}{j}\binom{n+1}{j}^{-s/2} + (A_1/A)^{n+1}(2/n(n + 1)) \leq 1$. In fact we have

$$\sum_{j=1}^n \binom{n}{j}\binom{n + 1}{j}^{-s/2} = \sum_{j=1}^n \frac{n + 1 - j}{n + 1}\binom{n + 1}{j}^{1-s/2} \leq \sum_{j=1}^n \binom{n + 1}{j}^{-\delta} \leq \frac{1}{2}$$

and hence the assertion for $w(x)$ holds. To check the assertion for $w^{-1}(x)$ we assume that $|\partial_x^\alpha w^{-1}| \leq w^{-1}\tilde{A}^{|\alpha|}|\alpha|!^{s/2}(|\alpha|^{s/2} + B)^{|\alpha|}$ holds for $|\alpha| \leq n$. From $ww^{-1} = 1$

taking into account the estimate (7.10) for $w(x)$ we have for $|\alpha| = n + 1$

$$|\partial_x^\alpha w^{-1}| \leq w^{-1} \tilde{A}^{n+1} (n+1)!^{s/2} ((n+1)^{s/2} + B)^{n+1}$$

$$\times \sum_{j=1}^{n+1} \binom{n+1}{j} \binom{n+1}{j}^{-s/2} (A/\tilde{A})^j.$$

Thus it suffices to choose $\tilde{A}$ so that $A/\tilde{A} \leq 1/2$ to obtain the desired estimate for $|\partial_x^\alpha w^{-1}|$ and hence by induction on $|\alpha|$ we can prove (7.10) with $A = \tilde{A}$. $\quad\square$

Corollary 7.1 *Assume that $s > 2$ then for any $m \in \mathbb{Z}$ there is $A > 0$ such that*

$$|\partial_x^\alpha w^{-m}| \leq w^{-m} A^{|\alpha|} |\alpha|!^{s/2} (|\alpha|^{s/2} + B)^{|\alpha|}.$$

Proof It is clear from Lemma 7.7 and the proof of Lemma 7.5. $\quad\square$

Proof of Lemma 7.6 Note that one can write

$$\partial_{\xi'}^\alpha w = \partial_{\xi'}^\alpha (w^2)^{1/2} = \sum C_{\alpha_1,\ldots,\alpha_k} w^{1-2k} (\partial_{\xi'}^{\alpha_1} w^2) \cdots (\partial_{\xi'}^{\alpha_k} w^2) \qquad (7.12)$$

where the sum is taken over $\alpha_1 + \cdots + \alpha_k = \alpha$ with $|\alpha_j| \geq 1$. Remarking that $|\partial_{\xi'}^\alpha w^2| \leq C_\alpha w \langle \xi' \rangle_\mu^{-1} \leq C_\alpha w^2 \langle \mu \xi' \rangle^\delta \langle \xi' \rangle_\mu^{-1}$ for $|\alpha| = 1$ and noting $w^2 \langle \mu \xi' \rangle^{2\delta} \geq 1$

$$|\partial_x^\beta \partial_{\xi'}^\alpha w^2| \leq C_\alpha \langle \xi' \rangle_\mu^{-|\alpha|} A^{|\beta|} |\beta|!^s \leq C_\alpha w^2 \langle \mu \xi' \rangle^{2\delta} \langle \xi' \rangle_\mu^{-|\alpha|} A^{|\beta|} |\beta|!^s$$

$$\leq C_\alpha w^2 (\langle \mu \xi' \rangle^\delta \langle \xi' \rangle_\mu^{-1})^{|\alpha|} A_1^{|\beta|} |\beta|!^{s/2} (|\beta|^{s/2} + \langle \mu \xi' \rangle^\delta)^{|\beta|} \qquad (7.13)$$

for $|\alpha + \beta| \geq 2$ ($|\alpha| \geq 1$). Applying Corollary 7.1 to w^{1-2k} with $B = \langle \mu \xi' \rangle^\delta$ we get

$$|\partial_x^\beta w^{1-2k}| \leq C w^{1-2k} A_1^{|\beta|} |\beta|!^{s/2} (|\beta|^{s/2} + \langle \mu \xi' \rangle^\delta)^{|\beta|}. \qquad (7.14)$$

Then the assertion follows immediately from (7.13), (7.14) and (7.12). $\quad\square$

Lemma 7.8 *We have $\phi_1 \bar{k} \in \tilde{S}_\delta^{(s)}(w^2 \langle \mu \xi' \rangle)$ and $\partial_x^\beta \partial_{\xi'}^\alpha (\phi_1 k) \in \tilde{S}_\delta^{(s)}(\langle \mu \xi' \rangle \langle \xi' \rangle_\mu^{-|\alpha|})$ for $|\alpha + \beta| \leq 2$.*

Proof Since $\phi_1 \in \tilde{S}_\delta^{(s)}(w \langle \mu \xi' \rangle)$ the first assertion is clear. To check the second assertion it is enough to note that $\partial_x^\beta \partial_{\xi'}^\alpha k \in \tilde{S}_\delta^{(s)}(\langle \xi' \rangle_\mu^{-|\alpha|})$ for $|\alpha + \beta| = 1$ and $\phi_1 \in \tilde{S}_\delta^{(s)}(w \langle \mu \xi' \rangle) \cap \tilde{S}_0^{(s)}(\langle \mu \xi' \rangle)$. $\quad\square$

Definition 7.5 Let $0 \leq \delta < 1/2$. We denote by $\tilde{S}_\delta(m)$ and $\tilde{S}_w(m)$ the symbol class defined by the metric

$$g_\delta = \langle \mu \xi' \rangle^{2\delta} (|dx'|^2 + \langle \xi' \rangle_\mu^{-2} |d\xi'|^2),$$

$$g_w = w^{-2} (|dx'|^2 + \langle \xi' \rangle_\mu^{-2} |d\xi'|^2) = w^{-2} g_0$$

respectively (see Definition 7.1), that is the set of all $a(x, \xi', \mu) \in C^\infty(\mathbb{R}^{n+1} \times \mathbb{R}^n)$ such that for all α, β one has

$$|\partial_x^\beta \partial_{\xi'}^\alpha a(x, \xi', \mu)| \leq C_{\alpha\beta} m(x, \xi', \mu) \langle \mu\xi' \rangle^{\delta|\beta|} (\langle \mu\xi' \rangle^\delta \langle \xi' \rangle_\mu^{-1})^{|\alpha|},$$

$$|\partial_x^\beta \partial_{\xi'}^\alpha a(x, \xi', \mu)| \leq C_{\alpha\beta} m(x, \xi', \mu) w^{-|\alpha+\beta|} \langle \xi' \rangle_\mu^{-|\alpha|}$$

respectively, where x_0 is regarded as a parameter and also $\mu > 0$ is a parameter and the constants $C_{\alpha\beta}$ do not depend on μ.

It is easy to check that g_δ is an admissible metric and $\langle \xi' \rangle_\mu^k$, $\langle \mu\xi' \rangle^k$ with $k \in \mathbb{R}$ are g_δ-admissible weights. We note that $g_w \leq g_\delta$ since $w \geq \langle \mu\xi' \rangle^{-\delta}$ and hence $\tilde{S}_w(m) \subset \tilde{S}_\delta(m)$. We are now going to prove that g_w is an admissible metric.

Lemma 7.9 *Write $X = (x', \xi'), Y = (y', \eta')$. Then there are $c, C > 0$ such that*

$$(g_w)_X(Y) < c \implies w(X + Y)/C \leq w(X) \leq Cw(X + Y).$$

Proof Since $w(X) \leq C$ one can choose $c > 0$ so that $(g_w)_X(Y) < c$ implies $|\eta'| \leq \langle \xi' \rangle_\mu/2$. Assume $|\eta'| \leq \langle \xi' \rangle_\mu/2$ and hence $\langle \mu\xi' \rangle^{-\delta}/C \leq \langle \mu(\xi' + \eta') \rangle^{-\delta} \leq C \langle \mu\xi' \rangle^{-\delta}$. Write $F = \langle \mu\xi' \rangle^{\delta-1} \phi_1$ and $W = \langle \mu\xi' \rangle^\delta w$ so that $W = (F^2 + 1)^{1/2}$. It suffices to show $W(X)/C \leq W(X + Y) \leq CW(X)$. Note that

$$|W(X + Y) - W(X)| = |F(X + Y) - F(X)|$$

$$\times |F(X + Y) + F(X)|/(W(X + Y) + W(X))$$

$$\leq 2|F(X + Y) - F(X)|.$$

Since $F \in \tilde{S}_0(\langle \mu\xi' \rangle^\delta)$ we have $|F(X + Y) - F(X)| \leq C \langle \mu\xi' \rangle^\delta (|y'| + \langle \xi' \rangle_\mu^{-1} |\eta'|)$ because $|\eta'| \leq \langle \xi' \rangle_\mu/2$. Therefore one obtains

$$|W(X + Y)/W(X) - 1| \leq C \langle \mu\xi' \rangle^\delta (|y'| + \langle \xi' \rangle_\mu^{-1} |\eta'|)/W(X)$$

$$\leq Cw^{-1}(X)(|y'| + \langle \xi' \rangle_\mu^{-1} |\eta'|) \leq C'(g_w)_X^{1/2}(Y)$$

which proves the assertion. $\square$

Lemma 7.10 *There exist $C > 0$ and $N > 0$ such that*

$$w(X + Y)^{\pm 1} \leq Cw(X)^{\pm 1}(1 + (g_w)_X^\sigma(Y))^N,$$

$$\langle \xi' + \eta' \rangle_\mu^{\pm 1} \leq C \langle \xi' \rangle_\mu^{\pm 1}(1 + (g_w)_X^\sigma(Y))^N.$$

Proof We first prove the assertion for w^{-1}. If $(g_w)_X(Y) < c$ then the assertion is clear from Lemma 7.9. Assume $(g_w)_X(Y) \geq c$ and note

$$(g_w)^\sigma_X(Y) = w^4(X)\langle\xi'\rangle^2_\mu(g_w)_X(Y) \geq (cw^4(X)\langle\xi'\rangle^2_\mu + w^2(X)|\eta'|^2)/2.$$

If $|\eta'| \leq \langle\xi'\rangle_\mu/2$ then $1 \leq w^4(X+Y)\langle\mu(\xi'+\eta')\rangle^{4\delta} \leq w^4(X+Y)\langle\xi'\rangle^2_\mu$ and hence $Cw^4(X+Y)(g_w)^\sigma_X(Y) \geq w^4(X)$. If $|\eta'| \geq \langle\xi'\rangle_\mu/2$ then one has

$$w(X+Y)^{-1} \leq C(1+|\eta'|)^\delta \leq Cw(X)^{-\delta}(1+(g_w)^\sigma_X)^\delta.$$

We turn to the assertion for w. If $(g_w)_{X+Y}(Y) < c$ then the assertion follows from Lemma 7.9. Assume $(g_w)_{X+Y}(Y) \geq c$. If $|\eta'| \leq \langle\xi'\rangle_\mu/2$ we have

$$cw(X+Y)^2 \leq w(X+Y)^2(g_w)_{X+Y}(Y) \leq C(|y'|^2 + \langle\xi'\rangle^{-2}_\mu|\eta'|^2)$$

$$\leq Cw(X)^2(g_w)_X(Y)$$

$$= Cw(X)(w(X)^{-3}\langle\xi'\rangle^{-2}_\mu)(g_w)^\sigma_X(Y).$$

From $w(X)^{-3}\langle\xi'\rangle^{-2}_\mu \leq \langle\mu\xi'\rangle^{3\delta}\langle\xi'\rangle^{-2}_\mu \leq 1$ the assertion follows. If $|\eta'| \geq \langle\xi'\rangle_\mu/2$ noting that $\langle\xi'\rangle^{1-\delta}_\mu \leq 2|\eta'|w(X) \leq \{2(g_w)^\sigma_X(Y)\}^{1/2}$ we conclude

$$w(X+Y) \leq C \leq C\langle\xi'\rangle^\delta_\mu w(X) \leq C'w(X)(1+(g_w)^\sigma_X(Y))^{\delta/(2(1-\delta))}$$

which proves the assertion. The proof for $\langle\xi'+\eta'\rangle^{\pm1}_\mu$ is easier.   □

Proposition 7.3 *The metric g_w is admissible and $w^{\pm1}$ are g_w-admissible weights.*

Proof The proof follows from Lemmas 7.9 and 7.10 immediately.   □

Lemma 7.11 *We have $w^{\pm1} \in \tilde{S}_w(w^{\pm1})$ and $\phi_1 \in \tilde{S}_w(\langle\mu\xi'\rangle w)$ and moreover we have $\partial^\beta_x\partial^\alpha_{\xi'}(\phi_1 k) \in \tilde{S}_w(\langle\mu\xi'\rangle\langle\xi'\rangle^{-|\alpha|}_\mu)$ for $|\alpha+\beta| \leq 2$.*

Proof It suffices to show $|\partial^\beta_x\partial^\alpha_{\xi'}w^{\pm1}|/w^{\pm1} \leq C_{\alpha\beta}w^{-|\beta|}(\langle\xi'\rangle_\mu w)^{-|\alpha|}$ with some $C_{\alpha,\beta}$ independent of $0 < \mu < \mu_0$ for any α, β. We first show the assertion for w. For $|\alpha+\beta| = 1$ the assertion is clear because $|\partial^\beta_x\partial^\alpha_{\xi'}(\langle\mu\xi'\rangle^{-1}\phi_1)| \leq C\langle\xi'\rangle^{-|\alpha|}_\mu$. For $|\alpha+\beta| \geq 2$ noting $w^2 = \langle\mu\xi'\rangle^{-2}\phi^2_1 + \langle\mu\xi'\rangle^{-2\delta}$ we have

$$|\partial^\beta_x\partial^\alpha_{\xi'}w^2| \leq C_{\alpha\beta}\langle\xi'\rangle^{-|\alpha|}_\mu \leq CC_{\alpha\beta}\langle\xi'\rangle^{-|\alpha|}_\mu w^{-(|\alpha+\beta|-2)}$$

$$= CC_{\alpha\beta}w^2w^{-|\beta|}(\langle\xi'\rangle_\mu w)^{-|\alpha|}.$$

Therefore noting

$$(\partial^\beta_x\partial^\alpha_{\xi'}w)w = -\sum_{\alpha'+\beta'<\alpha+\beta}C_{\alpha'\beta'}(\partial^{\beta'}_x\partial^{\alpha'}_{\xi'}w)\partial^{\beta-\beta'}_x\partial^{\alpha-\alpha'}_{\xi'}w + \partial^\beta_{x'}\partial^\alpha_{\xi'}w^2$$

the assertion follows from the induction on $|\alpha + \beta|$. As for w^{-1} taking $w^{-1}w = 1$ into account the proof is similar. $\square$

Notation 7.1 We write $a \in \mu^k \tilde{S}_w(m)$ or $a \in \mu^k \tilde{S}_\delta(m)$ if $\mu^{-k}a \in \tilde{S}_w(m)$ or $\mu^{-k}a \in \tilde{S}_\delta(m)$.
Note $w^{-1/2} \in \tilde{S}_w(\langle\mu\xi'\rangle^\kappa)$. Since $g_w = w^{-2}g_0$ and $g_w = w^{-2}\langle\xi'\rangle_\mu^{-2}g_0^\sigma \le g_0^\sigma$ from [34, Theorem 18.5.4] we have

Lemma 7.12 *Let* $a \in \tilde{S}_w(m_1)$ *and* $b \in \tilde{S}_0(m_2)$. *Then we have*

$$a \# a - a^2 \in \mu^2 \tilde{S}_w(m_1^2 w^{-4}\langle\mu\xi'\rangle^{-2}),$$

$$a \# b - b \# a - \frac{1}{i}\{a, b\} \in \mu^3 \tilde{S}_w(m_1 m_2 w^{-3}\langle\mu\xi'\rangle^{-3}),$$

$$a \# b + b \# a - 2ab \in \mu^2 \tilde{S}_w(m_1 m_2 w^{-2}\langle\mu\xi'\rangle^{-2}).$$

Corollary 7.2 *Assume that* $a \in \tilde{S}_w(m_1)$ *and* $b \in \tilde{S}_0(m_2)$ *are real. Then there is* $T \in \mu^2 \tilde{S}_w(m_1 m_2 w^{-2}\langle\mu\xi'\rangle^{-2})$ *such that* $(\mathrm{Op}(ab)u, u) = \mathrm{Re}(bu, au) + (Tu, u)$.

7.3 A Lemma on Composition with $e^{\pm\langle D\rangle^\kappa}$

In this section, to simplify notations we use ξ, x for ξ' and x'. Consider

$$e^{\phi(D,\mu)}b(x, D, \mu)e^{-\phi(D,\mu)}$$

where we assume that ϕ verifies

$$|\partial_\xi^\alpha \phi(\xi, \mu)| \le C_\alpha \langle\mu\xi\rangle^\kappa \langle\xi\rangle_\mu^{-|\alpha|} \tag{7.15}$$

with $0 < \kappa < 1$ for any α where C_α is independent of μ. Since

$$e^{\phi(D,\mu)}b(x, D, \mu)v = \int e^{i(x\xi - z\xi + (z-y)\eta)}e^{\phi(\xi,\mu)}b((z+y)/2, \eta, \mu)v(y)dy d\eta dz d\xi$$

inserting

$$e^{-\phi(D,\mu)}u(y) = \int e^{iy\zeta - \phi(\zeta,\mu)}\hat{u}(\zeta)d\zeta$$

into v we have

$$e^{\phi(D,\mu)}b(x, D, \mu)e^{-\phi(D,\mu)}u = \int e^{ix\zeta}I(x, \zeta, \mu)\hat{u}(\zeta)d\zeta$$

where

$$I = \int e^{i(x\xi - \tilde{z}\xi + (z-y)\eta + y\zeta - x\zeta)} e^{\phi(\xi,\mu)} b((z+y)/2, \eta, \mu) e^{-\phi(\zeta,\mu)} dy d\eta dz d\xi.$$

Make a change of variables: $\tilde{z} = (y+z)/2$, $\tilde{y} = (y-z)/2$ and hence we have

$$I = 2^n \int e^{i(-\tilde{z}(\xi-\zeta) + \tilde{y}(\xi - 2\eta + \zeta) + x(\xi-\zeta))} e^{\phi(\xi,\mu)} b(\tilde{z}, \eta, \mu) e^{-\phi(\zeta,\mu)} d\tilde{y} d\eta d\tilde{z} d\xi$$

$$= 2^n \int e^{i\tilde{y}(\xi - 2\eta + \zeta)} d\tilde{y} \int e^{-i(\tilde{z}-x)(\xi - \zeta)} e^{\phi(\xi,\mu)} b(\tilde{z}, \eta, \mu) e^{-\phi(\zeta,\mu)} d\eta d\tilde{z} d\xi$$

$$= 2^n \int e^{-2i(\tilde{z}-x)(\eta-\zeta)} e^{\phi(2\eta - \zeta,\mu)} b(\tilde{z}, \eta, \mu) e^{-\phi(\zeta,\mu)} d\eta d\tilde{z}$$

$$= \int e^{-i\tilde{z}\eta} e^{\phi(\sqrt{2}\eta + \zeta,\mu) - \phi(\zeta,\mu)} b(x + \tilde{z}/\sqrt{2}, \zeta + \eta/\sqrt{2}, \mu) d\eta d\tilde{z}.$$

Thus we conclude that

$$e^{\phi(D,\mu)} b(x, D, \mu) e^{-\phi(D,\mu)} u = \int e^{i(x-y)\xi} a(x, \xi, \mu) u(y) dy d\xi = \mathrm{Op}^0(a) u$$

with

$$a(x, \xi, \mu) = \int e^{-iy\eta} e^{\phi(\xi + \sqrt{2}\eta,\mu) - \phi(\xi,\mu)} b(x + y/\sqrt{2}, \xi + \eta/\sqrt{2}, \mu) dy d\eta.$$

By Lemma 1.2 one sees that $c(x, D, \mu) = \mathrm{Op}^0(a)$ with

$$c(x, \xi, \mu) = (2\pi)^{-n} \int e^{iz\zeta} a(x + z/\sqrt{2}, \xi + \zeta/\sqrt{2}, \mu) dz d\zeta. \tag{7.16}$$

We now insert the expression of $a(x, \xi, \mu)$ into (7.16) to get

$$c(x, \xi, \mu) = (2\pi)^{-n} \int e^{i(z\zeta - y\eta)} e^{\phi(\sqrt{2}\eta + \xi + \frac{\zeta}{\sqrt{2}},\mu) - \phi(\xi + \frac{\zeta}{\sqrt{2}},\mu)}$$

$$\times b(x + (z+y)/\sqrt{2}, \xi + (\eta + \zeta)/\sqrt{2}, \mu) dy d\eta dz d\zeta.$$

The change of variables $\tilde{z} = (z+y)/\sqrt{2}$, $\tilde{y} = (y-z)/\sqrt{2}$, $\tilde{\zeta} = (\zeta + \eta)/\sqrt{2}$, $\tilde{\eta} = (\eta - \zeta)/\sqrt{2}$ gives

$$c(x, \xi, \mu) = (2\pi)^{-n} \int e^{-i(\tilde{z}\tilde{\eta} + \tilde{y}\tilde{\zeta})} e^{\phi(\frac{3\tilde{\zeta}}{2} + \xi + \frac{\tilde{\eta}}{2},\mu) - \phi(\xi + \frac{\tilde{\zeta}}{2} - \frac{\tilde{\eta}}{2},\mu)}$$

$$\times b(x + \tilde{z}, \xi + \tilde{\zeta}, \mu) d\tilde{y} d\tilde{\eta} d\tilde{z} d\tilde{\zeta}$$

$$= (2\pi)^{-n} \int e^{-i\tilde{z}\tilde{\eta}} e^{\phi(\xi + \frac{\tilde{\eta}}{2},\mu) - \phi(\xi - \frac{\tilde{\eta}}{2},\mu)} b(x + \tilde{z}, \xi, \mu) d\tilde{z} d\tilde{\eta}.$$

Thus we obtain

Lemma 7.13 *We have* $e^{\phi(D,\mu)}b(x, D, \mu)e^{-\phi(D,\mu)} = c(x, D, \mu)$ *where*

$$c(x, \xi, \mu) = (2\pi)^{-n} \int e^{-iy\eta} e^{\phi(\xi + \frac{\eta}{2}, \mu) - \phi(\xi - \frac{\eta}{2}, \mu)} b(x + y, \xi, \mu) dy d\eta.$$

Proposition 7.4 *Assume* $2\delta + \kappa \le 1$ *and* (7.15). *Then for* $b(x, \xi, \mu) \in \tilde{S}_\delta^{(1/\kappa)}(\langle \mu \xi \rangle^\ell)$ *we have* $e^{\phi(D,\mu)}b(x, D, \mu)e^{-\phi(D,\mu)} = c(x, D, \mu)$ *of which symbol has the form for any* $N \in \mathbb{N}$

$$c(x, \xi, \mu) = \sum_{j=0}^{N-1} c_j(x, \xi, \mu) + R_N(x, \xi, \mu)$$

where $R_N(x, \xi, \mu) \in \mu^N \tilde{S}_\delta(\langle \mu \xi \rangle^{\ell - N(1 - \kappa - \delta) + n\delta})$ *and* $c_j(x, \xi, \mu)$ *is given by*

$$c_j = \sum_{|\alpha| = j} \frac{(-i)^{|\alpha|}}{\alpha!} \partial_\eta^\alpha e^{\phi(\xi + \frac{\eta}{2}, \mu) - \phi(\xi - \frac{\eta}{2}, \mu)} \Big|_{\eta = 0} \partial_x^\alpha b(x, \xi, \mu). \tag{7.17}$$

We divide the proof into two steps. In the first step we prove (7.17). Recall

$$c(x, \xi, \mu) = (2\pi)^{-n} \int e^{-iy\eta} e^{\phi(\xi + \frac{\eta}{2}, \mu) - \phi(\xi - \frac{\eta}{2}, \mu)} b(x + y, \xi, \mu) dy d\eta. \tag{7.18}$$

With $b_{(\alpha)}(x, \xi, \mu) = (-i)^{|\alpha|} \partial_x^\alpha b(x, \xi, \mu)$ we write

$$b(x + y, \xi, \mu) = \sum_{|\alpha| < N} \frac{1}{\alpha!} b_{(\alpha)}(x, \xi, \mu)(iy)^\alpha$$

$$+ \sum_{|\alpha| = N} \frac{N}{\alpha!} (iy)^\alpha \int_0^1 (1 - s)^{N-1} b_{(\alpha)}(x + sy, \xi, \mu) ds$$

and insert this expression into (7.18) to conclude that $c(x, \xi, \mu)$ is

$$(2\pi)^{-n} \sum_{|\alpha| < N} \frac{1}{\alpha!} \int e^{-iy\eta} e^{\phi(\xi + \frac{\eta}{2}, \mu) - \phi(\xi - \frac{\eta}{2}, \mu)} b_{(\alpha)}(x, \xi, \mu)(iy)^\alpha dy d\eta$$

$$+ (2\pi)^{-n} \sum_{|\alpha| = N} \frac{N}{\alpha!} \int e^{-iy\eta} e^{\phi(\xi + \frac{\eta}{2}, \mu) - \phi(\xi - \frac{\eta}{2}, \mu)} (iy)^\alpha dy d\eta \tag{7.19}$$

$$\times \int_0^1 (1 - s)^{N-1} b_{(\alpha)}(x + sy, \xi, \mu) ds.$$

The first term on the right-hand side is

$$\sum_{|\alpha|<N} \frac{1}{\alpha!} \partial_\eta^\alpha e^{\phi(\xi+\frac{\eta}{2},\mu)-\phi(\xi-\frac{\eta}{2},\mu)}\Big|_{\eta=0} b_{(\alpha)}(x,\xi,\mu) \tag{7.20}$$

because $e^{-iy\eta}(iy)^\alpha = (-\partial_\eta)^\alpha e^{-iy\eta}$ and $\delta = (2\pi)^{-n}\int e^{-iy\eta}dy$. Denote

$$\partial_\eta^\alpha e^{\phi(\xi+\frac{\eta}{2},\mu)-\phi(\xi-\frac{\eta}{2},\mu)} = \sum_{\beta+\gamma=\alpha} \frac{\alpha!}{\beta!\gamma!} \partial_\eta^\beta e^{\phi(\xi+\frac{\eta}{2},\mu)} \partial_\eta^\gamma e^{-\phi(\xi-\frac{\eta}{2},\mu)}$$

$$= 2^{-|\alpha|} \sum_{\beta+\gamma=\alpha} \frac{\alpha!}{\beta!\gamma!} \partial_\xi^\beta e^{\phi(\xi+\frac{\eta}{2},\mu)} (-\partial_\xi)^\gamma e^{-\phi(\xi-\frac{\eta}{2},\mu)}$$

$$= 2^{-|\alpha|} \alpha! H_\alpha(\xi,\eta,\mu)$$

that is

$$H_\alpha(\xi,\eta,\mu) = \frac{2^{|\alpha|}}{\alpha!} \partial_\eta^\alpha e^{\phi(\xi+\eta/2,\mu)-\phi(\xi-\eta/2,\mu)}.$$

Then the second term on the right-hand side of (7.19) yields, up to the factor $2^{-N}(2\pi)^{-n}N$

$$\sum_{|\alpha|=N} \int e^{-iy\eta} H_\alpha(\xi,\eta,\mu)dyd\eta \int_0^1 (1-s)^{N-1} b_{(\alpha)}(x+sy,\xi,\mu)ds$$

$$= \sum_{|\alpha|=N} \int \int_0^1 e^{ix\eta}(1-s)^{N-1} H_\alpha(\xi,s\eta,\mu)d\eta ds \int e^{-iy\eta} b_{(\alpha)}(y,\xi,\mu)dy.$$

With $B_\alpha(\eta,\xi,\mu) = \int e^{-iy\eta} b_{(\alpha)}(y,\xi,\mu)dy$ this can be written as

$$\sum_{|\alpha|=N} \int \int_0^1 e^{ix\eta}(1-s)^{N-1} H_\alpha(\xi,s\eta,\mu)B_\alpha(\eta,\xi,\mu)d\eta ds. \tag{7.21}$$

It is easy to see that $\partial_\eta^\alpha e^{\phi(\xi+\frac{\eta}{2},\mu)-\phi(\xi-\frac{\eta}{2},\mu)}\big|_{\eta=0}$ is a linear combination of terms such as

$$\partial_\xi^{\alpha_1}\phi(\xi,\mu)\cdots\partial_\xi^{\alpha_s}\phi(\xi,\mu), \quad \sum \alpha_p = j, \ |\alpha_p| \geq 1$$

which are in $\tilde{S}_0(\langle\xi\rangle_\mu^{j\kappa-j})$. Therefore we conclude (7.17).

In the second step we prove the assertion for $R_N = 2^{-N}(2\pi)^{-n}N\tilde{R}_N$. We consider

$$\tilde{R}_N = \sum_{|\alpha|=N} \int \int_0^1 e^{ix\eta}(1-s)^{N-1} H_\alpha(\xi,s\eta,\mu)B_\alpha(\eta,\xi,\mu)d\eta ds.$$

Lemma 7.14 *There exists $M > 0$ and $c > 0$ such that we have*

$$|\partial_\xi^\gamma B_\alpha(\eta,\xi,\mu)| \leq C_{\alpha,\gamma}\langle\mu\xi\rangle^\ell (\langle\mu\xi\rangle^\delta \langle\xi\rangle_\mu^{-1})^{|\gamma|} e^{-c|\eta|^\kappa}, \quad |\eta| \geq M\langle\mu\xi\rangle^{2\delta},$$

$$|\partial_\xi^\gamma B_\alpha(\eta,\xi,\mu)| \leq C_{\alpha,\delta}(\langle\mu\xi\rangle^\delta \langle\xi\rangle_\mu^{-1})^{|\gamma|} \langle\mu\xi\rangle^{\ell+\delta|\alpha|} e^{-c((\mu\xi)^{-\delta}|\eta|)^\kappa}, \quad |\eta| \leq M\langle\mu\xi\rangle^{2\delta}.$$

Proof Recall that we have

$$\eta^\nu \partial_\xi^\gamma B_\alpha(\eta,\xi,\mu) = \int e^{-iy\eta} \partial_\xi^\gamma b_{(\alpha+\nu)}(y,\xi,\mu)dy$$

and hence one gets

$$|\partial_\xi^\gamma B_\alpha(\eta,\xi,\mu)| \leq C_\gamma \langle\mu\xi\rangle^\ell (\langle\mu\xi\rangle^\delta \langle\xi\rangle_\mu^{-1})^{|\gamma|}$$

$$\times A^{|\alpha+\nu|}|\alpha+\nu|!^{s/2}(|\alpha+\nu|^{s/2} + \langle\mu\xi\rangle^\delta)^{|\alpha+\nu|}|\eta|^{-|\nu|}$$

$$\leq C_\gamma \langle\mu\xi\rangle^\ell (\langle\mu\xi\rangle^\delta \langle\xi\rangle_\mu^{-1})^{|\gamma|} C_1 C_2^{|\nu|}|\nu|!^{s/2}(|\nu|^{s/2} + \langle\mu\xi\rangle^\delta)^{|\alpha+\nu|}|\eta|^{-|\nu|}$$

where $C_i = C_i(|\alpha|)$. We minimize $C_2^{|\nu|}|\nu|!^{s/2}(|\nu|^{s/2} + \langle\mu\xi\rangle^\delta)^{|\nu|}|\eta|^{-|\nu|}$. Note that

$$C_2^{|\nu|}|\nu|!^{s/2}(|\nu|^{s/2} + \langle\mu\xi\rangle^\delta)^{|\nu|} \leq (2C_2)^{|\nu|}(|\nu|^s + \langle\mu\xi\rangle^{2\delta})^{|\nu|}.$$

If $|\eta| \geq 4C_2 e^s \langle\mu\xi\rangle^{2\delta}$ we choose ν so that $|\nu| = [e^{-s}(2C_2)^{-1}|\eta| - \langle\mu\xi\rangle^{2\delta}]^{1/s}$ then we have

$$(2C_2)^{|\nu|}(|\nu|^s + \langle\mu\xi\rangle^{2\delta})^{|\nu|}|\eta|^{-|\nu|} \leq e^{-c|\eta|^{1/s}}$$

with some $c > 0$. Therefore there is $c' > 0$ such that

$$|\partial_\xi^\gamma B_\alpha(\eta,\xi,\mu)| \leq C_{\alpha,\gamma}\langle\mu\xi\rangle^\ell (\langle\mu\xi\rangle^\delta \langle\xi\rangle_\mu^{-1})^{|\gamma|} (C|\eta|^{1/2} + \langle\mu\xi\rangle^\delta)^{|\alpha|} e^{-c|\eta|^\kappa}$$

$$\leq C_{\alpha,\gamma}\langle\mu\xi\rangle^\ell (\langle\mu\xi\rangle^\delta \langle\xi\rangle_\mu^{-1})^{|\gamma|} e^{-c'|\eta|^\kappa}.$$

If $|\eta| \leq 4C_2 e^s \langle\mu\xi\rangle^{2\delta}$ noting that

$$C_2^{|\nu|}|\nu|!^{s/2}(|\nu|^{s/2} + \langle\mu\xi\rangle^\delta)^{|\nu|} \leq (2C_2)^{|\nu|}(|\nu|^s \langle\mu\xi\rangle^\delta)^{|\nu|}$$

we choose ν so that $|\nu| = e^{-1}(2C_2)^{-1/s}(|\eta|\langle\mu\xi\rangle^{-\delta})^{1/s}$ then we have

$$(2C_2)^{|\nu|}(|\nu|^s \langle\mu\xi\rangle^\delta)^{|\nu|}|\eta|^{-|\nu|} \leq e^{-c(|\eta|\langle\mu\xi\rangle^{-\delta})^{1/s}}$$

with some $c > 0$. Thus we have

$$|\partial_\xi^\gamma B_\alpha(\eta, \xi, \mu)| \le C_{\alpha,\gamma} \langle \mu\xi \rangle^\ell ((\langle \mu\xi \rangle^\delta \langle \xi \rangle_\mu^{-1})^{|\gamma|} (C(|\eta| \langle \mu\xi \rangle^{-\delta})^{1/2}$$

$$+ \langle \mu\xi \rangle^\delta)^{|\alpha|} e^{-c(|\eta| \langle \mu\xi \rangle^{-\delta})^\kappa}$$

$$\le C_{\alpha,\gamma} (\langle \mu\xi \rangle^\delta \langle \xi \rangle_\mu^{-1})^{|\gamma|} \langle \mu\xi \rangle^{\ell + \delta|\alpha|} e^{-c'(|\eta| \langle \mu\xi \rangle^{-\delta})^\kappa}$$

with some $c' > 0$ which completes the proof.   □

Note that $H_\alpha(\xi, \eta, \mu)$ is a linear combination of terms;

$$\partial_\xi^{\beta_1} \phi(\xi + \eta/2, \mu) \cdots \partial_\xi^{\beta_s} \phi(\xi + \eta/2, \mu) \partial_\xi^{\gamma_1} \phi(\xi - \eta/2, \mu) \cdots \partial_\xi^{\gamma_t} \phi(\xi - \eta/2, \mu)$$

$$\times e^{\phi(\xi + \frac{\eta}{2}, \mu) - \phi(\xi - \frac{\eta}{2}, \mu)} = h_{\beta_1, \dots, \beta_s, \gamma_1, \dots, \gamma_t}(\xi, \eta, \mu) e^{\phi(\xi + \frac{\eta}{2}, \mu) - \phi(\xi - \frac{\eta}{2}, \mu)}$$

where $\sum \beta_j = \beta$, $\sum \gamma_j = \gamma$ and $|\beta_j| \ge 1$, $|\gamma_j| \ge 1$, $\beta + \gamma = \alpha$. It is easy to examine that

$$|\partial_\xi^\nu h_{\beta_1, \dots, \beta_s, \gamma_1, \dots, \gamma_t}(\xi, \eta, \mu)| \le C_\nu \mu^{|\alpha|} \langle \xi \rangle_\mu^{-|\alpha|(1-\kappa)-|\nu|} \langle \mu\eta \rangle^{|\alpha|+|\nu|}. \tag{7.22}$$

On the other hand noting that

$$\partial_\xi^\alpha \phi(\xi + \eta/2, \mu) - \partial_\xi^\alpha \phi(\xi - \eta/2, \mu)$$

$$= \sum_{k=1}^n \frac{1}{2} \eta_k (\partial_\xi^\alpha \partial_{\xi_k} \phi(\xi + \theta\eta/2, \mu) + \partial_\xi^\alpha \partial_{\xi_k} \phi(\xi - \theta\eta/2, \mu)) \tag{7.23}$$

we see that

$$|\partial_\xi^\nu e^{\phi(\xi + \frac{\eta}{2}, \mu) - \phi(\xi - \frac{\eta}{2}, \mu)}| \le C_\nu \langle \xi \rangle_\mu^{-|\nu|} \langle \mu\eta \rangle^{2|\nu|} \langle \eta \rangle^{|\nu|} e^{C\langle \mu\eta \rangle^\kappa}. \tag{7.24}$$

From Lemma 7.14 and (7.22), (7.24) there exists $c'' > 0$ such that for $|\eta| \ge M\langle \mu\xi \rangle^{2\delta}$, $|\alpha| = N$

$$|\partial_\xi^\nu (H_\alpha(\xi, \eta, \mu) B_\alpha(\eta, \xi, \mu))| \le C_\nu \mu^N ((\langle \mu\xi \rangle^{-\delta} \langle \xi \rangle_\mu)^{-|\nu|} \langle \mu\xi \rangle^{\ell - N(1-\kappa)} e^{-c'' \langle \mu\eta \rangle^\kappa}.$$

We consider the case $|\eta| \le M\langle \mu\xi \rangle^{2\delta}$. Since $2\delta < 1$, taking μ small, we have $\langle \mu\xi \rangle/C \le \langle \mu(\xi + \theta\eta) \rangle \le C\langle \mu\xi \rangle$ and $\langle \xi \rangle_\mu/C \le \langle \xi + \theta\eta \rangle_\mu \le C\langle \xi \rangle_\mu$ for $|\theta| \le 1$ and hence we have

$$|\partial_\xi^\nu h_{\beta_1, \dots, \beta_s, \gamma_1, \dots, \gamma_t}(\xi, \eta, \mu)| \le C_\nu \mu^{|\alpha|} \langle \mu\xi \rangle^{-|\alpha|(1-\kappa)} \langle \xi \rangle_\mu^{-|\nu|}.$$

On the other hand, since $\kappa + 2\delta \leq 1$ one sees from (7.23) and $\langle \mu\eta \rangle \leq C\langle \mu\xi \rangle^{2\delta}$ that $|\phi(\xi + \eta/2, \mu) - \phi(\xi - \eta/2, \mu)| \leq C\langle \mu\eta \rangle \langle \mu\xi \rangle^{\kappa-1} \leq C'$. Then we have

$$|\partial_\xi^\nu e^{\phi(\xi+\frac{\eta}{2},\mu) - \phi(\xi-\frac{\eta}{2},\mu)}| \leq C_\nu \langle \xi \rangle_\mu^{-|\nu|}$$

using (7.23) again. Thus we conclude with some $c'' > 0$

$$|\partial_\xi^\nu (H_\alpha(\xi, \eta, \mu) B_\alpha(\eta, \xi, \mu))| \leq C_\nu \mu^N (\langle \mu\xi \rangle^\delta \langle \xi \rangle_\mu^{-1})^{|\nu|}$$

$$\times \langle \mu\xi \rangle^{\ell + \delta N - N(1-\kappa)} e^{-c''(|\eta| \langle \mu\xi \rangle^{-\delta})^\kappa}.$$

Since $|\eta^\beta e^{-c''(|\eta| \langle \mu\xi \rangle^{-\delta})^\kappa}| \leq C_\beta \langle \mu\xi \rangle^{\delta|\beta|} e^{-c'''(|\eta| \langle \mu\xi \rangle^{-\delta})^\kappa}$ we have

$$|\partial_x^\alpha \partial_\xi^\nu \tilde{R}_N(x, \xi, \mu)| \leq C_{\alpha,\nu} \mu^N \langle \mu\xi \rangle^{\ell - N(1-\kappa-\delta)+\delta n} \langle \mu\xi \rangle^{\delta|\alpha|} (\langle \xi \rangle^\delta \langle \xi \rangle_\mu^{-1})^{|\nu|}$$

because $\int e^{-c'''(|\eta| \langle \xi \rangle_\mu^{-\delta})^\kappa} d\eta \leq C\langle \mu\xi \rangle^{\delta n}$. This completes the proof of Proposition 7.4.

7.4 Weighted Energy Estimates

We now prepare several lemmas to derive weighted energy estimates for P. From now on we fix κ and δ and define w so that

$$\kappa = 1/5, \quad \delta = 2/5, \quad w = (\langle \mu\xi' \rangle^{-2} \phi_1^2 + \langle \mu\xi' \rangle^{-4/5})^{1/2}.$$

Notation 7.2 To simplify notations, we denote $\lambda = \langle \mu\xi' \rangle$ and $\lambda_\mu = \langle \xi' \rangle_\mu$. Recall that $\mathrm{Op}(a_1 a_2 \cdots a_l)$ is abbreviated to $[a_1 a_2 \cdots a_l]$ (see Notation 4.1).

Lemma 7.15 *Let* $a \in \mu\tilde{S}_w(1)$. *Then we have*

$$\mathrm{Re}([a\phi_1^2 w]u, u) \leq C\mu \mathrm{Re}([\phi_1^2 w]u, u) + C\mu^3 \|\lambda^\kappa u\|^2,$$

$$\mathrm{Re}([a\phi_j^2]u, u) \leq C\mu \mathrm{Re}([\phi_j^2]u, u) + C\mu^3 \|\lambda^{2\kappa} u\|^2, \quad j \geq 2.$$

Let $a \in \mu\tilde{S}_w(\lambda^\kappa)$. *Then we have*

$$\mathrm{Re}([a\phi_1^2 w]u, u) \leq C\mu \mathrm{Re}([\lambda^\kappa \phi_1^2 w]u, u) + C\mu^3 \|\lambda^{3\kappa/2} u\|^2,$$

$$\mathrm{Re}([a\phi_j^2]u, u) \leq C\mu \mathrm{Re}([\phi_j^2 \lambda^\kappa]u, u) + C\mu^3 \|\lambda^{1/2} u\|^2, \quad j \geq 2.$$

Let $a \in \mu\tilde{S}_w(1)$ *then we have*

$$\|[a\lambda^{\kappa/2}\phi_j]u\|^2 \leq C\mu^2 \mathrm{Re}([\lambda^\kappa \phi_j^2]u, u) + C\mu^4 \|\lambda^{1/2} u\|^2,$$

$$\|[a\lambda^{\kappa/2}\sqrt{w}\phi_1]u\|^2 \leq C\mu^2 \mathrm{Re}([\lambda^\kappa w\phi_1^2]u, u) + C\mu^4 \|\lambda^{3\kappa/2} u\|^2.$$

Proof It suffices to prove the case $a \in \mu \tilde{S}_w(\lambda^\kappa)$. Since $\mathsf{Re}(au, u) = ((\mathsf{Re}\, a)u, u)$ we may assume that a is real. Let us consider

$$C\mu\lambda^\kappa \phi_1^2 w - a\phi_1^2 w = C\mu\lambda^\kappa \phi_1^2 w(1 - C^{-1}\mu^{-1}a\lambda^{-\kappa})$$
$$= C\mu\lambda^\kappa \phi_1^2 w\psi^2 = C\mu(\lambda^{\kappa/2}\phi_1 \sqrt{w}\psi)\#(\lambda^{\kappa/2}\phi_1 \sqrt{w}\psi) + R$$

with $R \in \mu^3 \tilde{S}_w(\lambda^{3\kappa})$ and $\psi = (1 - C^{-1}\mu^{-1}a\lambda^{-\kappa})^{1/2} \in \tilde{S}_w(1)$. Hence we have

$$C\mu\mathsf{Re}([\lambda^\kappa \phi_1^2 w]u, u) - \mathsf{Re}([a\phi_1^2 w]u, u) \geq -C\mu^3\|\lambda^{3\kappa/2}u\|^2$$

which shows the first assertion. From Corollary 7.2 we have

$$([a\phi_j^2]u, u) = \mathsf{Re}(\mu^{1/2}[\lambda^{\kappa/2}\phi_j]u, \mu^{-1/2}[\lambda^{-\kappa/2}a\phi_j]u) + (Tu, u)$$

with $T \in \mu^3\tilde{S}_w(\lambda^\kappa w^{-2}) \subset \mu^3\tilde{S}_w(\lambda)$. Since one can write

$$\mu^{-1/2}\lambda^{-\kappa/2}a\phi_j - \mu^{-1/2}(a\lambda^{-\kappa})\#(\lambda^{\kappa/2}\phi_j) \in \mu^{3/2}\tilde{S}_w(\lambda^{\kappa/2}w^{-1}, g)$$

and noting $\tilde{S}_w(\lambda^{\kappa/2}w^{-1}) \subset \tilde{S}_w(\lambda^{1/2})$ we have

$$\|\mu^{-1/2}[\lambda^{-\kappa/2}a\phi_j]u\|^2 \leq C\mu\|[\lambda^{\kappa/2}\phi_j]u\|^2 + C\mu^3\|\lambda^{1/2}u\|^2$$

which proves the assertion.

We turn to the next assertion. Note that $a\lambda^{\kappa/2}\phi_j - a\#(\lambda^{\kappa/2}\phi_j) \in \mu^2\tilde{S}_w(\lambda^{1/2})$ and hence one has $\|[a\lambda^{\kappa/2}\phi_j]u\|^2 \leq C\mu^2\|[\lambda^{\kappa/2}\phi_j]u\|^2 + C\mu^4\|\lambda^{1/2}u\|^2$. Since it is clear $(\lambda^{\kappa/2}\phi_j)\#(\lambda^{\kappa/2}\phi_j) - \lambda^\kappa \phi_j^2 \in \mu^2\tilde{S}_w(\lambda^\kappa)$ we apply the assertion just proven to get the assertion. Finally we note

$$(a\sqrt{w}\lambda^{\kappa/2}\phi_1)\#(a\sqrt{w}\lambda^{\kappa/2}\phi_1) - a^2 w\lambda^\kappa \phi_1^2 \in \mu^4\tilde{S}_w(\lambda^{3\kappa})$$

which, together with the above assertion, proves the desired assertion.   □

Lemma 7.16 *There exist $c > 0$, $C > 0$ such that*

$$c(\mu[\lambda^{1+\kappa}\sqrt{w}]u, u) \leq \mathsf{Re}([\lambda^\kappa \phi_1^2 w]u, u)$$
$$+ \mathsf{Re}([\lambda^\kappa \phi_2^2]u, u) + \mathsf{Re}([\phi_{r+1}^2]u, u) + C\mu\|\lambda^{3\kappa/2}u\|^2.$$

Proof Denote $A = \lambda^{\kappa/2}\phi_1 \sqrt{w}$ and $B = \lambda^{\kappa/2}\phi_2$ and note that $A \in \tilde{S}_w(\lambda^{1+\kappa/2}w^{3/2})$ and $B \in \tilde{S}_0(\lambda^{1+\kappa/2})$. Remark that $|([A, B]u, u)| \leq (\mathrm{Op}(A\#A)u, u) + (\mathrm{Op}(B\#B)u, u)$ and recall

$$i\,[A, B] - \mathrm{Op}(\{A, B\}) \in \mu^3\tilde{S}_w(\lambda^{-1+\kappa}w^{-3/2}) \subset \mu^3\tilde{S}_w(\lambda^{-\kappa}).$$

Here write

$$\{A, B\} = \{\phi_1 \sqrt{w}, \lambda^{\kappa/2}\} \phi_2 \lambda^{\kappa/2} + \{\lambda^{\kappa/2}, \phi_2\} \sqrt{w} \phi_1 \lambda^{\kappa/2}$$
$$+ \{\sqrt{w}, \phi_2\} \phi_1 \lambda^{\kappa} + \{\phi_1, \phi_2\} \sqrt{w} \lambda^{\kappa} = K_1 + K_2 + K_3 + K_4.$$

Since $\{\phi_1 \sqrt{w}, \lambda^{\kappa/2}\} \in \mu \tilde{S}_w(\lambda^{\kappa/2} \sqrt{w})$ and $\{\lambda^{\kappa/2}, \phi_2\} \in \mu \tilde{S}_0(\lambda^{\kappa/2})$ we see that the first and the second term can be written as $T_1 \# B + R_1$ and $T_2 \# A + R_2$ respectively with $T_i \in \mu \tilde{S}_w(\lambda^{\kappa/2})$ and $R_i \in \mu^2 \tilde{S}_w(\lambda^{2\kappa})$. Then we have

$$\mathsf{Re}(K_1 u, u) \geq -C\mu\big(\|\lambda^{\kappa/2} u\|^2 + \|Bu\|^2\big) - C\mu^2 \|\lambda^{\kappa} u\|^2,$$
$$\mathsf{Re}(K_2 u, u) \geq -C\mu\big(\|\lambda^{\kappa/2} u\|^2 + \|Au\|^2\big) - C\mu^2 \|\lambda^{\kappa} u\|^2.$$

Consider K_3. Note that

$$\{\sqrt{w}, \phi_2\} = \frac{1}{4} w^{-3/2} \big(2\{\phi_1, \phi_2\} \phi_1 \lambda^{-2} + \{\lambda^{-2}, \phi_2\} \phi_1^2 + \{\lambda^{-2\delta}, \phi_2\}\big).$$

Noting that $w^{-5/2} \{\phi_1, \phi_2\} \lambda^{-2} \in \mu \tilde{S}_w(1)$ and $w^{-5/2} \{\lambda^{-2}, \phi_2\} \phi_1 \in \mu \tilde{S}_w(1)$ and that $w^{-3/2} \{\lambda^{-\delta}, \phi_2\} \in \mu \tilde{S}_w(w^{-3/2} \lambda^{-2\delta})$ one can write $K_3 = a\phi_1^2 w \lambda^{\kappa} + b\phi_1$ with $a \in \mu \tilde{S}_w(1)$ and $b \in \mu \tilde{S}_w(1)$. We first consider $\mathsf{Re}\,([b\phi_1]u, u)$. Note that

$$2\mathsf{Re}([b\phi_1]u, u) \geq -\mu^{-1} \|[\lambda^{-\kappa/2} \# (b\phi_1)]u\|^2 - \mu \|\lambda^{-\kappa/2} u\|^2$$

and $\lambda^{-\kappa/2} \# (b\phi_1) - b\lambda^{-\kappa/2} \phi_1 \in \mu^2 \tilde{S}_w(\lambda^{-\kappa/2})$. Remarking that

$$(b\lambda^{-\kappa/2}) \# (b\lambda^{-\kappa/2}) - b^2 \lambda^{-\kappa} \phi_1^2 \in \mu^4 \tilde{S}_w(\lambda^{3\kappa})$$

and $b^2 \lambda^{-\kappa} \phi_1^2 = a\lambda^{\kappa} w \phi_1^2$ with $a \in \mu^2 \tilde{S}_w(1)$ we conclude that, applying Lemma 7.15

$$2\mathsf{Re}([b\phi_1]u, u) \geq -C\mu \mathsf{Re}([\lambda^{\kappa} w \phi_1^2]u, u) - C\mu \|\lambda^{3\kappa/2} u\|^2.$$

From Lemma 7.15 again we see that

$$\mathsf{Re}(K_3 u, u) \geq -C\mu \mathsf{Re}([\phi_1^2 w \lambda^{\kappa}]u, u) - C\mu \|\lambda^{3\kappa/2} u\|^2.$$

From the assumption we may assume $C\mu\lambda \geq \{\phi_1, \phi_2\} + \mu \phi_{r+1}^2 \langle \mu \xi' \rangle^{-1} = T \geq c\mu\lambda$ with some $C > 0, c > 0$. Write

$$C(K_4 + \mu \phi_{r+1}^2 \sqrt{w} \langle \mu \xi' \rangle^{-1+\kappa}) - \mu \sqrt{w} \lambda^{1+\kappa}$$
$$= C\sqrt{w} \lambda^{\kappa} T\big(1 - C^{-1} \mu T^{-1} \lambda\big) = f \# f + R$$

with $f = C^{1/2}w^{1/4}\lambda^{\kappa/2}T^{1/2}(1 - C^{-1}\mu T^{-1}\lambda)^{1/2}$ and $R \in \mu^3\tilde{S}_w(\lambda^{3\kappa})$ from which we see that

$$
\begin{aligned}
C\mathrm{Re}(K_4 u, u) &\geq \mathrm{Re}([\mu\sqrt{w}\lambda^{1+\kappa}]u, u) \\
&\quad -C\mu\mathrm{Re}([\sqrt{w}\phi_{r+1}^2\langle\mu\xi'\rangle^{-1+\kappa}]u, u) - C\mu^3\|\lambda^{3\kappa/2}u\|^2.
\end{aligned}
$$

The above estimates show

$$
\begin{aligned}
\mathrm{Re}(\mathrm{Op}(\{A, B\})u, u) &\geq c\,\mathrm{Re}([\mu\sqrt{w}\lambda^{1+\kappa}]u, u) \\
&\quad -C\mu\big(\|Au\| + \|Bu\|^2 + \|\lambda^{3\kappa/2}u\|^2\big) \\
&\quad -C\mu\mathrm{Re}([\phi_1^2 w\lambda^\kappa]u, u) - C\mu\|[\phi_{r+1}^2]u\|^2
\end{aligned}
$$

because $\phi_{r+1}^2\sqrt{w}\langle\mu\xi'\rangle^{-1+\kappa} - \phi_{r+1}\#(\phi_{r+1}\sqrt{w}\langle\mu\xi'\rangle^{-1+\kappa}) \in \mu^{4/5}\tilde{S}_w(\langle\mu\xi'\rangle^{2\kappa})$. We turn to $A\#A$ and $B\#B$. Since $A \in \tilde{S}_w(w^{3/2}\lambda^{1+\kappa/2})$ and $B \in \tilde{S}_0(\lambda^{1+\kappa/2})$ we see

$$
A\#A - w\phi_1^2\lambda^\kappa \in \mu^2\tilde{S}_w(\lambda^{3\kappa}), \quad B\#B - \phi_2^2\lambda^\kappa \in \mu^2\tilde{S}_0(\lambda^\kappa)
$$

and hence

$$
\begin{aligned}
\|Au\|^2 &= ([A\#A]u, u) \leq \mathrm{Re}([\phi_1^2 w\lambda^\kappa]u, u) + C\mu^2\|\lambda^{3\kappa/2}u\|^2, \\
\|Bu\|^2 &= ([B\#B]u, u, u) \leq \mathrm{Re}([\phi_2^2\lambda^\kappa]u, u) + C\mu^2\|\lambda^{\kappa/2}u\|^2.
\end{aligned}
$$

These prove the assertion.   □

Corollary 7.3 *We have*

$$
\begin{aligned}
\mu\|\lambda^{1/2}u\|^2 &\leq C\mathrm{Re}([\phi_1^2 w\lambda^\kappa]u, u) + C\mathrm{Re}([\phi_2^2\lambda^\kappa]u, u) \\
&\quad + C\mathrm{Re}([\phi_{r+1}^2]u, u) + C\mu\|\lambda^{3\kappa/2}u\|^2.
\end{aligned}
$$

Proof Write $C\mu\lambda^{1+\kappa}\sqrt{w} - \mu\lambda$ as

$$
C\mu\lambda^{1+\kappa}\sqrt{w}(1 - C^{-1}w^{-1/2}\lambda^{-\kappa}) - C\mu f\#f \in \mu^3\tilde{S}_w(\lambda^{3\kappa})
$$

with $f = \langle\mu\xi'\rangle^{(1+\kappa)/2}w^{1/4}(1 - C^{-1}w^{-1/2}\langle\mu\xi'\rangle^{-\kappa})^{1/2}$ and hence

$$
C\mu\mathrm{Re}([\lambda^{1+\kappa}\sqrt{w}]u, u) - \mu\|\lambda^{1/2}u\|^2 \geq -C\mu^3\|\lambda^{3\kappa/2}u\|^2
$$

which proves the assertion.   □

Lemma 7.17 *Assume $a \in \mu\tilde{S}_w(1)$. Then for $j \neq 1$ we have*

$$
\begin{aligned}
\mathrm{Re}([a\phi_1\phi_j]u, u) &\leq C\mu\mathrm{Re}([\phi_j^2\lambda^\kappa]u, u) \\
&\quad + C\mu\mathrm{Re}([\phi_1^2 w\lambda^\kappa]u, u) + C\mu^3\|\lambda^{3\kappa/2}u\|^2.
\end{aligned}
$$

Proof We may assume that a is real. Consider

$$a\phi_1\phi_j - \mathsf{Re}((\mu^{1/2}\lambda^{\kappa/2}\phi_j)\#(\mu^{-1/2}a\lambda^{-\kappa/2}\phi_1)) \in \mu^3\tilde{S}_w(\lambda^{2\kappa}).$$

Since $(\mu^{1/2}\lambda^{\kappa/2}\phi_j)\#(\mu^{1/2}\lambda^{\kappa/2}\phi_j) - \mu\lambda^\kappa\phi_j^2 \in \mu^3\tilde{S}_0(\lambda^\kappa)$ and

$$(\mu^{-1/2}a\lambda^{-\kappa/2}\phi_1)\#(\mu^{-1/2}a\lambda^{-\kappa/2}\phi_1)$$
$$-(\mu^{-1}a^2w^{-1}\lambda^{-2\kappa})\phi_1^2 w\lambda^\kappa \in \mu^3\tilde{S}_w(\lambda^{3\kappa})$$

we see that, noting $\mu^{-1}a^2w^{-1}\lambda^{-2\kappa} \in \mu\tilde{S}_w(1)$,

$$\mathsf{Re}([a\phi_1\phi_j]u, u) \le \frac{1}{2}\left(\mu\|[\phi_j\lambda^{\kappa/2}]u\|^2 + \mu^{-1}\|[a\lambda^{-\kappa/2}\phi_1]u\|^2\right) + C\mu^3\|\lambda^\kappa u\|^2$$
$$\le C\mu\mathsf{Re}([\lambda^\kappa\phi_j^2]u, u) + C\mu\mathsf{Re}([\phi_1^2 w\lambda^\kappa]u, u) + C\mu^3\|\lambda^{3\kappa/2}u\|^2.$$

This is the assertion.   □

Lemma 7.18 *Let* $a \in \mu\tilde{S}_w(\lambda w)$. *Then for* $j \ne 1$ *we have*

$$\mathsf{Re}([a\phi_j]u, u) \le C\mu^{1/2}\mathsf{Re}([\phi_j^2\lambda^\kappa]u, u) + C\mu^{1/2}\mathsf{Re}([\phi_1^2 w\lambda^\kappa]u, u)$$
$$+C\mu^{1/2}\mathsf{Re}([\phi_2^2\lambda^\kappa]u, u) + C\mu^{1/2}\mathsf{Re}([\phi_{r+1}^2]u, u) + C\mu^{3/2}\|\lambda^{3\kappa/2}u\|^2.$$

Proof Write $a\phi_j - \mathsf{Re}((\mu^{1/4}\lambda^{\kappa/2}\phi_j)\#(\mu^{-1/4}\lambda^{-\kappa/2}a)) \in \mu^3\tilde{S}_w(\lambda^{2\kappa})$ and hence

$$\mathsf{Re}([a\phi_j]u, u) \le \mu^{1/2}\|[\lambda^{\kappa/2}\phi_j]u\|^2 + \mu^{-1/2}\|[\lambda^{-\kappa/2}a]u\|^2 + C\mu^3\|\lambda^\kappa u\|^2.$$

Note that $(\lambda^{-\kappa/2}a)\#(\lambda^{-\kappa/2}a) - \lambda^{-\kappa}a^2 \in \mu^4 S_w(\lambda^{3\kappa})$ and write

$$\lambda^{-\kappa}a^2 = (w^{-2}a^2\lambda^{-2})w^2\lambda^{2-\kappa} = b(\lambda^{-2}\phi_1^2 + \lambda^{-2\delta})\lambda^{2-\kappa}$$
$$= b\lambda^{-2\kappa}w^{-1}(\phi_1^2 w\lambda^\kappa) + b\lambda^{2-2\delta-\kappa}$$

where $b = w^{-2}a^2\lambda^{-2} \in \mu^2\tilde{S}_w(1)$. Since $2 - 2\delta - \kappa = 1$ thanks to Lemma 7.15 we get

$$\mu^{-1/2}\|[\lambda^{-\kappa/2}a]u\|^2 \le C\mu^{3/2}\mathsf{Re}([\phi_1^2 w\lambda^\kappa]u, u)$$
$$+C\mu^{5/2}\|\lambda^{3\kappa/2}u\|^2 + C\mu^{3/2}\|\lambda^{1/2}u\|^2.$$

Then applying Corollary 7.3 we obtain the assertion.   □

We now estimate $\{\xi_0 - \phi_1 k, \phi_j^2\}$ and $\{\xi_0 - \phi_1 k, w\phi_1^2\}$. Recall that

$$\{\xi_0 - \phi_1 k, \phi_j^2\} = \{\xi_0 - \phi_1, \phi_j^2\} + \{\phi_1\bar{k}, \phi_j^2\}.$$

From Lemma 7.2 we have $\{\xi_0 - \phi_1, \phi_j^2\} = 2\{\xi_0 - \phi_1, \phi_j\}\phi_j = \sum_{i=1}^{r+1} C_{ji}\phi_i\phi_j$ where $C_{ji} \in \mu \tilde{S}_0(1)$. Note that for $j, i \geq 2$ one has

$$\mathrm{Re}([C_{ji}\phi_i\phi_j]u, u) \leq C\mu\mathrm{Re}([\phi_i^2]u, u) + C\mu\mathrm{Re}([\phi_j^2]u, u) + C\mu^3\|u\|^2.$$

For $C_{j1}\phi_1\phi_j$ we apply Lemma 7.17 to get

$$\mathrm{Re}([C_{j1}\phi_1\phi_j]u, u) \leq C\mu\mathrm{Re}([\phi_j^2\lambda^\kappa]u, u)$$
$$+ C\mu\mathrm{Re}([\phi_1^2 w\lambda^\kappa]u, u) + C\mu^3\|\lambda^{3\kappa/2}u\|^2.$$

Consider $\{\phi_1\bar{k}, \phi_j^2\} = 2\{\phi_1, \phi_j\}\phi_j\bar{k} + 2\{\bar{k}, \phi_j\}\phi_j\phi_1$ for $j \geq 2$. Write

$$\{\phi_1, \phi_j\}\phi_j\bar{k} - \mathrm{Re}((\lambda^{\kappa/2}\phi_j)\#(\{\phi_1, \phi_j\}\bar{k}\lambda^{-\kappa/2})) \in \mu^2\tilde{S}_w(\lambda^{2\kappa})$$

and note that with $T = \{\phi_1, \phi_j\}^2\bar{k}^2\lambda^{-2}w^{-2}$

$$\{\phi_1, \phi_j\}^2\bar{k}^2\lambda^{-\kappa} = (\{\phi_1, \phi_j\}^2\bar{k}^2\lambda^{-2}w^{-2})w^2\lambda^{2-\kappa}$$
$$= T(\lambda^{-2}\phi_1^2 + \lambda^{-\delta})\lambda^{2-\kappa} = (Tw^{-1}\lambda^{-2\kappa})w\lambda^\kappa\phi_1^2 + T\lambda^{2-\delta-\kappa}$$

where $Tw^{-1}\lambda^{-2\kappa} \in \mu^2\tilde{S}_w(1)$ and $T\lambda^{2-\delta-\kappa} \in \mu^2\tilde{S}_w(\lambda)$. We now apply Lemma 7.17 and Corollary 7.3 to get the desired estimate.

We turn to estimate $\{\xi_0 - \phi_1 k, w\phi_1^2\}$. Recalling $\xi_0 - \phi_1 k = \xi_0 - \phi_1 + \phi_1\bar{k}$ consider

$$\{\xi_0 - \phi_1, w\phi_1^2\} = 2\{\xi_0 - \phi_1, \phi_1\}\phi_1 w + \{\xi_0 - \phi_1, w\}\phi_1^2.$$

For the first term on the right-hand side remark that $\{\xi_0 - \phi_1, \phi_1\} = \sum_{i=1}^{r+1} C_{1i}\phi_i$ and apply Lemmas 7.17 and 7.15 to get the estimates. We next study

$$\{\xi_0 - \phi_1, w\} = \frac{1}{2}w^{-1}\{\xi_0 - \phi_1, \lambda^{-2}\phi_1^2 + \lambda^{-\delta}\}$$
$$= -\frac{1}{2}w^{-1}\{\phi_1, \lambda^{-2}\}\phi_1^2 + w^{-1}\{\xi_0 - \phi_1, \phi_1\}\phi_1\lambda^{-2} - \frac{1}{2}w^{-1}\{\phi_1, \lambda^{-\delta}\}.$$

Note that $w^{-1}\{\phi_1, \lambda^{-2}\}\phi_1^2$, $w^{-1}\{\phi_1, \lambda^{-\delta}\} \in \mu\tilde{S}_w(w)$ and apply Lemma 7.15 to $w^{-1}\{\phi_1, \lambda^{-2}\}\phi_1^4$ and $w^{-1}\{\phi_1, \lambda^{-\delta}\}\phi_1^2$. As for the second term on the right-hand side it is enough to note that $w^{-1}\{\xi_0 - \phi_1, \phi_1\}\phi_1^3\lambda^{-2} = \sum_{i=1}^r T_i\phi_i\phi_1$ with $T_i \in \mu\tilde{S}_w(w)$. We finally consider

$$\{\phi_1\bar{k}, \phi_1^2 w\} = \{\phi_1, w\}\phi_1^2\bar{k} + \{\bar{k}, w\}\phi_1^3 + 2\{\bar{k}, \phi_1\}\phi_1^2 w.$$

Note that $\{\phi_1, w\}, \{\bar{k}, \phi_1\} \in \mu\tilde{S}_w(1)$ and apply Lemma 7.15. Recall that

$$\{w, \bar{k}\} = \frac{1}{2}w^{-1}\big(2\{\phi_1, \bar{k}\}\phi_1\lambda^{-2} + \{\lambda^{-2}, \bar{k}\}\phi_1^2 + \{\lambda^{-\delta}, \bar{k}\}\big)$$

so that one can write $\{w, \bar{k}\}\phi_1^3 = Tw\phi_1^2$ with $T \in \mu\tilde{S}_w(1)$ and then we apply Lemmas 7.17 and 7.15 again to obtain the desired estimate.

Proposition 7.5 *We have*

$$\sum_{j=2}^{r+1} \left|\text{Re}\,(\text{Op}(\{\xi_0 - \phi_1 k, \phi_j^2\})u, u)\right| + \left|\text{Re}\,(\text{Op}(\{\xi_0 - \phi_1 k, w\phi_1^2\})u, u)\right|$$

$$\leq C\mu\Big(\sum_{j=2}^{r+1} \text{Re}\,([\lambda^\kappa \phi_j^2]u, u) + \text{Re}\,([w\lambda^\kappa \phi_1^2]u, u)\Big) + C\mu^2\|\lambda^{1/2}u\|^2.$$

7.5 Well-Posedness in the Gevrey Classes

We prove the well-posedness of the Cauchy problem for P in (7.5) in the Gevrey class of order 5.

Theorem 7.3 ([6]) *Assume* (1.5) *and that p is of spectral type 2 on Σ and $P_{sub} = 0$ everywhere on Σ. Then the Cauchy problem for P is well-posed in $\gamma^{(5)}$ near the origin, that is for any $f(x) \in C^\infty(\mathbb{R}; \gamma_0^{(5)}(\mathbb{R}^n))$ vanishing in $x_0 \leq 0$ there is $u(x)$ which is C^∞, vanishing in $x_0 \leq 0$ satisfying $Pu = f$ near the origin.*

Remark 7.1 It is clear from the proof that $u(x)$ that we obtain in the above theorem is indeed in $C^\infty((-\tau, \tau); \gamma^{(5)}(\mathbb{R}^n))$ for small $\tau > 0$.

The proof of Theorem 7.3 goes as follows: To avoid notational confusions, we denote by $\hat{P}$ the original operator. By a compactness argument we can assume that there are finite number of $h_\alpha(x', \xi') \in \tilde{S}_0^{(1/\kappa)}(1)$ such that $\sum_\alpha h_\alpha = 1$ in a neighborhood of the origin $x' = 0$ and second order operators $\hat{P}_\alpha$, differential in D_0 with pseudodifferential coefficients in $\tilde{S}_0^{(1/\kappa)}$, which coincides with $\hat{P}$ in a conic neighborhood of the support of h_α. From the discussions after Corollary 7.4 and Proposition 7.10 below there exists a parametrix G_α with finite propagation speed of micro supports for $P_\alpha = e^{(\tau-x_0)\lambda^\kappa}\hat{P}_\alpha e^{-(\tau-x_0)\lambda^\kappa}$. We define

$$\hat{G} = \sum_\alpha \hat{G}_\alpha h_\alpha = \sum_\alpha e^{-(\tau-x_0)\lambda^\kappa} G_\alpha e^{(\tau-x_0)\lambda^\kappa} h_\alpha.$$

Assume that f has small support with respect to x' and consider

$$\hat{P}\hat{G}f = \sum_\alpha \hat{P}\hat{G}_\alpha h_\alpha f = \sum_\alpha \hat{P}_\alpha \hat{G}_\alpha h_\alpha f + \sum_\alpha (\hat{P} - \hat{P}_\alpha)\hat{G}_\alpha h_\alpha f$$

which is equal to $\sum_\alpha h_\alpha f - Rf = f - Rf$ where $Rf = \sum_\alpha (\hat{P}_\alpha - \hat{P})\hat{G}_\alpha h_\alpha f$. From Lemma 7.24 below we see that

$$\|e^{(\tau-t)\lambda^\kappa} Rf(t)\| \le C \int^t \|e^{(\tau-x_0)\lambda^\kappa} f(x_0)\| dx_0.$$

Take $\tau > 0$ small so that $C\tau < 1$ and define the norm $[[f]]$ by

$$[[f]] = \sup_{0 \le x_0 \le \tau} \|e^{(\tau-x_0)\lambda^\kappa} f(x_0)\|$$

so that $[[Rf]] \le C\tau[[f]]$. Since $\sum_{j=0}^\infty [[R^j f]] \le \sum_{j=0}^\infty (C\tau)^j [[f]]$ then $(1 - R)^{-1} f = \sum_{j=0}^\infty R^j f$ exists and $[[(1 - R)^{-1} f]] < \infty$ for any f with $[[f]] < +\infty$ vanishing in $x_0 \le 0$. Therefore we conclude $\hat{P}\hat{G}(1 - R)^{-1} f = f$. Since $\hat{G}(1 - R)^{-1} f = 0$ for $x_0 \le 0$ we get a desired solution. It is clear that

$$e^{(\tau_1-x_0)\lambda^\kappa} (1 - R)^{-1} f \in C_+^0 ([-\tau_1, \tau_1]; H^\infty(\mathbb{R}^n))$$

for small $0 < \tau_1 < \tau$ and hence we have $u = \hat{G}(1 - R)^{-1} f \in C_+^2([-\tau_1, \tau_1]; H^\infty(\mathbb{R}^n))$. From $\hat{P}u = f$ it follows that $u \in C_+^\infty([-\tau_1, \tau_1]; H^\infty(\mathbb{R}^n))$ (see [34, Appendix B]). This proves the assertion.

We go into details. Denote $\psi = -\phi_1 k$ and $\Lambda = D_0 - \psi$, $M = D_0 + \psi$. Thanks to Lemma 7.8 we see that

$$(\xi_0 - \psi)\#(\xi_0 + \psi) = \xi_0^2 - \psi^2 + \{\xi_0 - \psi, \xi_0 + \psi\}/i + T$$

with $T \in \mu^2 \tilde{S}_w(1)$. Note that from Lemmas 7.2 and 7.4 we have

$$\{\xi_0 - \phi_1, \phi_j\} = \sum_{j=1}^{r+1} C_{jk}\phi_k, \quad \{\phi_2, \phi_1\} + \mu\phi_{r+1}^2 \langle\mu\xi'\rangle^{-1} \ge c\,\mu\langle\mu\xi'\rangle$$

with $C_{jk} \in \mu\tilde{S}_0^{(s)}(1)$. From Lemma 7.2 one can write $\{\xi_0, \phi_1\} = \mu\sum_{j=1}^{r+1} C_j\phi_j$ with some $C_j \in \tilde{S}_0^{(s)}(1)$ and hence $\{\xi_0-\psi, \xi_0+\psi\} = \mu\sum_{j=1}^{r+1} C_j\phi_j$ with some $C_j \in \tilde{S}_\delta^{(s)}(1)$. On the other hand Lemma 7.3 shows

$$P_{sub} = C_0(\xi_0 + \psi) + \sum_{j=1}^{r+1} C_j\phi_j$$

with some $C_j \in \tilde{S}_\delta^{(s)}(1)$. Noting that $P = \text{Op}(p + P_{sub}) + R$, $R \in \mu^2 \tilde{S}_0^{(s)}(1)$ one can write $P = -M\Lambda + B_0\Lambda + Q$ where $B_0 \in \mu\tilde{S}_\delta^{(s)}(1)$ and

$$Q = \text{Op}\Big(\sum_{j=2}^{r+1} \phi_j^2 + w\phi_1^2 + R\Big) + R_0, \quad R_0 \in \mu^2 \tilde{S}_\delta^{(s)}((\langle\mu\xi'\rangle)^{2\kappa})$$

where $R = \sum_{j=1}^{r+1} c_j \phi_j$ with $c_j \in \mu \tilde{S}_\delta^{(s)}(1)$. We now conjugate

$$e^\phi = \mathrm{Op}(e^{(\tau-x_0)\langle \mu \xi'\rangle^\kappa}) = \mathrm{Op}(e^{(\tau-x_0)\lambda^\kappa})$$

to P so that $e^\phi P e^{-\phi} = -e^\phi M e^{-\phi} e^\phi \Lambda e^{-\phi} + e^\phi B_0 e^{-\phi} e^{-\phi} e^\phi \Lambda e^{-\phi} + e^\phi Q e^{-\phi}$ where τ is a constant. Denote $e^\phi M e^{-\phi}$, $e^\phi \Lambda e^{-\phi}$, $e^\phi B_0 e^{-\phi}$, $e^\phi Q e^{-\phi}$ by M, Λ, B_0, Q again:

$$e^\phi M e^{-\phi} \ \text{by}\ M, \quad e^\phi \Lambda e^{-\phi} \ \text{by}\ \Lambda, \quad e^\phi Q e^{-\phi} \ \text{by}\ Q, \quad e^\phi B_0 e^{-\phi} \ \text{by}\ B_0.$$

We first consider $M = e^\phi (D_0 + \psi) e^{-\phi} = D_0 - i\lambda^\kappa + e^\phi \psi e^{-\phi}$. Since $\psi \in \tilde{S}_w^{(s)}(w\lambda) \subset \tilde{S}_\delta^{(s)}(\lambda)$ we apply Proposition 7.4 with $\delta = 2/5 = 2\kappa$ to get

$$e^\phi \psi e^{-\phi} = \mathrm{Op}(\psi_0 + \psi_1 + \psi_2), \quad \psi_0 = -\phi_1 k \tag{7.25}$$

where $\psi_1 \in \mu \tilde{S}_w(\lambda^\kappa)$ is pure imaginary and $\psi_2 \in \mu^2 \tilde{S}_\delta(\lambda^{-\kappa})$.

Next consider $e^\phi Q e^{-\phi}$ and we note that $e^\phi \mathrm{Op}(\phi_j^2) e^{-\phi} = \mathrm{Op}(\phi_j^2 + a_j \phi_j + r_j)$ where $a_j \in \mu \tilde{S}_w(\lambda^\kappa)$ is pure imaginary and $r_j \in \mu^2 \tilde{S}_\delta(\lambda^{2\kappa})$. Note that

$$e^\phi \mathrm{Op}(w\phi_1^2) e^{-\phi} = \mathrm{Op}(w\phi_1^2 + a_1 w\phi_1 + r_1)$$

where $a_1 \in \mu \tilde{S}_w(\lambda^\kappa)$ is pure imaginary and $r_1 \in \mu^2 \tilde{S}_\delta(w\lambda^{2\kappa})$. Remark that $e^\phi R e^{-\phi} = \mathrm{Op}(\sum_{j=1}^{r+1} c_j \phi_j + r)$ where $c_j \in \mu \tilde{S}_\delta(1)$ and $r \in \mu^2 \tilde{S}_\delta(\langle \mu \xi'\rangle^{2\kappa})$. Hence one can write

$$e^\phi Q e^{-\phi} = \mathrm{Op}\Big(\sum_{j=2}^{r+1} \phi_j^2 + w\phi_1^2 + \sum_{j=2}^{r+1} a_j \phi_j + a_1 w\phi_1 + \sum_{j=1}^{r+1} c_j \phi_j + r \Big) \tag{7.26}$$

where $a_j \in \mu \tilde{S}_w(\lambda^\kappa)$ and $r \in \mu^2 \tilde{S}_\delta(\lambda^{2\kappa})$. We summarize what we have proved in

Proposition 7.6 *We can write* $e^\phi P e^{-\phi} = -M\Lambda + B_0\Lambda + Q$ *with* $B_0 \in \mu \tilde{S}_\delta(1)$ *and*

$$M = D_0 - i\lambda^\kappa + (\psi_0 + \psi_1 + \psi_2) = D_0 - i\lambda^\kappa + \tilde{\psi},$$
$$\Lambda = D_0 - i\lambda^\kappa - (\psi_0 + \psi_1 + \psi_2) = D_0 - i\lambda^\kappa - \tilde{\psi}$$

with $\tilde{\psi} = \psi_0 + \psi_1 + \psi_2$ *where* $\psi_1 \in \mu \tilde{S}_w(\lambda^\kappa)$ *is pure imaginary* $\psi_2 \in \mu^2 \tilde{S}_\delta(\lambda^{-\kappa})$ *and* $Q = q + q_1 + q_2 + r$ *where*

$$q = \sum_{j=2}^{r+1} \phi_j^2 + w\phi_1^2, \quad q_1 = \sum_{j=2}^{r+1} a_j \phi_j + a_1 w\phi_1, \quad q_2 = \sum_{j=1}^{r+1} c_j \phi_j \tag{7.27}$$

and $a_j \in \mu \tilde{S}_w(\lambda^\kappa)$ *are pure imaginary and* $c_j \in \mu \tilde{S}_\delta(1)$ *and* $r \in \mu^2 \tilde{S}_\delta(\lambda^{2\kappa})$.

Writing ψ for $\tilde{\psi}$, to simplify notation, we write $P = -M\Lambda + B_0\Lambda + Q$ where $M = D_0 - i\lambda^\kappa + \psi$ and $\Lambda = D_0 - i\lambda^\kappa - \psi$. Repeating the proof of Proposition 5.1 we get

Proposition 7.7 *We have*

$$2\mathsf{Im}(Pu, \Lambda u) = \frac{d}{dx_0}(\|\Lambda u\|^2 + ((\mathsf{Re}\,Q)u, u)) + 2\|\lambda^{\kappa/2}\Lambda u\|$$

$$+2((\mathsf{Im}B_0)\Lambda u, \Lambda u) + 2\mathsf{Re}(\lambda^\kappa(\mathsf{Re}\,Q)u, u)$$

$$-2((\mathsf{Im}\,\psi)\Lambda u, \Lambda u) + 2\mathsf{Re}(\Lambda u, (\mathsf{Im}\,Q)u)$$

$$+\mathsf{Im}([D_0 - \mathsf{Re}\,\psi, \mathsf{Re}\,Q]u, u) + 2\mathsf{Re}((\mathsf{Re}\,Q)u, (\mathsf{Im}\,\psi)u).$$

On the other hand from (5.5) we have

$$-2\mathsf{Im}(\Lambda v, v) = \frac{d}{dx_0}\|v\|^2 + 2\|\lambda^{\kappa/2}v\|^2 + 2((\mathsf{Im}\,\psi)v, v).$$

From this it follows that for small $0 < \epsilon < 2$

$$\epsilon^{-1}\|\lambda^{-\kappa/2}\Lambda\lambda^\kappa u\|^2 \geq \frac{d}{dx_0}\|\lambda^\kappa u\|^2$$

$$+ (2 - \epsilon)\|\lambda^{3\kappa/2}u\|^2 + 2((\mathsf{Im}\,\psi)\lambda^\kappa u, \lambda^\kappa u). \tag{7.28}$$

Since $\mathsf{Im}\,\psi \in \mu\tilde{S}_w(\lambda^\kappa)$ one sees

$$\epsilon^{-1}\|\lambda^{-\kappa/2}\Lambda\lambda^\kappa u\|^2 \geq \frac{d}{dx_0}\|\lambda^\kappa u\|^2 + (2 - \epsilon - C\mu)\|\lambda^{3\kappa/2}u\|^2.$$

Since $\lambda^{-\kappa/2}\Lambda\lambda^\kappa = \lambda^{\kappa/2}\Lambda + \lambda^{-\kappa/2}[\Lambda, \lambda^\kappa]$ and $\psi_0 \in \tilde{S}_w(w\lambda)$ we have $[\Lambda, \lambda^\kappa] \in \mu\tilde{S}_\delta(\lambda^\kappa)$. Therefore one obtains $\|\lambda^{-\kappa/2}[\Lambda, \lambda^\kappa]u\|^2 \leq C\mu\|\lambda^{\kappa/2}u\|^2$.

Lemma 7.19 *We have*

$$\|\lambda^{\kappa/2}\Lambda u\|^2 \geq \frac{d}{dx_0}\|\lambda^\kappa u\|^2 + (1 - C\mu)\|\lambda^{3\kappa/2}u\|^2. \tag{7.29}$$

Since $\mathsf{Im}\,\psi \in \mu\tilde{S}_\delta(\lambda^\kappa)$ it follows that

$$|2((\mathsf{Im}\,\psi)\Lambda u, \Lambda u)| \leq C\mu\|\lambda^{\kappa/2}\Lambda u\|^2. \tag{7.30}$$

Consider $2\mathsf{Re}(\Lambda u, (\mathsf{Im}\,Q)u)$. Recall $\mathsf{Im}\,Q = q_1 + \mathsf{Im}\,q_2 + r_1$ with $r_1 \in \mu^2\tilde{S}_\delta(\lambda^{2\kappa})$. Then one can estimate

$$|2\mathsf{Re}(\Lambda u, (\mathsf{Im}\,Q)u)| \leq \mu\|\lambda^{\kappa/2}\Lambda u\|^2 + \mu^{-1}\|\lambda^{-\kappa/2}(\mathsf{Im}\,Q)u\|^2.$$

Note that $\lambda^{-\kappa/2}\#(q_1 + q_2' + r_1) = \lambda^{-\kappa/2}(q_1 + q_2') + T$ with $q_2' = \operatorname{Im} q_2$ and $T \in \mu^2 \tilde{S}_\delta(\lambda^{1/2})$ because $q_1 \in \mu \tilde{S}_w(\lambda^{1+\kappa})$. Here we remark that

$$\langle \mu \xi' \rangle^{-\kappa/2} c_1 \phi_1 = (c_1 \langle \mu \xi' \rangle^{-\kappa} w^{-1/2})(\langle \mu \xi' \rangle^{\kappa/2} w^{1/2} \phi_1)$$

where $c_1 \langle \mu \xi' \rangle^{-\kappa} w^{-1/2} \in \mu \tilde{S}_w(1)$. Recalling (7.27) and applying Lemma 7.15 to $\|[\lambda^{-\kappa/2}(q_1 + q_2')]u\|^2$ we get

Lemma 7.20 *We have*

$$|2\operatorname{Re}(\Lambda u, (\operatorname{Im} Q)u)| \le \mu\|\lambda^{\kappa/2}\Lambda u\|^2 + C\mu\Big\{\sum_{j=2}^{r+1} \operatorname{Re}([\lambda^\kappa \phi_j^2]u, u)$$

$$+ \operatorname{Re}([\lambda^\kappa w\phi_1^2]u, u)\Big\} + C\mu^3\|\lambda^{1/2}u\|^2.$$

We turn to $\operatorname{Re}((\operatorname{Re} Q)u, (\operatorname{Im} \psi)u)$. From Proposition 7.6 we have $\operatorname{Re} Q = q + q_2'' + r$ where $q_2'' = \sum_{j=1}^{r+1} c_j \phi_j$ with $c_j \in \mu \tilde{S}_\delta(1)$, $r \in \mu^2 \tilde{S}_\delta(\lambda^{2\kappa})$ and $\operatorname{Im} \psi \in \mu \tilde{S}_\delta(\lambda^\kappa)$. From Lemma 7.15 it is clear that $|(q_2'' u, (\operatorname{Im} \psi)u)|$ is bounded by a constant times

$$\mu \sum_{j=2}^{r+1} \operatorname{Re}([\lambda^\kappa \phi_j^2]u, u) + \operatorname{Re}([\lambda^\kappa w\phi_1^2]u, u) + \mu^3\|\lambda^{3\kappa/2}u\|^2 \qquad (7.31)$$

because $\langle \mu \xi' \rangle^{-\kappa/2} w^{-1/2} \operatorname{Im} \psi \in \mu \tilde{S}_\delta(\langle \mu \xi' \rangle^{3\kappa/2})$. Thus it suffices to study the term $\operatorname{Re}(q\, u, (\operatorname{Im} \psi)u)$ modulo (7.31). Since one can write

$$\operatorname{Im} \psi = \psi_1 + R_N, \quad \psi_1 \in \mu \tilde{S}_w(\lambda^\kappa), \quad R_N \in \mu^N \tilde{S}_\delta(\lambda^{1-2N\kappa+2n\kappa})$$

for any N we may assume $\operatorname{Im} \psi = \psi_1 \in \mu S(\lambda^\kappa, g)$ modulo $\mu^3\|\lambda^{3\kappa/2}u\|^2$. Note that $\operatorname{Re}(\psi_1 \# q) - \psi_1 q \in \mu^3 \tilde{S}_w(\lambda)$ because $5\kappa = 1$. Applying Lemma 7.15 we get

Lemma 7.21 *We have*

$$|2\operatorname{Re}((\operatorname{Re} Q)u, (\operatorname{Im} \psi)u)| \le C\mu\Big\{\sum_{j=2}^{r+1} \operatorname{Re}([\lambda^\kappa \phi_j^2]u, u)$$

$$+ \operatorname{Re}([\lambda^\kappa w\phi_1^2]u, u)\Big\} + C\mu^3\|\lambda^{1/2}u\|^2.$$

We now estimate $\operatorname{Im}([D_0 - \operatorname{Re} \psi, \operatorname{Re} Q]u, u)$. Note that one can write

$$\operatorname{Re} Q = q + q_2'' + r + R_N, \quad r \in \mu^2 \tilde{S}_w(\lambda^{2\kappa}),$$

$$\operatorname{Re} \psi = -\psi_0 - \psi_2 + R_N', \quad \psi_2 \in \mu^2 \tilde{S}_w(\lambda^{-\kappa})$$

where $R_N \in \mu^N \tilde{S}_\delta(\lambda^{2-2N\kappa+2n\kappa})$ and $R_N' \in \mu^N \tilde{S}_\delta(\lambda^{1-2N\kappa+2n\kappa})$. From this it follows that $|\operatorname{Im}([D_0 - \operatorname{Re} \psi, \operatorname{Re} Q]u, u)| = |\operatorname{Re}(\operatorname{Op}(\{\xi_0 - \operatorname{Re} \psi, \operatorname{Re} Q\})u, u)|$ modulo

$O(\mu^3 \|\lambda^{2\kappa} u\|^2)$. Note that

$$\{\xi_0 - \operatorname{Re}\psi, \operatorname{Re}Q\} = \{\xi_0 - \psi_0, q + q_2''\} - \{\psi_2, q\} + T, \quad T \in \mu^3 \tilde{S}_\delta(\lambda^{4\kappa}).$$

Since one can write $\{\psi_2, q\} = \sum_{j=2}^{r+1} a_j \lambda^\kappa \phi_j + a_1 \lambda^\kappa w \phi_1$ with $a_j \in \mu^3 \tilde{S}_w(1)$ and $\{\xi_0 - \psi_0, q_2''\} - \sum_{j=1}^{r+1} c_j \phi_j \in \mu^2 \tilde{S}_\delta(\lambda)$ with $c_j \in \mu^2 \tilde{S}_\delta(\lambda^{2\kappa})$ we have

$$|\operatorname{Im}([D_0 - \operatorname{Re}\psi, \operatorname{Re}Q]u, u)| \leq |\operatorname{Re}(\operatorname{Op}(\{\xi_0 - \psi_0, q\})u, u)|$$

$$+ C\mu^3 \{\sum_{j=2}^{r+1} \|\phi_j u\|^2 + \|[\sqrt{w}\phi_1]u\|^2 + \|\lambda^{1/2} u\|^2\}$$

$$\leq |\operatorname{Re}(\operatorname{Op}(\{\xi_0 - \operatorname{Re}\psi_0, q\})u, u)|$$

$$+ C\mu^3 \{\sum_{j=2}^{r+1} \operatorname{Re}([\phi_j^2]u, u) + \operatorname{Re}([w\phi_1^2]u, u) + \|\lambda^{1/2} u\|^2\}.$$

Thanks to Proposition 7.5 we conclude that

Lemma 7.22 *We have*

$$|\operatorname{Im}([D_0 - \operatorname{Re}\psi, \operatorname{Re}Q]u, u)| \leq C\mu \{\sum_{j=2}^{r+1} \operatorname{Re}([\lambda^\kappa \phi_j^2]u, u)$$

$$+ \operatorname{Re}([w\lambda^\kappa \phi_1^2]u, u)\} + C\mu^2 \|\lambda^{1/2} u\|^2.$$

It remains to estimate $\operatorname{Re}(\lambda^\kappa (\operatorname{Re}Q)u, u)$. We first note that

$$|\operatorname{Re}(\lambda^\kappa q_2'' u, u)| \leq C\mu \{\sum_{j=2}^{r+1} \|\phi_j u\|^2 + \|[\lambda^{\kappa/2}\sqrt{w}\phi_1]u\|^2 + \mu \|\lambda^{3\kappa/2} u\|^2\}$$

and hence it suffices to estimate $\operatorname{Re}(\lambda^\kappa q u, u)$ modulo $O(\mu^2 \|\lambda^{3\kappa/2} u\|^2)$. Writing $\operatorname{Re}(\lambda^\kappa \# \phi_j^2) = \lambda^\kappa \phi_j^2 + T_1$ and $\operatorname{Re}(\lambda^\kappa \#(w\phi_1^2)) = \lambda^\kappa w\phi_1^2 + T_2$ with $T_1 \in \mu^2 \tilde{S}_w(\lambda^\kappa)$ and $T_2 \in \mu^2 \tilde{S}_w(w\lambda^\kappa)$ we have

$$\operatorname{Re}(\lambda^\kappa q u, u) \geq \sum_{j=2}^{r+1} \operatorname{Re}([\lambda^\kappa \phi_j^2]u, u) + \operatorname{Re}([\lambda^\kappa w\phi_1^2]u, u) - C\mu^2 \|\lambda^{\kappa/2} u\|^2.$$

Therefore $\operatorname{Re}(\lambda^\kappa (\operatorname{Re}Q)u, u)$ is bounded from below by

$$\sum_{j=2}^{r+1} \operatorname{Re}([\lambda^\kappa \phi_j^2]u, u) + \operatorname{Re}([\lambda^\kappa w\phi_1^2]u, u) - C\mu^2 \|\lambda^{3\kappa/2} u\|^2.$$

Note that thanks to Lemma 7.15 there exist $c > 0$ and $C > 0$ such that

$$\sum_{j=2}^{r+1} \text{Re}([\phi_j^2 \lambda^\kappa]u, u) + \text{Re}([\phi_1^2 w \lambda^\kappa]u, u) \geq c \left(\sum_{j=2}^{r+1} \|[\lambda^{\kappa/2}\phi_j]u\|^2 + \|[\sqrt{w}\lambda^{\kappa/2}\phi_1]u\|^2 \right)$$

$$- C\mu^2 \|\lambda^{1/2}u\|^2 - C\mu^2 \|\lambda^{3\kappa/2}u\|^2.$$

Then from Lemmas 7.19–7.22 and (7.30) we have

Proposition 7.8 *There exist $\mu_0 > 0$, $C > 0$, $c > 0$ such that for $0 < \mu < \mu_0$ one has*

$$2\text{Im}\,(Pu, \Lambda u) \geq \frac{d}{dx_0}\{\|\Lambda u\|^2 + ((\text{Re}\,Q)u, u) + \|\lambda^\kappa u\|^2\}$$

$$+ c\|\lambda^{\kappa/2}\Lambda u\|^2 + c\left\{\sum_{j=2}^{r+1}\|[\lambda^{\kappa/2}\phi_j]u\|^2 + \|[\sqrt{w}\lambda^{\kappa/2}\phi_1]u\|^2\right\} \qquad (7.32)$$

$$+ c\,\|\lambda^{3\kappa/2}u\|^2 + c\,\mu\|\lambda^{1/2}u\|^2.$$

Taking into account that $\phi_j^2 - \phi_j\#\phi_j \in \mu^2\tilde{S}_0(1)$, $w\phi_1^2 = (\sqrt{w}\phi_1)\#(\sqrt{w}\phi_1) + R$, $R \in \mu^2\tilde{S}_w(\tilde{w}^{-1}) \subset \mu^2\tilde{S}_w(\lambda^{2\kappa})$ it follows that there is $c > 0$ such that

$$((\text{Re}\,Q)u, u) + \|\lambda^\kappa u\|^2 \geq c \left(\sum_{j=2}^{r+1}\|\phi_j u\|^2 + \|[\sqrt{w}\phi_1]u\|^2 + \|\lambda^\kappa u\|^2\right)$$

for small μ.

Since $\{a, \lambda^s\} \in \mu\tilde{S}_\delta(\lambda^{s-2\kappa})$ if $a \in \tilde{S}_w(\lambda^\kappa) \subset \tilde{S}_\delta(\lambda^\kappa)$ then it follows from (7.27) and Lemma 7.11 that one can write $\{Q, \lambda^s\} = \sum_{j=1}^{r+1} b_j\phi_j\lambda^s + b_0\lambda^s$ and $\{\Lambda, \lambda^s\} = c\lambda^s$ where $b_j, c \in \mu^2\tilde{S}_\delta(1)$ and $b_0 \in \mu^2\tilde{S}_\delta(\lambda^\kappa)$. Then repeating the same arguments proving Proposition 7.8 one can conclude that $2\text{Im}\,(\lambda^s Pu, \Lambda\lambda^s u)$ is bounded from below by the right-hand side of (7.32) with $\lambda^s u$ in place of u. Then integrating the resulting estimate and denoting

$$N_s(u) = \|\lambda^s \Lambda u\|^2 + \sum_{j=2}^{r+1}\|\lambda^s\phi_j u\|^2 + \|\lambda^s\sqrt{w}\phi_1 u\|^2 + \|\lambda^{s+\kappa}u\|^2$$

we obtain

Proposition 7.9 *For any $s \in \mathbb{R}$ there exist $C_s > 0$ and μ_s such that for any $u \in \cap_{j=0}^2 C_+^j([-T, T]; H^{s+2-j})$ and any $0 < \mu < \mu_s$ one has*

$$C_s \text{Im} \int^t (\lambda^s Pu, \Lambda\lambda^s u)dx_0 \geq N_s(u(t)) + \int^t N_{s+\kappa/2}(u(x_0))dx_0.$$

Corollary 7.4 *For any $s \in \mathbb{R}$ there exist $C, C' > 0$ and μ_0 such that for $0 < \mu < \mu_0$ and $u \in \cap_{j=0}^{2} C_+^j([-T, T]; H^{s+2-j})$ one has*

$$\|\lambda^{s+\kappa-1} D_0 u(t)\|^2 + \|\lambda^{s+\kappa} u(t)\|^2 \leq C N_s(u(t)) \leq C' \int^t \|\lambda^s P u\|^2 dx_0. \tag{7.33}$$

Recalling $P^* = \mathrm{Op}(p + \bar{P}_{sub}) + R$, $R \in \tilde{S}_0(1)$ we have the same energy estimates with the reversed time direction for P^*. Hence from the same arguments proving Proposition 5.19 we conclude that for any given $f \in C_+^0([-T, T]; H^s)$ there exists $u \in \cap_{j=0}^{1} C_+^j([-T, T]; H^{s-j})$ verifying $Pu = f$ and (7.33).

Define G by $Gf = u$ and we show that G is a parametrix of P with finite propagation speed of micro supports (Definition 4.4). Let $\phi(x', \xi') \in S^0$ and we still denote $\phi(x', \mu\xi')$, dilated according to (7.7), by $\phi(x', \xi')$ so that $\phi(x', \xi') \in \tilde{S}_0(1)$.

Lemma 7.23 *Let ϕ being as above. Then there exists $C > 0$ such that*

$$|\mathrm{Im}(\lambda^s[\phi, P]u, \Lambda\lambda^s\phi u)| \leq C\mu N_s(u).$$

Proof Write $P = -\Lambda^2 + B\Lambda + Q$ where $\Lambda = D_0 - i\lambda^\kappa - \psi$ and $B = B_0 - 2\psi$ and we first consider $(\lambda^s[\Lambda^2, \phi]u, \Lambda\lambda^s\phi u)$. Recall $\psi = -\phi_1 k + \psi_1 + \psi_2$ where $\psi_1 \in \mu\tilde{S}_w(\lambda^\kappa)$ and $\psi_2 \in \mu^2\tilde{S}_\delta(\lambda^{-\kappa})$. Since $\phi_1 k \in \tilde{S}_w(w\lambda)$ by Lemma 7.11 it is clear that $\{\psi, \phi\} \in \mu\tilde{S}_\delta(1)$ and hence $\{\Lambda, \phi\}, \{B, \phi\} \in \mu\tilde{S}_\delta(1)$ from which it follows that

$$|(\lambda^s[\Lambda, \phi]\Lambda u, \Lambda\lambda^s\phi u)| + |(\lambda^s\Lambda[\Lambda, \phi]u, \Lambda\lambda^s\phi u)| \leq C\mu N_s(u),$$

$$|(\lambda^s[B, \phi]\Lambda u, \Lambda\lambda^s\phi u)| + |(\lambda^s B[\Lambda, \phi]u, \Lambda\lambda^s\phi u)| \leq C\mu N_s(u).$$

We next consider $(\lambda^s[Q, \phi]u, \lambda^s\Lambda\lambda^s\phi u)$. Recall (7.27) then thanks to Lemma 7.11 again one can write

$$\{Q, \phi\} = \sum_{j=2}^{r+1} b_j \phi_j + b_1 \sqrt{w}\phi_1 + b_0$$

with $b_j \in \mu\tilde{S}_\delta(1), j = 1, \ldots, r+1$ and $b_0 \in \mu^2\tilde{S}_\delta(\lambda^\kappa)$. Therefore we have

$$|(\lambda^s[Q, \phi]u, \Lambda\lambda^s\phi u)| \leq C\mu N_s(u)$$

and hence the assertion.   □

From Proposition 7.9 and Lemma 7.23 we obtain

$$C \int^t \|\lambda^{s-\kappa/2}\phi P u(x_0)\|^2 dx_0 + C\mu \int^t N_s(u(x_0))dx_0$$

$$\geq N_s(\phi u(t)) + \int^t N_{s+\kappa/2}(\phi u(x_0))dx_0. \tag{7.34}$$

Thus we conclude that the regularity of ϕu is improved by $\kappa/2$ compared with u. Then by a repetition of similar arguments proving Lemma 4.4 we obtain

Proposition 7.10 *Let Γ_i ($i = 0, 1, 2$) be open conic sets in $\mathbb{R}^n \times (\mathbb{R}^n \setminus \{0\})$ such that $\Gamma_0 \Subset \Gamma_1 \Subset \Gamma_2$ where $\Gamma_2 \cap \{|\xi'| = 1\}$ is relatively compact and let $h_i(x', \xi') \in S^0$ be such that $\operatorname{supp} h_1 \subset \Gamma_0$ and $\operatorname{supp} h_2 \subset \Gamma_2 \setminus \overline{\Gamma}_1$. Then there is $\delta = \delta(\Gamma_i) > 0$ such that for any $p, q \in \mathbb{R}$ and $0 \leq t \leq \delta$ we have for $f \in C^0_+(I; H^{q-j})$*

$$\|D_0^j h_2 G h_1 f(t)\|_{p-j} \leq C_{pq} \int^t \|f(x_0)\|_q dx_0, \quad j = 0, 1.$$

Denote $e^{(\tau - x_0)\lambda^\kappa} \hat{P} e^{-(\tau - x_0)\lambda^\kappa} = P$ where a constant $\tau > 0$ will be fixed below. Let $f \in C^0_+([-T, T]; \gamma_0^{(1/\kappa)}(\mathbb{R}^n))$ then it is clear $F = e^{(\tau - x_0)\lambda^\kappa} f \in C^0_+([-\tau, \tau]; H^\infty(\mathbb{R}^n))$ for small $\tau > 0$. Then we have $GF \in C^2_+([-\tau, \tau]; H^\infty(\mathbb{R}^n))$ such that

$$e^{(\tau - x_0)\lambda^\kappa} \hat{P} e^{-(\tau - x_0)\lambda^\kappa} GF = PGF = F = e^{(\tau - x_0)\lambda^\kappa} f.$$

This implies that $\hat{P}(e^{-(\tau - x_0)\lambda^\kappa} G e^{(\tau - x_0)\lambda^\kappa} f) = f$. Define $\hat{G}$ by

$$\hat{G} = e^{-(\tau - x_0)\lambda^\kappa} G e^{(\tau - x_0)\lambda^\kappa}$$

so that $\hat{P} \hat{G} f = f$. Since $GF \in C^2_+([-\tau, \tau]; H^\infty(\mathbb{R}^n))$ it is easy to check that $\hat{G} f \in C^2_+([-\tau, \tau]; H^\infty(\mathbb{R}^n))$.

Lemma 7.24 *Let Γ_i ($i = 0, 1, 2$) be open conic sets in Proposition 7.10 and let $h_i(x', \xi') \in \tilde{S}_0^{(1/\kappa)}(1)$ be such that $\operatorname{supp} h_1 \subset \Gamma_0$ and $\operatorname{supp} h_2 \subset \Gamma_2 \setminus \overline{\Gamma}_1$. Then for any $p, q \in \mathbb{R}$ and for any $f \in C^0_+([-\tau, \tau]; \gamma_0^{(1/\kappa)}(\mathbb{R}^n))$ and $0 \leq t \leq \tilde{t}$ with small $\tilde{t} > 0$ we have*

$$\|D_0^j e^{(\tau - t)\lambda^\kappa} h_2 \hat{G} h_1 f(t)\|_p \leq C_{pq} \int^t \|e^{(\tau - x_0)\lambda^\kappa} f(x_0)\|_q dx_0, \quad j = 0, 1.$$

Proof Applying Proposition 7.4 with $\delta = 0$ one can write

$$e^{(\tau - x_0)\lambda^\kappa} h_2 \hat{G} h_1 = \tilde{h}_2 G \tilde{h}_1 e^{(\tau - x_0)\lambda^\kappa}, \quad \tilde{h}_i = e^{(\tau - x_0)\lambda^\kappa} h_i e^{-(\tau - x_0)\lambda^\kappa} \in \tilde{S}_0(1)$$

where for any $N \in \mathbb{N}$ we have $\tilde{h}_i = \tilde{h}_{i0} + \tilde{h}_{iN}$ with $\tilde{h}_{iN} \in \mu^N \tilde{S}_0(\lambda^{-N(1-\kappa)})$ where $\operatorname{supp} \tilde{h}_{i0} \subset \operatorname{supp} h_i$. Since one can take N arbitrarily large to prove the assertion it suffices to consider $D_0^j \tilde{h}_{20} G \tilde{h}_{10} e^{(\tau - x_0)\lambda^\kappa} f$. Therefore from Proposition 7.10 it follows that

$$\|D_0^j \tilde{h}_{20} G \tilde{h}_{10} e^{(\tau - t)\lambda^\kappa} f(t)\|_{p-j} \leq C_{pq} \int^t \|e^{(\tau - x_0)\lambda^\kappa} f(x_0)\|_q dx_0, \quad j = 0, 1$$

which shows the desired assertion. $\qquad \square$

Chapter 8
Ill-Posed Cauchy Problem, Revisited

Abstract In Chap. 6 we exhibited a second order differential operator with polynomial coefficients for which the Cauchy problem is C^∞ ill-posed even though the
Levi condition is satisfied. The Levi condition would be the most strict condition
that one can impose on lower order terms on the double characteristics as far as
we know. In this chapter we confirm this by proving that the Cauchy problem for
this operator is ill-posed in the Gevrey class of order grater than 6 for any lower
order term. In particular the Cauchy problem is C^∞ ill-posed for any lower order
term. This phenomenon never occurs in the case of one spatial dimension. In the last
section we give an example of second order differential operator of spectral type 1
on Σ, which shows that the IPH condition is not sufficient in general for the Cauchy
problem to be C^∞ well-posed.

8.1 Preliminaries

For a second order differential operator P in $\mathbb{R}^2$

$$P = -D_0^2 + a(x_0, x_1)D_1^2$$

with nonnegative real analytic coefficient $a(x_0, x_1) \geq 0$ defined near the origin
the Cauchy problem is C^∞ well-posed near the origin [70]. Since then it has
been conjectured that the Cauchy problem is C^∞ well-posed for any second order
differential operator in divergence form with real analytic coefficients

$$Pu = -D_0^2 u + \sum_{i,j=1}^{n} D_{x_i}(a_{ij}(x)D_{x_j}u), \quad a_{ij}(x) = a_{ji}(x)$$

where $a_{ij}(x)$ are real analytic and $\sum_{i,j=1}^{n} a_{ij}(x)\xi_i\xi_j \geq 0$ for all $\xi' = (\xi_1, \ldots, \xi_n) \in \mathbb{R}^n$.
However in Sect. 6.1 we have seen that

$$P_{mod}u = -D_0^2 u + D_1(D_1 u) + D_1(x_0 D_n u) + D_n(x_0 D_1 u)$$
$$+ D_n((1 + x_1^2(1 + x_1))D_n u)$$

© Springer International Publishing AG 2017

T. Nishitani, *Cauchy Problem for Differential Operators with Double
Characteristics*, Lecture Notes in Mathematics 2202,
DOI 10.1007/978-3-319-67612-8_8

so that P_{mod} is in divergence form with polynomial coefficients for which the Cauchy problem is C^∞ ill-posed, yielding a counter example to this conjecture. In this chapter we show a somewhat stronger assertion on the well-posedness of the Cauchy problem for P_{mod}, that is the Cauchy problem for $P_{mod} + Q$ is ill-posed in $\gamma^{(s)}$ for any $s > 6$ whatever the lower order term Q is. On the other hand note that the Cauchy problem for $P_{mod} + Q$ is $\gamma^{(s)}$ well-posed for any lower order term Q if $1 < s < 3$ by Proposition 6.6. Consider again

$$P_{mod} = -D_0^2 + 2x_1 D_0 D_2 + D_1^2 + x_1^3 D_2^2, \quad x = (x_0, x_1, x_2)$$

in $\mathbb{R}^3$. Then we have

Theorem 8.1 ([81])   *The Cauchy problem for $P = P_{mod} + \sum_{j=0}^{2} b_j D_j$ is not locally solvable in $\gamma^{(s)}$ at the origin for any b_0, b_1, $b_2 \in \mathbb{C}$ if $s > 6$. In particular the Cauchy problem for P_{mod} is C^∞ ill-posed for any $b_0, b_1, b_2 \in \mathbb{C}$.*
We begin with

Proposition 8.1   *The Cauchy problem for $P = P_{mod} + \sum_{j=0}^{1} b_j D_j$ is not locally solvable in $\gamma^{(s)}$ at the origin if $s > 5$ for any b_0, $b_1 \in \mathbb{C}$.*

Proof Consider

$$U(x) = \exp\left(i\rho^5 x_2 + \frac{i}{2}\zeta\rho x_0 - \frac{i}{2}b_1 x_1\right) w(x_1 \rho^2), \quad \zeta \in \mathbb{C}, \ \rho > 0.$$

It is easy to see that if w verifies

$$w''(x) = (x^3 + \zeta x - \zeta^2 \rho^{-2}/4 + b_0 \zeta \rho^{-3}/2 - b_1^2 \rho^{-4}/4) w(x)$$

then $P(x, D)U = 0$. Now instead of (6.10) we consider the equation

$$C_0(\zeta, -\zeta^2 s^2/4 + b_0 \zeta s^3/2 - b_1^2 s^4/4) = 0. \tag{8.1}$$

Repeating the same arguments as in Sect. 6.2 one can find $\tilde{\zeta}(s)$, holomorphic at $s = 0$ and $p \in \mathbb{N}$ such that $\zeta(s^p) = \tilde{\zeta}(s)$ verifies Eq. (8.1) and satisfies (6.13). Choosing $\eta(s) = -\tilde{\zeta}(s)^2 s^{2p}/4 + b_0 \tilde{\zeta}(s) s^{3p}/2 - b_1^2 s^{4p}/4$ and $e^{ib_1 x_1/2} \phi(x_1)$ in place of $\phi(x_1)$ in (6.16) and repeating the same arguments as in Sect. 6.3 we arrive at (6.18). The rest of the proof is just the repetition.   □
Thus in order to prove Theorem 8.1 we may assume that $b_2 \neq 0$. Moreover, making a change of the coordinate system $(x_0, x_1, x_2) \mapsto (x_0, x_1, -x_2)$ if necessary, we may assume that $b_2 \in \mathbb{C} \setminus \mathbb{R}^+$. We construct U_λ following [32, 38] which contradicts the a priori estimates, derived from the $\gamma^{(s)}$ well-posedness assumption, as $\lambda \to \infty$, and hence finally we prove Theorem 8.1.

8.2 Asymptotic Solutions

Let us consider

$$P = -D_0^2 + 2x_1 D_0 D_2 + D_1^2 + x_1^3 D_2^2 + \sum_{j=0}^{2} b_j D_j, \quad b_j \in \mathbb{C}$$

where we assume $b_2 \in \mathbb{C} \setminus \mathbb{R}^+$. We make a change of coordinates system such that $x_0 = \lambda^{-1} y_0$, $x_1 = \lambda^{-2} y_1$, $x_2 = \lambda^{-4} y_2$ then we obtain (after dividing the resulting operator by λ^4)

$$P_\lambda = -\lambda^{-2} D_0^2 + 2\lambda^{-1} y_1 D_0 D_2 + D_1^2 + \lambda^{-2} y_1^3 D_2^2$$

$$+ b_2 D_2 + \lambda^{-2} b_1 D_1 + \lambda^{-3} b_0 D_0.$$

We switch the notation to x and set $b_2 = b$. Denote

$$E_\lambda = \exp\left(i\lambda^2 x_2 + i\lambda\phi(x)\right)$$

and compute $\lambda^{-1} E_\lambda^{-1} P_\lambda E_\lambda$ which yields

$$\lambda^{-1} E_\lambda^{-1} P_\lambda E_\lambda = \lambda\{2x_1\phi_{x_0} + \phi_{x_1}^2 + x_1^3 + b\}$$

$$+ \{2x_1 D_0 + 2\phi_{x_1} D_1 + 2x_1\phi_{x_0}\phi_{x_2} + b\phi_{x_2} + 2x_1^3\phi_{x_1} - i\phi_{x_1 x_1}\}$$

$$+ \lambda^{-1} h^{(1)} + \lambda^{-2} h^{(2)} + \lambda^{-3} h^{(3)}$$

where $\phi_{x_i} = \partial_{x_i}\phi$ and $h^{(i)}$, $i = 1, 2, 3$ are differential operators of order 2 with coefficients which are polynomials in derivatives of ϕ. We first assume that

$$\text{Im } b \neq 0.$$

Take $y_1 \in \mathbb{R}$ small so that

$$\text{Im } \frac{b}{2y_1} > 0$$

and work near the point $(x_0, x_1, x_2) = (t, y_1, 0) = x^*$. We solve the equation

$$2x_1\phi_{x_0} + \phi_{x_1}^2 + x_1^3 + b = 0 \tag{8.2}$$

imposing the condition

$$\phi = (x_1 - y_1) + i(x_1 - y_1)^2 + ix_2^2 \quad \text{on} \quad x_0 = t.$$

Since $x_1 \neq 0$ near y_1 there is a unique analytic solution $\phi(x)$ near x^* by the Cauchy-Kowalevsky theorem (see, for example [92]). Noting $\phi = (x_1 - y_1) + i(x_1 - y_1)^2 + ix_2^2 + \phi_{x_0}(t, x_1, x_2)(x_0 - t) + O((x_0 - t)^2)$ we conclude

$$\operatorname{Im} \phi = (x_1 - y_1)^2 + x_2^2 + \{\operatorname{Im} \phi_{x_0}(t, y_1, 0) + R(x)\}(x_0 - t)$$

where $R(x) = O(|x - x^*|)$. Note that $\phi_{x_0}(x^*) = (-1 - b)/(2y_1) - y_1^2/2$ and hence $\operatorname{Im} \phi_{x_0}(x^*) < 0$. Writing $\alpha = \operatorname{Im} \phi_{x_0}(x^*)$ we have with small $\epsilon > 0$

$$\operatorname{Im} \phi = (x_1 - y_1)^2 + x_2^2 + \alpha(x_0 - t) + \frac{1}{2}(\epsilon^{-1}(x_0 - t) + \epsilon R(x))^2$$

$$- \frac{\epsilon^{-2}}{2}(x_0 - t)^2 - \frac{\epsilon^2}{2}R(x)^2$$

$$= (x_1 - y_1)^2 + x_2^2 + (x_0 - t)^2 - \frac{\epsilon^2}{2}R(x)^2$$

$$+ \left\{\alpha - \left(\frac{\epsilon^{-2}}{2} + 1\right)(x_0 - t)\right\}(x_0 - t) + \frac{1}{2}(\epsilon^{-1}(x_0 - t) + \epsilon R(x))^2$$

$$= |x - x^*|^2 - \frac{\epsilon^2}{2}R(x)^2 + \frac{1}{2}(\epsilon^{-1}(x_0 - t) + \epsilon R(x))^2$$

$$+ \left\{\alpha - \left(\frac{\epsilon^{-2}}{2} + 1\right)(x_0 - t)\right\}(x_0 - t).$$

Thus choosing $\epsilon > 0$ so that $\epsilon^2 R(x)^2 \leq |x - x^*|^2$ near x^* we see that $-\operatorname{Im} \phi$ attains its strict maximum at x^* in the set $\{x; |x - x^*| < \delta, x_0 \leq t\}$ if $\delta > 0$ is small enough. Let L be a compact set in $\mathbb{R}^3$. For $t \in \mathbb{R}$ we denote

$$L^t = \{x \in L \mid x_0 \leq t\}.$$

Then we have

Lemma 8.1 *Let L be a small compact neighborhood of x^*. Then for any small $\tau > 0$ we have*

$$\sup_{x \in L^{t+\tau}} \{-\operatorname{Im} \phi(x)\} \leq 2|\operatorname{Im} \phi_{x_0}(x^*)|\tau.$$

For any small $\delta > 0$ there exist $\nu(\delta), \tau(\delta) > 0$ such that for any $0 \leq \tau \leq \tau(\delta)$

$$\sup_{x \in L^{t+\tau} \cap \{|x - x^*| \geq \delta\}} \{-\operatorname{Im} \phi(x)\} \leq -\nu(\delta).$$

We denote $\lambda^{-1}P_\lambda E_\lambda = E_\lambda Q_\lambda$ and $Q_\lambda = Q_0 + Q_{1\lambda}$ where

$$\begin{cases} Q_0 = 2x_1 D_0 + 2\phi_{x_1} D_1 + 2x_1\phi_{x_0}\phi_{x_2} + b\phi_{x_2} + 2x_1^3\phi_{x_1} - i\phi_{x_1 x_1}, \\ Q_{1\lambda} = \lambda^{-1}h^{(1)} + \lambda^{-2}h^{(2)} + \lambda^{-3}h^{(3)}. \end{cases}$$

We set $V_\lambda = \sum_{n=0}^{N} v_\lambda^{(n)}$ and determine $v_\lambda^{(n)}$ by solving the Cauchy problem

$$\begin{cases} Q_0 v_\lambda^{(n)} = -g_\lambda^{(n)} = -Q_{1\lambda} v_\lambda^{(n-1)}, \\ v_\lambda^{(0)}(t, x_1, x_2) = 1, \\ v_\lambda^{(n)}(t, x_1, x_2) = 0, \quad n \geq 1 \end{cases} \tag{8.3}$$

where $v_\lambda^{(-1)} = 0$ so that $Q_\lambda V_\lambda = Q_{1\lambda} v_\lambda^{(N)}$. Thanks to the Cauchy-Kowalevsky theorem (8.3) has a unique analytic solution near x^*. Hence we have

$$\lambda^{-1}P_\lambda E_\lambda V_\lambda = E_\lambda Q_{1\lambda} v_\lambda^{(N)}. \tag{8.4}$$

We turn to the case

$$b \in \mathbb{R}, \quad b < 0.$$

We follow the arguments in [32]. We write $b = -\gamma^2$, $\gamma > 0$ and solve Eq. (8.2) under the condition

$$\phi = -i(x_0 - t) + ix_2^2 \quad \text{on } x_1 = 0.$$

That is, one solves the equation $\phi_{x_1} = \sqrt{\gamma^2 - x_1^3 - 2x_1\phi_{x_0}}$. From the Cauchy-Kowalevsky theorem again there exists a unique analytic solution ϕ near x^*. Since it is clear that $\phi_{x_1} = (\gamma + ix_1/\gamma) + O(x_1^2)$ one can write

$$\phi = -i(x_0 - t) + ix_2^2 + (\gamma + ix_1/\gamma)x_1 + R(x)$$

where $R(x) = O(x_1^3)$. Note that

$$\begin{aligned} \text{Im}\,\phi &= -(x_0 - t) + x_2^2 + \gamma^{-1}x_1^2 + R(x) \\ &= (x_0 - t)^2 + \gamma^{-1}x_1^2 + x_2^2 + R(x) + \{-1 - (x_0 - t)\}(x_0 - t) \end{aligned}$$

and hence the same assertion as Lemma 8.1 holds for this case. Noting that ϕ_{x_1} is different from zero in an open neighborhood of $x^* = (t, 0, 0)$ we can solve the following Cauchy problem in x_1 direction near x^* by the Cauchy-Kowalevsky theorem:

$$
\begin{cases}
Q_0 v_\lambda^{(n)} = -g_\lambda^{(n)} = -Q_{1\lambda} v_\lambda^{(n-1)}, \\
v_\lambda^{(0)}(x_0, 0, x_2) = 1, \\
v_\lambda^{(n)}(x_0, 0, x_2) = 0, \quad n \geq 1.
\end{cases}
$$

8.3   Estimates of Asymptotic Solutions, Majorant

To estimate $U_\lambda = E_\lambda V_\lambda$, which is constructed in the previous section, we apply the method of majorant following Ivrii [38, 39] (see also [56]). We first recall the notion of majorant.

Definition 8.1 Let $\Phi_i(\tau, \eta) = \sum_{j,k \geq 0}^{\infty} C_{ijk} \tau^j \eta^k$, $i = 1, 2$ be two formal power series in (τ, η). Then we say that Φ_2 is a majorant of Φ_1 if $|C_{1jk}| \leq C_{2jk}$ for any $j, k \geq 0$ and we write $\Phi_1 \ll \Phi_2$.

We first make some general observations on majorants. Consider a first order differential operator

$$
Q = \sum_{|\alpha| \leq 1} b_\alpha(x) D^\alpha, \quad D = (D_0, D_1, \ldots, D_n)
$$

where we assume that $b_\alpha(x)$ are holomorphic at $x = x^*$ and $b_{(1,0,\ldots,0)}(x)$ is different from zero near $x = x^*$.

Lemma 8.2 *Let $Qv = g$ and denote*

$$
\Phi(\tau, \eta; v) = \sum_{\alpha = (\alpha_0, \alpha')} \frac{\tau^{\alpha_0} \eta^{|\alpha'|}}{\alpha!} |D^\alpha v(x^*)|.
$$

Then one can find $C(\tau, \eta) \gg 0$ which is holomorphic at $(0,0)$ and depends only on Q such that

$$
\frac{\partial}{\partial \tau} \Phi(\tau, \eta; v) \ll C(\tau, \eta) \frac{\partial}{\partial \eta} \Phi(\tau, \eta; v) + C(\tau, \eta) \Phi(\tau, \eta; g).
$$

Proof Note that

$$
\frac{\partial}{\partial \tau} \Phi(\tau, \eta; v) = \sum_\beta \frac{\tau^{\beta_0} \eta^{|\beta'|}}{\beta!} |D^\beta (D_0 v)(x^*)| = \Phi(\tau, \eta; D_0 v)
$$

and $D_0 v = \sum_{j=1}^n b_j D_j v + b_0 v + cg$ which follows from $Qv = g$. Since $\Phi(\tau, \eta; f_1 f_2) \ll \Phi(\tau, \eta; f_1)\Phi(\tau, \eta; f_2)$ we hence have

$$\frac{\partial}{\partial \tau}\Phi(\tau, \eta; v) \ll C(\tau, \eta)\left(\sum_{j=1}^n \Phi(\tau, \eta; D_j v) + \Phi(\tau, \eta; g)\right).$$

To conclude the assertion it is enough to note

$$\frac{\partial \Phi}{\partial \eta} \gg \sum_{\alpha_j \geq 1} \frac{|\alpha'|\tau^{\alpha_0}\eta^{|\alpha'|-1}}{\alpha!}|D^{\tilde\alpha}(D_j v)(x^*)|,$$

$$\frac{|\alpha'|\tau^{\alpha_0}\eta^{|\alpha'|-1}}{\alpha!} = \frac{|\alpha'|\tau^{\tilde\alpha_0}\eta^{|\tilde\alpha'|}}{\alpha_j\tilde\alpha!} \geq \frac{\tau^{\tilde\alpha_0}\eta^{|\tilde\alpha'|}}{\tilde\alpha!}$$

where $\tilde\alpha = (\alpha_0, \ldots, \alpha_{j-1}, \alpha_j - 1, \ldots, \alpha_n)$. $\qquad\square$

Lemma 8.3 *Assume $Qv = g$ and*

$$\begin{cases} \dfrac{\partial}{\partial \tau}\Phi^*(\tau, \eta) \gg C(\tau, \eta)\dfrac{\partial}{\partial \eta}\Phi^*(\tau, \eta) + C(\tau, \eta)\Phi(\tau, \eta; g), \\ \Phi^*(0, \eta) \gg \Phi(0, \eta; v). \end{cases} \tag{8.5}$$

Then we have $\Phi(\tau, \eta; v) \ll \Phi^*(\tau, \eta)$.

Proof Let $\tilde\Phi$ be a solution to the Cauchy problem

$$\begin{cases} \dfrac{\partial}{\partial \tau}\tilde\Phi(\tau, \eta) = C(\tau, \eta)\dfrac{\partial}{\partial \eta}\tilde\Phi(\tau, \eta) + C(\tau, \eta)\Phi(\tau, \eta; g), \\ \tilde\Phi(0, \eta) = \Phi^*(0, \eta). \end{cases}$$

Then it is clear that $\Phi(\tau, \eta; v) \ll \tilde\Phi(\tau, \eta) \ll \Phi^*(\tau, \eta)$. $\qquad\square$

Lemma 8.4 *Assume $0 < a \leq \delta a_1$ and $0 < b \leq \delta b_1$ with some $0 < \delta < 1$. Then we have*

(i) $\left(1 - \eta/b - \tau/a\right)^{-1}\left(1 - \eta/b_1 - \tau/a_1\right)^{-1} \ll (1 - \delta)^{-1}\left(1 - \eta/b - \tau/a\right)^{-1}$,

(ii) $\left(1 - \eta/b\right)^{-1}\left(1 - \tau/a\right)^{-1} \ll \left(1 - \eta/b - \tau/a\right)^{-1}$.

Proof The assertion (i) follows from

$$\left\{\sum(\eta/b + \tau/a)^n\right\}\left\{\sum(\eta/b_1 + \tau/a_1)^n\right\} = \sum_{n,m}(\eta/b + \tau/a)^n(\eta/b_1 + \tau/a_1)^m$$

$$\ll \sum_{n,m}\delta^m(\eta/b + \tau/a)^{n+m}$$

$$\ll \sum_m \delta^m \sum_n(\eta/b + \tau/a)^n.$$

To examine the second assertion it is enough to note that the coefficient of $\eta^n \tau^m$ in $\sum (\eta/b)^k \sum (\tau/a)^j$ is $b^{-n} a^{-m}$ while that of $\eta^n \tau^m$ in $\sum (\eta/b + \tau/a)^k$ is $b^{-n} a^{-m}(n + m)!/(n! m!)$. □

Here we recall that if $\phi(\tau, \eta)$ is holomorphic in a neighborhood of $\{(\tau, \eta) \mid |\eta| \leq b, |\tau| \leq a\}$ then we have

$$\phi(\tau, \eta) \ll (1 - \tau/a)^{-1}(1 - \eta/b)^{-1} \sup_{|\tau|=a, |\eta|=b} |\phi(\tau, \eta)|$$

which follows from Cauchy's integral formula. Assume that $C(\tau, \eta)$ in (8.5) satisfies

$$C(\tau, \eta) \ll (1 - \tau/a_1)^{-1}(1 - \eta/b_1)^{-1} B \ll (1 - \tau/a_1 - \eta/b_1)^{-1} B.$$

Then we have

Lemma 8.5 *Assume that* $Qv = g$ *and*

$$\Phi(0, \eta; v) \ll \omega^{-1}(1 - \eta/b)^{-k}, \quad \Phi(\tau, \eta; g) \ll L(1 - \tau/a - \eta/b)^{-k} e^{M \tau \omega}.$$

We also assume that $Ba/b \leq (1 - \delta)$ *and* $B \leq (1 - \delta)M$. *Then we have*

$$\Phi(\tau, \eta; v) \ll L \omega^{-1}(1 - \tau/a - \eta/b)^{-k} e^{M \tau \omega}.$$

Proof Denote $\Phi^* = L \omega^{-1}(1 - \tau/a - \eta/b)^{-k} e^{M \tau \omega}$. Then it is easy to see by Lemma 8.4 that

$$\frac{\partial \Phi^*}{\partial \tau} \gg C(\tau, \eta) \frac{\partial \Phi^*}{\partial \eta} + C(\tau, \eta)\Phi(\tau, \eta; g).$$

Then the assertion follows from Lemma 8.3. □

We now turn to our aim, that is to estimate $V_\lambda = \sum_{n=0}^{N} v_\lambda^{(n)}$. Denote $\Phi_\lambda^n(\tau, \eta) = \Phi(\tau, \eta; v_\lambda^{(n)})$ so that $\Phi_\lambda^0(0, \eta) = 1$ and $\Phi_\lambda^n(0, \eta) = 0$ for $n \geq 1$.

Lemma 8.6 *We have for any* $\omega \geq 1$

$$\Phi_\lambda^n(\tau, \eta) \ll A^{n+1} \lambda^{-n} \sum_{k=0}^{2n} \omega^{n-k} k! \left(1 - \tau/a - \eta/b\right)^{-k-1} e^{M \tau \omega}. \tag{8.6}$$

Proof It is clear that (8.6) holds for $n = 0$. Suppose that (8.6) holds for $n = 0, \ldots, n - 1$. Denoting

$$g = \left(\sum_{j=1}^{3} \lambda^{-j} h^{(j)}\right) v_\lambda^{(n-1)} = Q_{1\lambda} v_\lambda^{(n-1)}$$

we first estimate $\Phi(\tau, \eta; g)$. Note that for $c(x)D^\alpha v_\lambda^{(n-1)}$ with $|\alpha| \leq 2$ we have

$$\Phi(\tau, \eta; cD^\alpha v_\lambda^{(n-1)}) \ll C(1 - \tau/a_1 - \eta/b_1)^{-1} \Phi(\tau, \eta; D^\alpha v_\lambda^{(n-1)})$$

$$\ll C(1 - \tau/a_1 - \eta/b_1)^{-1} \Big(\frac{\partial^2}{\partial \tau^2} + \frac{\partial^2}{\partial \tau \partial \eta} + \frac{\partial^2}{\partial \eta^2}\Big) \Phi(\tau, \eta; v_\lambda^{(n-1)}).$$

We now estimate

$$\Big(\frac{\partial^2}{\partial \tau^2} + \frac{\partial^2}{\partial \tau \partial \eta} + \frac{\partial^2}{\partial \eta^2}\Big) \sum_{k=0}^{2(n-1)} \omega^{n-1-k} k! (1 - \tau/a - \eta/b)^{-k-1} e^{M\tau\omega}$$

which is bounded by, writing $\Theta = (1 - \tau/a - \eta/b)$

$$\sum_{k=0}^{2(n-1)} M^2 \omega^{n+1-k} k \Theta^{-k-1} + 2M\omega^{n-k}(k+1)! a^{-1} \Theta^{-k-2}$$

$$+ \omega^{n-1-k}(k+2)! a^{-2} \Theta^{-k-3}$$

$$+ M\omega^{n-k}(k+1)! b^{-1} \Theta^{-k-2} + \omega^{n-1-k}(k+2)! a^{-1} b^{-1} \Theta^{-k-3}$$

$$+ \omega^{n-1-k}(k+2)! b^{-2} \Theta^{-k-3}$$

$$\ll \omega\big(M^2 + 2Ma^{-1} + a^{-2} + Mb^{-1} + a^{-1}b^{-1} + b^{-2}\big) \sum_{k=0}^{2n} \omega^{n-k} k! \Theta^{-k-1}$$

up to the factor $e^{M\tau\omega}$. Taking A so that

$$A \geq M^2 + 2Ma^{-1} + a^{-2} + Mb^{-1} + a^{-1}b^{-1} + b^{-2}$$

we conclude that

$$\Phi(\tau, \eta; g) \ll A^{n+1} \lambda^{-n} \omega \sum_{k=0}^{2n} \omega^{n-k} k! (1 - \tau/a - \eta/b)^{-k-1} e^{M\tau\omega}.$$

Note that $\Phi(0, \eta; v_\lambda^{(n)}) = \Phi_\lambda^n(0, \eta) = 0 \ll \omega^{-1}(1 - \eta/b)^{-k-1}$ for any $k \geq 0$ and $n \geq 1$ and any $\omega > 0$. Since $Q_0 v_\lambda^{(n)} = -g$ applying Lemma 8.5 we obtain the assertion (8.6). Therefore by induction on n we conclude the assertion.   □

Lemma 8.7 *There are $h > 0$ and $\delta > 0$ such that*

$$\sum_\alpha \frac{h^{|\alpha|}}{\alpha!} \sup_{|x-x^*| \leq \delta} |D^\alpha v_\lambda^{(n)}(x)| \leq B^{n+1} \lambda^{-n} \sum_{k=0}^{2n} \omega^{n-k} k! e^{M_1\omega}.$$

Proof Note that

$$\sum_\alpha \frac{\eta^{|\alpha|}}{\alpha!} |D^\alpha v_\lambda^{(n)}(x^*)| \leq A^{n+1}\lambda^{-n} \sum_{k=0}^{2n} \omega^{n-k} k! \left(1 - \eta/a - \eta/b\right)^{-k-1} e^{M\eta\omega}$$

and hence for $0 < \eta \leq \eta_0$ we have

$$\sum_\alpha \frac{\eta^{|\alpha|}}{\alpha!} |D^\alpha v_\lambda^{(n)}(x^*)| \leq B^{n+1}\lambda^{-n} \sum_{k=0}^{2n} \omega^{n-k} k! e^{M\eta_0\omega}.$$

This shows that

$$|v_\lambda^{(n)}(x)| \leq \sum_\alpha \frac{|D^\alpha v_\lambda^{(n)}(x^*)|}{\alpha!} |(x - x^*)^\alpha| \leq B^{n+1}\lambda^{-n} \sum_{k=0}^{2n} \omega^{n-k} k! e^{M_1\omega}$$

for $|x - x^*| \leq \eta_0$. From the Cauchy's inequality it follows that

$$\sup_{|x-x^*| \leq \eta_0/2} |D^\alpha v_\lambda^{(n)}(x)| \leq (\eta_0/2)^{-|\alpha|} \alpha! B^{n+1}\lambda^{-n} \sum_{k=0}^{2n} \omega^{n-k} k! e^{M_1\omega}$$

and hence we have for $2h < \eta_0$ and $2\delta < \eta_0$

$$\sum_\alpha \frac{h^{|\alpha|}}{\alpha!} \sup_{|x-x^*| \leq \delta} |D^\alpha v_\lambda^{(n)}(x)| \leq B^{n+1}\lambda^{-n} \sum_{k=0}^{2n} \omega^{n-k} k! e^{M_1\omega} \qquad (8.7)$$

with a possibly different B.   □

Let $N \in \mathbb{N}$ and $\lambda = 4NBe^\ell$ and define

$$V_\lambda(x) = \sum_{n=0}^{N} v_\lambda^{(n)}(x)$$

where $\ell \geq 1$ will be determined later. Denote $\omega = 4N$ so that $\lambda = \omega Be^\ell$. We have for $n \leq N$

$$\sum_{k=0}^{2n} \omega^{n-k} k! e^{M_1\omega} \leq \omega^n e^{M_1\omega} \sum_{k=0}^{2n} \left(\frac{k}{\omega}\right)^k \leq \omega^n e^{M_1\omega} \sum_{k=0}^{2n} \left(\frac{1}{2}\right)^k$$

and hence from (8.7)

$$\sum_\alpha \frac{h^{|\alpha|}}{\alpha!} \sup_{|x-x^*|\leq\delta} |D^\alpha v_\lambda^{(n)}(x)| \leq B^{n+1}\lambda^{-n}\omega^n e^{M_1\omega}$$

$$\leq B^{n+1}(B^{-1}e^{-\ell})^n e^{M_1\omega} = Be^{-\ell n + M_1\omega}. \tag{8.8}$$

In particular one has

$$\sum_\alpha \frac{h^{|\alpha|}}{\alpha!} \sup_{|x-x^*|\leq\delta} |D^\alpha v_\lambda^{(N)}(x)| \leq Be^{-\ell N + 4M_1 N} = Be^{-e^{-\ell}(\ell - 4M_1)\lambda/4B}. \tag{8.9}$$

On the other hand, from (8.8) it follows

$$\sum_\alpha \frac{h^{|\alpha|}}{\alpha!} \sup_{|x-x^*|\leq\delta} |D^\alpha V_\lambda(x)| \leq \sum_{n=0}^N B^{n+1}\lambda^{-n}\omega^n e^{M_1\omega}$$

$$= e^{M_1\omega}B\sum_{n=0}^N \left(\frac{B\omega}{\lambda}\right)^n \leq e^{M_1\omega}B = Be^{e^{-\ell}M_1\lambda/B}. \tag{8.10}$$

8.4   A Priori Estimates in the Gevrey Classes

In this section assuming that the Cauchy problem for P is $\gamma^{(s)}$ well-posed near the origin we derive a priori estimates following [38, 43]. Let L be a compact set in $\mathbb{R}^3$ and recall the definition of $\gamma_0^{(s),h}(L)$ (Definition 6.3) which is a Banach space equipped with the norm

$$\sum_\alpha \sup_x \frac{h^{|\alpha|}|\partial_x^\alpha f(x)|}{(\alpha!)^s}.$$

From now on, we fix $h > 0$ and $\delta > 0$ so that Lemma 8.7 holds and hence we have (8.9) and (8.10). Making a change of coordinates system $x_0 = \lambda^{-\sigma_0}y_0$, $x_1 = \lambda^{-\sigma_1}y_1$, $x_2 = \lambda^{-\sigma_2}y_2$ we consider the resulting P_λ such that

$$(P_\lambda u)(\lambda^\sigma x) = Pv, \quad v(x) = u(\lambda^\sigma x), \quad u \in C_0^\infty(\mathbb{R}^3)$$

where $\lambda^\sigma x = (\lambda^{-\sigma_0}x_0, \lambda^{-\sigma_1}x_1, \lambda^{-\sigma_2}x_2)$ and $\sigma_j \geq 0$. Then we have

Lemma 8.8 *Assume that the Cauchy problem for P is $\gamma^{(s)}$ well-posed near the origin and let $h > 0$ and W be a compact set in $\mathbb{R}^3$. Then there are $c > 0$, $C > 0$ such that for any $u \in \gamma_0^{(s)}(W_0)$ with some $1 < \tilde{s} < s$, any $t > 0$, $\tau > 0$, any $1 < s' < s$ and any $1 < \kappa < s$ we have*

$$|u|_{C^0(W')} \le C \exp\left(c(\lambda^{\sigma_0}/\tau)^{1/(s-\kappa)}\right) \exp\left(\lambda^{\tilde{\sigma}/s'}\right) \sum_\alpha \sup_{x_0 \le t+\tau} \frac{h^{|\alpha|}|\partial_x^\alpha P_\lambda u|}{(\alpha!)^{s-s'}}$$

where $\bar{\sigma} = \max_j\{\sigma_j\}$ and $W_0 = W \cap \{x_0 \ge 0\}$ (the inequality is understood to be trivial if the right-hand side is divergent).

Proof Assume that the Cauchy problem for P is $\gamma^{(s)}$ well-posed near the origin. Let $h > 0$ and K be a compact neighborhood of the origin. From an analogue of Proposition 6.5, where the Cauchy data is replaced by the forcing term f, it follows that there exists a neighborhood D of the origin such that for any $f(x) \in \gamma_0^{(s),h}(K_0)$, $K_0 = K \cap \{x_0 \ge 0\}$, there is a unique $u \in C^2(D)$ satisfying $Pu = f$ in D vanishing in $x_0 \le 0$. Replacing K by a smaller one if necessary we assume that $K \subset D$. Since the graph of the linear map $f \in \gamma_0^{(s),h}(K_0) \mapsto u \in C^2(D)$ is closed then from Banach's closed graph theorem it follows that this map is continuous (see [66, Theorem 4.4] for example). Then for any compact set $L \subset D$ there is $C > 0$ such that

$$|u|_{C^0(L)} \le C \sum_\alpha \sup \frac{h^{|\alpha|}|\partial_x^\alpha f(x)|}{(\alpha!)^s}.$$

Applying this inequality with $L = K$ and $f = Pu \in \gamma_0^{(s),h}(K_0)$ for $u \in \gamma_0^{(s),h}(K_0)$ one has

$$|u|_{C^0(K)} \le C \sum_\alpha \sup \frac{h^{|\alpha|}|\partial_x^\alpha Pu|}{(\alpha!)^s}, \quad \forall u \in \gamma_0^{(s),h}(K_0).$$

Take $\chi(r) \in \gamma^{(\kappa)}(\mathbb{R})$, $1 < \kappa < s$ such that $\chi(r) = 1$ for $r \le 0$, $\chi(r) = 0$ for $r \ge 1$ and set $\chi_1(x_0) = \chi((x_0 - t)/\tau)$ so that

$$\begin{cases} \chi_1(x_0) = 1, & x_0 \le t, \\ \chi_1(x_0) = 0, & x_0 \ge t + \tau. \end{cases}$$

Let $u \in \gamma_0^{(s),h}(K_0)$ and consider $\chi_1 Pu \in \gamma_0^{(s),h}(K_0)$. As noted above there exists a unique solution $v \in C^2(D)$ to $Pv = \chi_1 Pu$ vanishing in $x_0 \le 0$. Since $Pv = Pu$ in $x_0 \le t$ by the uniqueness of solution one has $u = v$ in $x_0 \le t$ and hence

$$|u|_{C^0(K')} = |v|_{C^0(K')} \le C \sum_\alpha \sup \frac{h^{|\alpha|}|\partial_x^\alpha(\chi_1 Pu)|}{(\alpha!)^s}.$$

Since $\chi_1 = 0$ for $x_0 \geq t + \tau$ one has

$$\sum_\alpha \sup \frac{h^{|\alpha|}|\partial_x^\alpha(\chi_1 Pu)|}{(\alpha!)^s} \leq \sum \sup \frac{\alpha!}{\alpha_1!\alpha_2!} \frac{h^{|\alpha|}|\partial_x^{\alpha_1}\chi_1||\partial_x^{\alpha_2}Pu|}{(\alpha!)^s}$$

$$\leq \sum \sup \frac{h^{|\alpha|}|\partial_x^{\alpha_1}\chi_1||\partial_x^{\alpha_2}Pu|}{(\alpha_1!)^s(\alpha_2!)^s}$$

$$\leq \sum_{\alpha_1} \sup \frac{h^{|\alpha_1|}|\partial_x^{\alpha_1}\chi_1|}{(\alpha_1!)^s} \sum_{\alpha_2} \sup_{x_0 \leq t+\tau} \frac{h^{|\alpha_2|}|\partial_x^{\alpha_2}Pu|}{(\alpha_2!)^s}.$$

Note that $|\partial_x^\beta \chi_1(x)| \leq C^{|\beta|+1}(\beta!)^\kappa \tau^{-|\beta|}$ and then

$$\sum_{\alpha_1} \sup \frac{h^{|\alpha_1|}|\partial_x^{\alpha_1}\chi_1|}{(\alpha_1!)^s} \leq \sum_{\alpha_1} \frac{C^{|\alpha_1|+1}\tau^{-|\alpha_1|}h^{|\alpha_1|}}{(\alpha_1!)^{s-\kappa}}$$

$$\leq C \exp\left(c(1/\tau)^{1/(s-\kappa)}\right) \sum_{\alpha_1}(Ch)^{|\alpha_1|} \leq C_h \exp\left(c(1/\tau)^{1/(s-\kappa)}\right)$$

so that we have

$$|u|_{C^0(K^t)} \leq C \exp\left(c(1/\tau)^{1/(s-\kappa)}\right) \sum_\alpha \sup_{x_0 \leq t+\tau} \frac{h^{|\alpha|}|\partial_x^\alpha Pu|}{(\alpha!)^s}. \tag{8.11}$$

Let $u \in \gamma_0^{(s)}(W_0)$ then it is clear that $v(x) = u(\lambda^\sigma x) \in \gamma_0^{(s),h}(K_0)$ for large λ since

$$\sum_\alpha \sup \frac{h^{|\alpha|}|\partial_x^\alpha v(x)|}{(\alpha!)^s} \leq \sum_\alpha \sup \frac{h^{|\alpha|}\lambda^{\tilde\sigma|\alpha|}|(\partial_x^\alpha u)(\lambda^\sigma x)|}{(\alpha!)^{\tilde s}(\alpha!)^{s-\tilde s}}$$

$$\leq \sup_\alpha \left(\frac{\lambda^{\tilde\sigma|\alpha|/(s-\tilde s)}}{\alpha!}\right)^{s-\tilde s} \sum_\alpha \sup \frac{h^{|\alpha|}|(\partial_x^\alpha u)(\lambda^\sigma x)|}{(\alpha!)^{\tilde s}}$$

$$\leq \exp\left((s-\tilde s)\lambda^{\tilde\sigma/(s-\tilde s)}\right) \sum_\alpha \sup \frac{h^{|\alpha|}|\partial_x^\alpha u(x)|}{(\alpha!)^{\tilde s}}.$$

For $v(x) = u(\lambda^\sigma x)$ we apply the inequality (8.11) with $t = \lambda^{-\sigma_0}\hat t$, $\tau = \lambda^{-\sigma_0}\hat\tau$ and $K = \{\lambda^{-\sigma}x \mid x \in W\}$. Since $Pv = Pu(\lambda^\sigma x) = (P_\lambda u)(\lambda^\sigma x)$ so $\partial_x^\alpha[(P_\lambda u)(\lambda^\sigma u)] = \lambda^{\langle\sigma,\alpha\rangle}(\partial_x^\alpha P_\lambda u)(\lambda^\sigma x)$ we obtain

$$|u|_{C^0(W^{\hat t})} \leq C \exp\left(c(\lambda^{\sigma_0}/\hat t)^{1/(s-\kappa)}\right) \sum_\alpha \sup_{x_0 \leq \hat t+\hat\tau} \frac{h^{|\alpha|}\lambda^{\tilde\sigma|\alpha|}|\partial_x^\alpha(P_\lambda u)(x)|}{(\alpha!)^{s'}(\alpha!)^{s-s'}}$$

$$\leq C \exp\left(c(\lambda^{\sigma_0}/\hat t)^{1/(s-\kappa)}\right) \exp\left(c\lambda^{\tilde\sigma/s'}\right) \sum_\alpha \sup_{x_0 \leq \hat t+\hat\tau} \frac{h^{|\alpha|}|\partial_x^\alpha(P_\lambda u)(x)|}{(\alpha!)^{s-s'}}$$

which proves the assertion. $\qquad\qquad\square$

8.5   Proof of Ill-Posed Results

Let W be a compact neighborhood of x^*. Take $1 < \kappa < s$ which will be made precise later and choose $\chi(x) \in \gamma_0^{(\kappa)}(W_0)$ such that $\chi(x) = 1$ in a neighborhood of x^* and supported in $\{x \mid |x - x^*| \leq \delta\}$. We denote $\tilde{U}_\lambda = E_\lambda V_\lambda \chi \in \gamma_0^{(\kappa)}(W_0)$ and note that $|\tilde{U}_\lambda(x^*)| = 1$. Then we have from (8.4)

$$
\begin{aligned}
P_\lambda \tilde{U}_\lambda &= (P_\lambda E_\lambda V_\lambda)\chi + \sum_{|\alpha| \leq 1, 1 \leq |\beta| \leq 2} c_{\alpha\beta}(x, \lambda)\partial_x^\alpha (E_\lambda V_\lambda)\partial_x^\beta \chi \\
&= E_\lambda Q_{1\lambda} v_\lambda^{(N)} \chi + \sum_{|\alpha| \leq 1, 1 \leq |\beta| \leq 2} c_{\alpha\beta}(x, \lambda)\partial_x^\alpha (E_\lambda V_\lambda)\partial_x^\beta \chi.
\end{aligned}
\tag{8.12}
$$

To estimate the second term on the right-hand side we first remark

Lemma 8.9 *Assume that $f(x)$ satisfies $|\partial_x^\alpha f(x)| \leq C_1^{|\alpha|+1}|\alpha|!$ for any α and $x \in K$ and denote $\omega^\alpha(x) = e^{-\lambda f(x)}\partial_x^\alpha e^{\lambda f(x)}$. Then there is C independent of λ such that one has*

$$
|\partial_x^\beta \omega^\alpha(x)| \leq C^{|\alpha+\beta|+1} \sum_{j=0}^{|\alpha|} (C\lambda)^{|\alpha|-j}(|\beta|+j)!, \quad x \in K.
$$

In particular one has

$$
|\partial_x^\alpha e^{\lambda f(x)}| \leq C^{|\alpha|+1}(|\alpha|+\lambda)^{|\alpha|} e^{\lambda \mathrm{Re} f(x)}, \quad x \in K.
$$

Proof We prove the assertion by induction on $|\alpha|$. When $|\alpha| = 0$ the assertion is obvious. For $|\alpha| = \ell \geq 1$ assume that there exist $C > 0, A_1 > 0, A_2 > 0$ such that we have

$$
|\partial_x^\beta \omega^\alpha(x)| \leq C A_1^{|\beta|} A_2^{|\alpha|} \sum_{j=0}^{|\alpha|} (C\lambda)^{|\alpha|-j}(|\beta|+j)!.
\tag{8.13}
$$

For $|e| = 1$ note that $\omega^{\alpha+e} = \partial_x^e \omega^\alpha + \lambda(\partial_x^e f)\omega^\alpha$. Since we can assume $C \geq C_1 \geq 1$ it follows from (8.13) that

$$
|\partial_x^\beta \omega^{\alpha+e}| = \left| \partial_x^{\beta+e}\omega^\alpha + \lambda \sum \binom{\beta}{\beta'} \partial_x^{\beta'+e} f \partial_x^{\beta-\beta'}\omega^\alpha \right|
$$

$$
\leq C A_1^{|\beta|+1} A_2^{|\alpha|} \sum_{j=0}^{|\alpha|} (C\lambda)^{|\alpha|-j}(|\beta|+1+j)!
$$

$$
+ \sum \binom{\beta}{\beta'} C^{|\beta'|+1}(|\beta'|+1)! C A_1^{|\beta-\beta'|} A_2^{|\alpha|} \sum_{j=0}^{|\alpha|} (C\lambda)^{|\alpha|+1-j}(|\beta-\beta'|+j)!
$$

which is bounded by

$$A_1 A_2^{-1} C A_1^{\beta|} A_2^{|\alpha|+1} \sum_{j=0}^{|\alpha|+1} (C\lambda)^{|\alpha|+1-j} (|\beta| + j)!$$

$$+ 2C^2 A_1 (A_1 - 2C)^{-1} A_1^{|\beta|} A_2^{|\alpha|} \sum_{j=0}^{|\alpha|} (C\lambda)^{|\alpha|+1-j} (|\beta| + j)!$$

where we have used

$$\sum \binom{\beta}{\beta'} C^{|\beta'|+1} (|\beta'| + 1)! A_1^{|\beta-\beta'|} (|\beta - \beta'| + j)!$$

$$\leq \sum_{k=0}^{|\beta|} \binom{|\beta|}{k} (2C)^{k+1} k! A_1^{|\beta|-k} (|\beta| - k + j)! \leq 2CA_1 (A_1 - 2C)^{-1} A_1^{|\beta|} (|\beta| + j)!.$$

Choosing $A_1 > 2C$, A_2 so that $A_1 A_2^{-1} C (1 + 2C(A_1 - 2C)^{-1}) \leq 1$ the assertion holds for $|\alpha| = \ell + 1$ and hence for all α. The second assertion follows immediately because $\sum_{j=0}^{|\alpha|} \lambda^{|\alpha|-j} j! \leq (|\alpha| + \lambda)^{|\alpha|}$. □

Lemma 8.10 *There exists $c > 0$ such that*

$$\sum_\alpha \sup_{x \in K} \frac{h^{|\alpha|} |\partial_x^\alpha E_\lambda|}{(\alpha!)^s} \leq C \exp \left(c\lambda^{2/s} + \lambda \sup_{x \in K} \{ -\mathrm{Im}\, \phi(x) \} \right).$$

Proof Recall that $E_\lambda = \exp(i\lambda^2 x_2 + i\lambda \phi(x))$. Since $\phi(x)$ is real analytic in a neighborhood K of x^* then it follows from Lemma 8.9 that

$$|\partial_x^\alpha E_\lambda| \leq C^{|\alpha|+1} (\lambda^2 + |\alpha|)^{|\alpha|} e^{-\lambda \mathrm{Im}\, \phi(x)}, \quad x \in K. \tag{8.14}$$

Noting $h^{|\alpha|} (\lambda^2 + |\alpha|)^{|\alpha|} / (\alpha!)^s \leq C e^{c\lambda^{2/s}}$ we get the assertion from (8.14). □

From Lemma 8.1 there exist $\nu > 0$ and $\bar{\tau} > 0$ such that $-\mathrm{Im}\, \phi(x) \leq -\nu$ if $x \in \mathrm{supp}\, (\partial_x^\beta \chi) \cap \{x_0 \leq t + \tau\}$, $0 < \tau \leq \bar{\tau}$, $|\beta| \geq 1$. Then from Lemma 8.10 and (8.10) it follows that

$$\sum_\gamma \sup_{x_0 \leq t+\tau} \frac{h^{|\gamma|} |\partial_x^\gamma (\partial_x^\alpha (E_\lambda V_\lambda) \partial_x^\beta \chi)|}{(\gamma!)^{\tilde{s}}} \leq C \exp (c\lambda^{2/\tilde{s}} - \nu\lambda + e^{-\ell} M_1 B^{-1} \lambda). \tag{8.15}$$

We turn to estimate the first term on the right-hand side of (8.12); $E_\lambda Q_{1\lambda} v_\lambda^{(N)} \chi$. Thanks to Lemma 8.1 we have $-\mathrm{Im}\, \phi(x) \leq 2a\tau$ if $x \in \mathrm{supp}\, \chi \cap \{x_0 \leq t + \tau\}$ where

$a = |\mathrm{Im}\,\phi_{x_0}(x^*)|$. Thus from Lemma 8.10 and (8.9) it follows that

$$\sum_\alpha \sup_{x_0 \leq t + \tau} \frac{h^{|\alpha|} |\partial_x^\alpha (E_\lambda Q_{1\lambda} v_\lambda^{(N)} \chi)|}{(\alpha!)^{\tilde{s}}} \tag{8.16}$$

$$\leq C \exp \left(c\lambda^{2/\tilde{s}} + 2a\tau\lambda - e^{-\ell}(\ell - 4M_1)(4B)^{-1}\lambda \right).$$

Take ℓ large so that $e^{-\ell} M_1 B^{-1} < \nu$ and $\ell > 4M_1$ then choose $\tau > 0$ such that

$$2a\tau - e^{-\ell}(\ell - 4M_1)(4B)^{-1} < 0.$$

It is clear that the right-hand side of (8.16) decays exponentially in λ if $\tilde{s} > 2$. Therefore if $s - s' > 2$ then from (8.15) and (8.16) there is $\nu_1 > 0$ such that

$$\sum_\alpha \sup_{x_0 \leq t + \tau} \frac{h^{|\alpha|} |\partial_x^\alpha (P_\lambda \tilde{U}_\lambda)|}{(\alpha!)^{s-s'}} \leq Ce^{-\nu_1 \lambda}.$$

We now assume $s > 6$. Recalling $\sigma_0 = 1$, $\sigma_1 = 2$, $\sigma_2 = 4$ and hence $\bar{\sigma} = 4$ then we can choose $s' > 4$ such that $s - s' > 2$ and $\bar{\sigma}/s' < 1$. Taking $1 < \kappa < s$ so that $\sigma_0/(s - \kappa) < 1$ we now apply Lemma 8.8 to conclude with some $c > 0$ that

$$|\tilde{U}_\lambda|_{C^0(W')} \leq Ce^{-c\lambda + o(\lambda)}, \quad \lambda \to \infty.$$

This gives a contradiction because $|\tilde{U}_\lambda(x^*)| = 1$. This completes the proof of Theorem 8.1.

The method we employed here can be applied to prove Proposition 6.9 so we give a sketch of the proof. Without restrictions we can assume $n = 2$ so that we consider

$$P = -D_0^2 + 2x_1 D_0 D_2 + D_1^2 + AD_2, \quad x = (x_0, x_1, x_2)$$

with $A \in \mathbb{C} \setminus \mathbb{R}_+$. We make a change of variables; $x_0 = y_0, x_1 = \lambda^{-1} y_1, x_2 = \lambda^{-2} y_2$. Dividing the resulting operator by λ^2 we obtain

$$P_\lambda = -\lambda^{-2} D_0^2 + 2\lambda^{-1} x_1 D_0 D_2 + D_1^2 + AD_2.$$

With $E_\lambda = \exp(i\lambda^2 x_2 + i\lambda\phi(x))$ we have

$$\lambda^{-1} E_\lambda^{-1} P_\lambda E_\lambda = \lambda \{ 2x_1 \phi_{x_0} + \phi_{x_1}^2 + A \}$$

$$+ \{ 2x_1 D_0 + 2\phi_{x_1} D_1 + 2x_1 \phi_{x_0} \phi_{x_1} + A\phi_{x_2} - i\phi_{x_1 x_1} \}$$

$$+ \lambda^{-1} h^{(1)} + \lambda^{-2} h^{(2)} + \lambda^{-3} h^{(3)}.$$

One can construct a family V_λ such that $E_\lambda^{-1} P E_\lambda V_\lambda = O(\lambda^{-\infty})$ as exactly the same way in Sect. 8.2. With $\tilde{U}_\lambda = E_\lambda V_\lambda \chi$ from Lemma 8.8 one concludes

$$|\tilde{U}_\lambda|_{C^0(W')} \le C \exp\left(c\tau^{-1/(s-\kappa)}\right) \exp\left(\lambda^{\bar{\sigma}/s'}\right) \sum_\alpha \sup_{x_0 \le t+\tau} \frac{h^{|\alpha|} |\partial_x^\alpha P_\lambda \tilde{U}_\lambda|}{(\alpha!)^{s-s'}}$$

where $\bar{\sigma} = 2$. Let $s > 4$. Then we can choose $s' > 2$ so that $s - s' > 2$ and $\bar{\sigma}/s' < 1$ and hence we conclude that

$$|\tilde{U}_\lambda|_{C^0(W')} \le C \exp\left(-c\lambda + o(\lambda)\right)$$

as $\lambda \to \infty$. This is a contradiction.

8.6 Non Strict IPH Condition, An Example

In this section, exhibiting an example, we show that the IPH condition is not sufficient *in general* for the Cauchy problem to be C^∞ well-posed for differential operators verifying (1.5). Consider the following differential operator

$$P(x, D) = -D_0^2 + \sum_{j=1}^{\ell} \mu_j(x_j^2 D_n^2 + D_j^2) + b(x)D_n \tag{8.17}$$

where $\ell \le n - 1$ and μ_j are positive constants and $b(x)$ is a C^∞ function defined near the origin which will be made precise below. We consider the Cauchy problem for P near the origin of $\mathbb{R}^{n+1}$;

$$\begin{cases} P(x, D)u(x) = 0 & \text{in} \quad \omega \cap \{x_0 > \tau\}, \\ u(\tau, x') = 0, \quad D_0 u(\tau, x') = \phi(x') & \text{on} \quad \omega \cap \{x_0 = \tau\} \end{cases} \tag{8.18}$$

for small $|\tau|$ where ω is a neighborhood of the origin. It is easy to check that P satisfies the IPH condition on Σ if and only if the following condition holds;

$$b(x) \in \mathbb{R}, \quad |b(x)| \le \sum_{j=1}^{k} \mu_j \quad \text{near} \quad x = 0. \tag{8.19}$$

Our aim in this section is to prove

Proposition 8.2 *There exist a C^∞ function $b(x) = b(x_0)$, depending only on x_0 defined near $x_0 = 0$ satisfying (8.19) such that the Cauchy problem for P is C^∞ ill-posed near the origin.*

On the other hand in [89] the C^∞ well-posedness of the Cauchy problem (8.18) is proved assuming that the set $\{|b(x)| = \sum_{j=1}^{k} \mu_j\}$ is not empty but satisfies some conditions. In our case $\sum_{j=1}^{k} \mu_j - |b(x_0)|$ vanishes of infinite order at some point and does not satisfy the assumption of [89, Theorem B] (well-posedness is proved in a more general setting, see [89, Theorem 4.5]).

For the proof of Proposition 8.2 we note

Lemma 8.11 *Let $K_1 \Subset K_2 \subset \omega$ be two compact sets. Then there are $\delta > 0$, $\epsilon > 0$ such that for any $f(x) \in C_0^\infty(K_1 \cap \{x_0 \geq \tau\})$, $|\tau| < \epsilon$ and $u \in C^2(\omega)$, vanishing in $x_0 \leq \tau$ satisfying $Pu = f$ in ω we have $\operatorname{supp} u \cap \{x \mid \tau \leq x_0 \leq \tau + \delta\} \subset K_2$.*

Proof Note that $P^* = -D_0^2 + \sum_{j=1}^{\ell} \mu_j(x_j^2 D_n^2 + D_j^2) - b(x_0)D_n$ is a second order differential operator with coefficients which are analytic in x' and C^∞ in x_0. Then from [36] we know that for any compact sets $\tilde{K}_1 \Subset \tilde{K}_2 \subset \omega$ there is $\tilde{\delta} > 0$ such that for any $g \in \gamma_0^{(2)}(\tilde{K}_1 \cap \{x_0 \leq \tilde{\tau}\})$ there exists $v \in C^\infty$, vanishing in $x_0 \geq \tilde{\tau}$, and satisfying $P^*v = g$ in ω and $\operatorname{supp} v \cap \{x \mid x_0 - \tilde{\delta} \leq x_0 \leq \tilde{\tau}\} \subset \tilde{K}_2$. With the use of the standard Holmgren's arguments (arguments we have employed in the end of Sect. 5.7) we conclude the assertion.   □

Lemma 8.12 *Assume that the Cauchy problem for P is C^∞ well-posed near the origin. Then there exist $\delta > 0$ and $r \in \mathbb{N}$ such that for any small $\epsilon > 0$ there is $C > 0$ such that if $u_\tau(x)$ satisfies (8.18) in $\omega \cap \{\tau < x_0 < \epsilon\}$ with $\phi(x') \in C_0^\infty(\{|x'| \leq \delta\})$ and $|\tau| < \epsilon$ then we have*

$$|u_\tau|_{C^1(\omega \cap \{\tau < x_0 < \epsilon\})} \leq C|\phi|_{C^r(\{|x'| \leq \delta\})} \tag{8.20}$$

where $|u|_{C^r(B)} = \sup_{x \in B, |\alpha| \leq r} |\partial_x^\alpha u(x)|$.
Postponing the proof of this lemma we give a proof of Proposition 8.2.

Proof of Proposition 8.2 In [14] Colombini and Spagnolo have constructed a C^∞ function $a(x_0)$ on $(-\infty, \rho]$ (here we take $a(x_0) = \tilde{a}(\rho - x_0)$ where $\tilde{a}(x_0)$ is given in [14]) vanishing in $(-\infty, 0]$, strictly positive on $(0, \rho]$, where ρ is any positive constant given beforehand, and a sequence of solutions $v_k(x_0)$ to the ordinary differential equations

$$\frac{d^2 v_k}{dx_0^2} + h_k^2 a(x_0) v_k = 0$$

and a sequence of positive numbers t_k tending to 0 such that for every $\epsilon > 0$ and $p \in \mathbb{N}$ there is $C(\epsilon, p)$ verifying

$$\begin{cases} |v_k(\epsilon)|, \quad |D_0 v_k(\epsilon)| \leq C(\epsilon, p) h_k^{-p}, \quad k = 1, 2, \ldots, \\ |v_k(t_k)| h_k^{-p} \to \infty \quad \text{as} \quad k \to \infty. \end{cases} \tag{8.21}$$

Here $h_k \in \mathbb{N}$ such that $h_k \to \infty$ as $k \to \infty$ and $\sum_{k=1}^{\infty} h_k^{-1}$ is convergent. Define $b(x_0)$ by

$$b(x_0) = \sum_{j=1}^{\ell} \mu_j - a(x_0)$$

then by virtue of the non-negativity of $a(x_0)$ it is clear that $b(x_0)$ verifies (8.19) in $(-\bar{\rho}, \infty)$ with some positive $\bar{\rho} > 0$. Define

$$V_k(x) = v_k(x_0) e^{i x_n h_k^2} \prod_{j=1}^{\ell} e^{-x_j^2 h_k^2/2}$$

then it is clear that $P^* V_k = 0$.

Now we suppose that the Cauchy problem for P is C^∞ well-posed near the origin. In view of Lemma 1.1 there exists a solution $u_k(x) \in C^2(\omega)$ to the Cauchy problem (8.18) with $\tau = t_k$ and $\phi \in C_0^\infty(\{|x'| \leq \delta\})$. If $\delta > 0$ and $\epsilon > 0$ is small then thanks to Lemma 8.11 one can assume that supp $u_k \cap \{t_k \leq x_0 \leq \epsilon\} \subset \omega$ and Lemma 8.12 holds. We now observe the integral

$$\int_{t_k}^{\epsilon} (P u_k, V_k) dx_0 = 0.$$

From integration by parts it follows that

$$(\phi, V_k(t_k)) = (D_0 u_k(\epsilon), V_k(\epsilon)) - (u_k(\epsilon), D_0 V_k(\epsilon)). \tag{8.22}$$

From (8.20) and (8.21) with $p = -1$ it follows that the right-hand side of (8.22) converges to 0 as $k \to \infty$ and hence so does $(\phi, V_k(t_k))$. We show that one can choose ϕ contradicting this fact. Choose $\phi(x') = \theta(x_n) \psi(x'') \prod_{j=1}^{\ell} \theta_1(x_j)$ with θ, $\theta_1 \in C_0^\infty(\mathbb{R})$ where $\theta_1(0) = 1$ and $\psi(x'') \in C_0^\infty(\mathbb{R}^{n-\ell-1})$, $x'' = (x_{\ell+1}, \dots, x_{n-1})$ such that $\int_{\mathbb{R}^{n-\ell-1}} \psi(x'') dx'' = 1$ and the support of ϕ is contained in $\{|x'| \leq \delta\}$. Then $(\phi, V_k(t_k))$ turns to be

$$v_k(t_k) \hat{\theta}(h_k^2) \prod_{j=1}^{\ell} \int_{\mathbb{R}} \theta_1(x_j) e^{-x_j^2 h_k^2/2} dx_j$$

where $\hat{\theta}$ is the Fourier transform of θ. Remarking $h_k \int_{\mathbb{R}} \theta_1(t) e^{-t^2 h_k^2/2} dt \to (2\pi)^{1/2}$ as $k \to \infty$ we would have

$$|v_k(t_k)| |\hat{\theta}(h_k^2)| h_k^{-\ell} \to 0 \quad \text{as } k \to \infty. \tag{8.23}$$

Since $|v_k(t_k)|h_k^{-p} \to \infty$ as $k \to \infty$ for any $p \in \mathbb{N}$ it is clear that we can choose $\theta \in C_0^\infty(\mathbb{R})$ of arbitrarily small support which does not satisfy (8.23). In fact it is enough to take

$$\theta(s) = \sum_k |v_k(t_k)|^{-1} h_k^\ell e^{ish_k^2} \alpha(s), \quad \alpha(s) = \beta(s) * \bar{\beta}(s)$$

where $\beta \in C_0^\infty(\mathbb{R})$ has support in a small neighborhood of 0 such that $\int \beta(s)ds = 1$. It is clear from (8.21) that θ is $C^\infty(\mathbb{R})$ and supported in a small neighborhood of 0. Noting $\hat{\alpha}(\zeta) = |\hat{\beta}(\zeta)|^2 \geq 0$ and $\hat{\alpha}(0) = 1$ we have

$$\hat{\theta}(h_k^2) \geq |v_k(t_k)|^{-1} h_k^\ell$$

which contradicts (8.23).   □

It remains to prove Lemma 8.12. We first prove the following lemma

Lemma 8.13 *Assume that the Cauchy problem for P is C^∞ well-posed near the origin. Then there exist a compact neighborhood K of the origin and $p \in \mathbb{N}$ such that for any $\epsilon > 0$ there is $C > 0$ such that for any $u \in C_0^\infty(K \cap \{x_0 \geq -\epsilon\})$ one has*

$$|u|_{C^1(K \cap \{x_0 < \epsilon\})} \leq C|Pu|_{C^p(K \cap \{x_0 < \epsilon\})}.$$

Proof In view of Lemma 8.11 we can take compact neighborhoods of the origin $K \Subset \tilde{K} \subset \omega$ and $\epsilon_1 > 0$ such that for the solution u to $Pu = f$ in ω vanishing in $x_0 \leq -\epsilon_1$ with $f \in C_0^\infty(\tilde{K} \cap \{x_0 \geq -\epsilon_1\})$ we have $\operatorname{supp} u \cap \{-\epsilon_1 \leq x_0 \leq \epsilon_1\} \subset \omega$. We consider the mapping $T : C_0^\infty(\tilde{K} \cap \{x_0 \geq -\epsilon_1\}) \ni f \mapsto u \in C^\infty(\omega)$ where u is the uniquely determined solution to $Pu = f$ vanishing in $x_0 < -\epsilon_1$. It is easy to check that the graph of T is closed. Then from Banach's closed graph theorem it follows that T is continuous. This proves that there exist $q \in \mathbb{N}$ and $C > 0$ such that for $u \in C^\infty(\omega)$ vanishing in $x_0 < -\epsilon_1$ and satisfying $Pu = f$ in ω with $f \in C_0^\infty(\tilde{K} \cap \{x_0 \geq -\epsilon_1\})$ one has

$$|u|_{C^2(\tilde{K})} \leq C|f|_{C^q(\tilde{K} \cap \{x_0 \geq -\epsilon_1\})}. \tag{8.24}$$

Note that from the continuity it follows that for any $f \in C_0^q(\tilde{K} \cap \{x_0 \geq -\epsilon_1\})$ there exists $u \in C^2(\tilde{K})$ vanishing in $x_0 < -\epsilon_1$ and satisfies $Pu = f$ in the interior of $\tilde{K}$ and (8.24) remains valid for this solution. Let $0 < \epsilon < \epsilon_1$ be fixed. Then there exists $B > 0$ such that for any $g \in C_0^\infty(K)$ one can find $\tilde{g} \in C_0^q(\tilde{K})$ which coincides with g in $x_0 < \epsilon$ and $|\tilde{g}|_{C^q(\tilde{K})} \leq B|g|_{C^q(K \cap \{x_0 \leq \epsilon\})}$. Indeed it suffices to replace g in $x_0 \geq \epsilon$ by

$$\tilde{g}(x_0, x') = \chi(x_0) \sum_{\nu=1}^{q+1} a_\nu g(-\nu(x_0 - \epsilon) + \epsilon, x')$$

where $(a_1, \ldots, a_{q+1})$ verifies $1 = \sum_{\nu=1}^{q+1}(-\nu)^j a_j$ for $j = 0, 1, \ldots, q$ and $\chi \in C^\infty(\mathbb{R})$ is 1 in $x_0 \leq \epsilon$ and 0 for $x_0 \geq \epsilon + \epsilon'$ where we choose $\epsilon' > 0$ small so that $\tilde{g} \in C_0^q(\tilde{K})$. Let $u \in C_0^\infty(K \cap \{x_0 \geq -\epsilon_1\})$ and denote by $\widetilde{Pu}$ such extension of Pu. Recall that there exists $w \in C^2(\tilde{K})$ such that $Pw = \widetilde{Pu}$ on K vanishing in $x_0 < -\epsilon_1$. Since $u = w$ in $x_0 < \epsilon$ we conclude that

$$|u|_{C^1(K \cap \{x_0 < \epsilon\})} = |w|_{C^1(K \cap \{x_0 < \epsilon\})} \leq |w|_{C^2(\tilde{K})}$$
$$\leq C|\widetilde{Pu}|_{C^q(\tilde{K})} \leq CB|Pu|_{C^q(K \cap \{x_0 \leq \epsilon\})}$$

which proves the assertion. □

Proof of Lemma 8.12 Let $K \subset \omega$ be the compact neighborhood of the origin in Lemma 8.13. Then choosing $\epsilon_1 > 0$ and $\delta > 0$ small, in view of Lemma 8.11, one can assume that $\text{supp} \, u_\tau \cap \{\tau \leq x_0 \leq \epsilon_1\} \subset K$ for the solution u_τ to (8.18) with $\phi \in C_0^\infty(\{|x'| \leq \delta\})$ and $|\tau| < \epsilon_1$. Let $0 < \epsilon < \epsilon_1$ be small and take $\chi(x_0) \in C^\infty(\mathbb{R})$ such that $\chi = 1$ in $x_0 \leq \epsilon$ and 0 for $x_0 \geq \epsilon_1$. Denote $w = u_\tau - i(x_0 - \tau)\phi(x')$ then it is easy to check that $D_0^j w(\tau, x') = 0$ for all $j \in \mathbb{N}$. Hence extending w in $x_0 < \tau$ to be 0 it is clear that $\chi w \in C_0^\infty(K)$. We now consider $P(\chi w) = [P, \chi]w + \chi Pw$. Thanks to Lemma 8.13 we have

$$|u_\tau - i(x_0 - \tau)\phi|_{C^1(K \cap \{\tau < x_0 < \epsilon\})} \leq C|P(x_0 - \tau)\phi|_{C^p(K \cap \{\tau < x_0 < \epsilon\})}$$

since $[P, \chi]w = 0$ and $Pw = -iP(x_0 - \tau)\phi$ in $\tau \leq x_0 \leq \epsilon$. This proves the assertion immediately. □

References

1. A. Barbagallo, V. Esposito, A global existence and uniqueness result for a class of hyperbolic operators. Ricerche Mat. **63**, 25–40 (2014)
2. R. Beals, A general calculus of pseudodifferential operators. Duke Math. J. **42**, 1–42 (1975)
3. E. Bernardi, A. Bove, Geometric results for a class of hyperbolic operators with double characteristics. Commun. Partial Differ. Equ. **13**, 61–86 (1988)
4. E. Bernardi, A. Bove, Geometric transition for a class of hyperbolic operators with double characteristics. Jpn. J. Math. **23**, 1–87 (1997)
5. E. Bernardi, A. Bove, A remark on the Cauchy problem for a model hyperbolic operator, in *Hyperbolic Differential Operators and Related Problems.* Lecture Notes in Pure and Applied Mathematics, vol. 233 (Dekker, New York, 2003), pp. 41–51
6. E. Bernardi, T. Nishitani, On the Cauchy problem for noneffectively hyperbolic operators, the Gevrey 5 well posedness. J. Anal. Math. **105**, 197–240 (2008)
7. E. Bernardi, T. Nishitani, On the Cauchy problem for noneffectively hyperbolic operators. The Gevrey 4 well-posedness. Kyoto J. Math. **51**, 767–810 (2011)
8. E. Bernardi, T. Nishitani, On the Cauchy problem for noneffectively hyperbolic operators. The Gevrey 3 well-posedness. J. Hyperbolic Differ. Equ. **8**, 615–650 (2011)
9. E. Bernardi, A. Bove, C. Parenti, Geometric results for a class of hyperbolic operators with double characteristics. II. J. Funct. Anal. **116**, 62–82 (1993)
10. E. Bernardi, C. Parenti, A. Parmeggiani, The Cauchy problem for hyperbolic operators with double characteristics in presence of transition. Commun. Partial Differ. Equ. **37**, 1315–1356 (2012)
11. C. Briot, J.C. Bouquet, Recherches sur les propriétés des fonctions définies par des équations différentielles. J. Ec. Imp. Polytech. **36**, 133–198 (1856)
12. M.D. Bronshtein, Cauchy problem for hyperbolic operators with characteristics of variable multiplicity. Tr. Mosk. Mat. Obs. **41**, 83–99 (1980)
13. J. Chazarain, Opérateurs hyperboliques à caractéristiques de multiplicité constant. Ann. Inst. Fourier (Grenoble) **24**, 173–202 (1974)
14. F. Colombini, S. Spagnolo, An example of a weakly hyperbolic Cauchy problem not well posed in C^∞. Acta Math. **148**, 243–253 (1982)
15. F. Colombini, E. De Giorgi, S. Spagnolo, Sur les équations hyperboliques avec des coefficients qui ne dépendent que du temps. Ann. Sc. Norm. Super. Pisa **4**, 511–559 (1979)
16. F. Colombini, E. Jannelli, S. Spagnolo, Well posedness in the Gevrey class of the Cauchy problem for a non strictly hyperbolic equation with coefficients depending on time. Ann. Sc. Norm. Super. Pisa **10**, 291–312 (1983)

© Springer International Publishing AG 2017

T. Nishitani, *Cauchy Problem for Differential Operators with Double Characteristics*, Lecture Notes in Mathematics 2202,
DOI 10.1007/978-3-319-67612-8

17. F. Colombini, T. Nishitani, N. Orrù, Some well-posed Cauchy problem for second order hyperbolic equations with two independent variables. Osaka J. Math. **48**, 647–673 (2011)
18. P. D'Ancona, Well posedness in C^∞ for a weakly hyperbolic second order equation. Rend. Sem. Mat. Univ. Padova **91**, 65–83 (1994)
19. F.R. de Hoog, R. Weiss, On the boundary value problem for systems of ordinary differential equations with a singularity of the second kind. SIAM J. Math. Anal. **11**, 41–60 (1980)
20. J.J. Duistermaat, *Fourier Integral Operators*. Progress in Mathematics, vol. 130 (Birkhäuser, Basel, 1994)
21. Yu.V. Egorov, Canonical transformations and pseudodifferential operators. Trans. Moscow Math. Soc. **24**, 3–28 (1971)
22. V. Esposito, On the well posedness of the Cauchy problem for a class of hyperbolic operators with double characteristics. Ricerche Mat. **49**, 221–239 (2000)
23. H. Flaschka, G. Strang, The correctness of the Cauchy problem. Adv. Math. **6**, 347–379 (1971)
24. L. Gårding, Solution directe du problème de Cauchy pour les équations hyperboliques, in *La théorie des équations aux dérivées partielles, Nancy, 1956*. Coll. Int. CNRS, vol. 71 (CNRS, Nancy, 1956), pp. 71–90
25. L. Gårding, *Cauchy's Problem for Hyperbolic Equations* (University of Chicago, Chicago, 1957)
26. L. Gårding, Hyperbolic differential operators, in *Perspectives in Mathematics* (Birkhäuser, Basel, 1984), pp. 215–247
27. L. Gårding, Hyperbolic equations in the twentieth century, in *Matériaux pour l'histoire des Mathématiques au XXe*, Papers from the Colloquium, Nice, 1996, Semin. Congr., vol. 3 (Soc. Math. France, Paris, 1998), pp. 37–68
28. M. Gevrey, Sur la nature analytiques des solutions des équations aux dérivées partielles. Ann. Sc. Norm. super. **35**, 129–189 (1917)
29. A. Grigis, J. Sjöstrand, *Microlocal Analysis for Differential Operators*. London Mathematical Society Lecture Note Series, vol. 196 (Cambridge University Press, Cambridge, 1994)
30. J. Hadamard, *Lectures on Cauchy's Problem in Linear Partial Differential Equations* (Dover Publications, New York, 1953)
31. Q. Han, Energy estimates for a class of degenerate hyperbolic equations. Math. Ann. **347**, 339–364 (2010)
32. L. Hörmander, The Cauchy problem for differential equations with double characteristics. J. Anal. Math. **32**, 118–196 (1977)
33. L. Hörmander, *The Analysis of Linear Partial Differential Operators. I*. Grundlehren Math. Wiss., vol. 274 (Springer, Berlin, 1983)
34. L. Hörmander, *The Analysis of Linear Partial Differential Operators. III*. Grundlehren Math. Wiss., vol. 274 (Springer, Berlin, 1985)
35. L. Hörmander, Quadratic hyperbolic operators, in *Microlocal Analysis and Applications*, ed. by L. Cattabriga, L. Rodino (Springer, Berlin, 1989), pp. 118–160
36. V.Ja. Ivrii, Correctness of the Cauchy problem in Gevrey classes for nonstrictly hyperbolic operators. Uspehi Math. USSR Sbornik **25**, 365–387 (1975)
37. V.Ja. Ivrii, Sufficient conditions for regular and completely regular hyperbolicity. Tr. Mosk. Mat. Obs. **33**, 3–65 (1975); English transl., Trans. Moscow Math. Soc. **33**, 1–65 (1978)
38. V.Ja. Ivrii, Conditions for correctness in Gevrey classes of the Cauchy problem for hyperbolic operators with characteristics of variable multiplicity. Sib. Math. J. **17**, 422–435 (1976)
39. V.Ja. Ivrii, Conditions for correctness in Gevrey classes of the Cauchy problem for not strictly hyperbolic operators. Sib. Math. J. **17**, 921–931 (1976)
40. V.Ja. Ivrii, The wellposed Cauchy problem for non-strictly hyperbolic operators, III. The energy integral. Trans. Moscow Math. Soc. (English transl.) **34**, 149–168 (1978)
41. V.Ja. Ivrii, Wave fronts of solutions of some hyperbolic pseudodifferential equations. Trans. Moscow Math. Soc. (English transl.) **39**, 87–119 (1981)
42. V.Ja. Ivrii, Linear hyperbolic equations, in *Partial Differential Equations IV*, ed. by Yu.V. Egorov, M.A. Shubin. Encyclopaedia of Mathematical Sciences, vol. 33 (Springer, New York, 1988), pp. 149–235

43. V.Ja. Ivrii, V.M. Petkov, Necessary conditions for the correctness of the Cauchy problem for non-strictly hyperbolic equations. Uspehi Mat. Nauk **29**, 3–70 (1974); English transl., Russ. Math. Surv. **29**, 1–70 (1974)
44. N. Iwasaki, The Cauchy problem for effectively hyperbolic equations (a special case). J. Math. Kyoto Univ. **23**, 503–562 (1983)
45. N. Iwasaki, The Cauchy problem for effectively hyperbolic equations (a standard type). Publ. Res. Inst. Math. Sci. **20**, 551–592 (1984)
46. N. Iwasaki, The Cauchy problem for effectively hyperbolic equations. Sugaku Exposition **36**, 227–238 (1984)
47. N. Iwasaki, The Cauchy problem for effectively hyperbolic equations (general case). J. Math. Kyoto Univ. **25**, 727–743 (1985)
48. N. Iwasaki, The Cauchy problem for effectively hyperbolic equations (remarks), in *Hyperbolic Equations and Related Topics* (Academic, Boston, 1986), pp. 89–100
49. N. Iwasaki, Bicharacteristic curves and well-posedness for hyperbolic equations with noninvolutive multiple characteristics. J. Math. Kyoto Univ. **34**, 41–46 (1994)
50. K. Kajitani, T. Nishitani, *The Hyperbolic Cauchy Problem*. Lecture Notes in Mathematics, vol. 1505 (Springer, Berlin, 1991)
51. K. Kasahara, M. Yamaguti, Strongly hyperbolic systems of linear partial differential equations with constant coefficients. Mem. Coll. Sci. Univ. Kyoto, Ser. A. Math. **33**, 1–23 (1960/1961)
52. G. Komatsu, T. Nishitani, Continuation of bicharacteristics for effectively hyperbolic operators. Publ. Res. Inst. Math. Sci. **28**, 885–911 (1992)
53. A. Lax, On Cauchy's problem for partial differential equations with multiple characteristics. Commun. Pure Appl. Math. **9**, 135–169 (1956)
54. P.D. Lax, Asymptotic solutions of oscillatory initial value problem. Duke Math. J. **24**, 627–646 (1957)
55. J. Leray, *Hyperbolic Differential Equations* (The Institute for Advanced Study, Princeton, NJ, 1953)
56. J. Leray, Y. Ohya, Équations et systèmes non-linéaires, hyperboliques nonstrict. Math. Ann. **170**, 167–205 (1967)
57. N. Lerner, *Metrics on the Phase Space and Non-selfadjoint Pseudo-Differential Operators* (Birkhäuser, Basel, 2010)
58. E.E. Levi, Carateristiche multiple e problema di Cauchy. Ann. Mat. Pura Appl. **16**, 161–201 (1909)
59. A. Martinez, *An Introduction to Semiclassical and Microlocal Analysis*. Universitext (Springer, New York, 2002)
60. A. Melin, Lower bounds for pseudo-differential operators. Ark. Mat. **9**, 117–140 (1971)
61. R. Melrose, The Cauchy problem for effectively hyperbolic operators. Hokkaido Math. J. **12**, 371–391 (1983)
62. R. Melrose, The Cauchy problem and propagation of singularities, in *Seminar on Nonlinear Partial Differential Equations*, Papers from the Seminar, Berkeley, CA, 1983, ed. by S.S.Chern. Mathematical Sciences Research Institute Publications, vol. 2 (Springer, Berlin, 1984), pp. 185–201
63. S. Mizohata, Systèmes hyperboliques. J. Math. Soc. Jpn. **11**, 205–233 (1959)
64. S. Mizohata, Note sur le traitement par les opérateurs d'intégrale singulière du problème de Cauchy. J. Math. Soc. Jpn. **11**, 234–240 (1959)
65. S. Mizohata, Some remarks on the Cauchy problem. J. Math. Kyoto Univ. **1**, 109–127 (1961)
66. S. Mizohata, *The Theory of Partial Differential Equations* (Cambridge University Press, Cambridge, 1973)
67. S. Mizohata, Y. Ohya, Sur la condition de E.E.Levi concernant des équations hyperboliques. Publ. Res. Inst. Math. Sci. **4**, 511–526 (1968/1969)
68. S. Mizohata, Y. Ohya, Sur la condition d'hyperbolicité pour les équations à caractéristiques multiples. II. Jpn. J. Math. **40**, 63–104 (1971)
69. S. Mizohata, Y. Ohya, M. Ikawa, Comments on the development of hyperbolic analysis, in *Hyperbolic Equations and Related Topics (Katata/Kyoto, 1984)* (Academic, Boston, MA, 1986), pp. ix–xxxiv

70. T. Nishitani, The Cauchy problem for weakly hyperbolic equations of second order. Commun. Partial Differ. Equ. **5**, 1273–1296 (1980)
71. T. Nishitani, On the finite propagation speed of wave front sets for effectively hyperbolic operators. Sci. Rep. College Gen. Ed. Osaka Univ. **32**(1), 1–7 (1983)
72. T. Nishitani, Note on some non effectively hyperbolic operators. Sci. Rep. College Gen. Ed. Osaka Univ. **32**(2), 9–17 (1983)
73. T. Nishitani, A necessary and sufficient condition for the hyperbolicity of second order equations with two independent variables. J. Math. Kyoto Univ. **24**, 91–104 (1984)
74. T. Nishitani, Local energy integrals for effectively hyperbolic operators. I, II. J. Math. Kyoto Univ. **24**, 623–658, 659–666 (1984)
75. T. Nishitani, Microlocal energy estimates for hyperbolic operators with double characteristics, in *Hyperbolic Equations and Related Topics* (Academic, Boston, 1986), pp. 235–255
76. T. Nishitani, Note on wave front set of solutions to non effectively hyperbolic operators. J. Math. Kyoto Univ. **27**, 657–662 (1987)
77. T. Nishitani, Note on a paper of N.Iwasaki: "Bicharacteristic curves and well-posedness for hyperbolic equations with noninvolutive multiple characteristics". J. Math. Kyoto Univ. **38**, 415–418 (1998)
78. T. Nishitani, Non effectively hyperbolic operators, Hamilton map and bicharacteristics. J. Math. Kyoto Univ. **44**, 55–98 (2004)
79. T. Nishitani, Effectively hyperbolic Cauchy problem, in *Phase Space Analysis of Partial Differential Equations*, vol. II, ed. F. Colombini, L. Pernazza. Publ. Cent. Ric. Mat. Ennio Giorgi (Scuola Norm., Pisa, 2004), pp. 363–449
80. T. Nishitani, On the Cauchy problem for non-effectively hyperbolic operators, the Ivrii-Petkov-Hörmander condition and the Gevrey well posedness. Serdica Math. J. **34**, 155–178 (2008)
81. T. Nishitani, On Gevrey well-posedness of the Cauchy problem for some noneffectively hyperbolic operators, in *Advances in Phase Space Analysis of PDEs*. Progress in Nonlinear Differential Equations and Their Application, vol. 78 (Birkhäuser, Boston, 2009), pp. 217–233
82. T. Nishitani, A note on zero free regions of the Stokes multipliers for second order ordinary differential equations with cubic polynomial coefficients. Funkcialaj Ekvac. **54**, 473–483 (2011)
83. T. Nishitani, On the Cauchy problem for noneffectively hyperbolic operators, a transition casse, in *Studies in Phase Space Analysis with Applications to PDEs*, ed. by M. Cicognani, F. Colombini, D. Del Santo (Birkhäuser, Basel, 2013), pp. 259–290
84. T. Nishitani, *Cauchy Problem for Noneffectively Hyperbolic Operators*. MSJ Memoirs, vol. 30 (Mathematical Society of Japan, Tokyo, 2013)
85. T. Nishitani, Local and microlocal Cauchy problem for non-effectively hyperbolic operators. J. Hyperbolic Differ. Equ. **11**, 185–213 (2014)
86. T. Nishitani, On the Cauchy problem for hyperbolic operators with double characteristics, a transition case, in *Fourier Analysis*. Trends in Mathematics (Birkhäuser, Basel, 2014), pp. 311–334
87. T. Nishitani, A simple proof of the existence of tangent bicharacteristics for noneffectively hyperbolic operators. Kyoto J. Math. **55**, 281–297 (2015)
88. O.A. Oleinik, On the Cauchy problem for weakly hyperbolic equations. Commun. Pure Appl. Math. **23**, 569–586 (1970)
89. C. Parenti, A. Parmeggiani, On the Cauchy problem for hyperbolic operators with double characteristics. Commun. Partial Differ. Equ. **34**, 837–888 (2009)
90. C. Parenti, A. Parmeggiani, On the Cauchy problem for hyperbolic operators with double characteristics, in *Evolution Equations of Hyperbolic and Schrödinger Type*. Progress in Mathematics, vol. 301 (Birkäuser/Springer, Basel, 2012), pp. 247–266
91. I.G. Petrovsky, Über das Cauchysche Problem für Systeme von partiellen Differentialgleichungen. Mat. Sb. N.S. **2**(44), 815–870 (1937)
92. I.G. Petrovsky, *Lectures on Partial Differential Equations* (Interscience, New York, 1950)

93. Y. Sibuya, *Global Theory of a Second Order Linear Ordinary Differential Equation with a Polynomial Coefficient*. North-Holland Mathematical Studies, vol. 18 (Elsevier, Amsterdam, 1975)
94. G. Strang, On strong hyperbolicity. J. Math. Kyoto Univ. **6**, 397–417 (1967)
95. D.T. Trinh, On the simpleness of zeros of Stokes multipliers. J. Differ. Equ. **223**, 351–366 (2006)
96. S. Wakabayashi, On the Cauchy problem for hyperbolic operators of second order whose coefficients depend on the time variable. J. Math. Soc. Jpn. **62**, 95–133 (2010)

Index

© Springer International Publishing AG 2017
T. Nishitani, *Cauchy Problem for Differential Operators with Double Characteristics*, Lecture Notes in Mathematics 2202,
DOI 10.1007/978-3-319-67612-8

LECTURE NOTES IN MATHEMATICS   ◢ Springer

Editors in Chief: J.-M. Morel, B. Teissier;

Editorial Policy

1. Lecture Notes aim to report new developments in all areas of mathematics and their applications – quickly, informally and at a high level. Mathematical texts analysing new developments in modelling and numerical simulation are welcome.

   Manuscripts should be reasonably self-contained and rounded off. Thus they may, and often will, present not only results of the author but also related work by other people. They may be based on specialised lecture courses. Furthermore, the manuscripts should provide sufficient motivation, examples and applications. This clearly distinguishes Lecture Notes from journal articles or technical reports which normally are very concise. Articles intended for a journal but too long to be accepted by most journals, usually do not have this "lecture notes" character. For similar reasons it is unusual for doctoral theses to be accepted for the Lecture Notes series, though habilitation theses may be appropriate.

2. Besides monographs, multi-author manuscripts resulting from SUMMER SCHOOLS or similar INTENSIVE COURSES are welcome, provided their objective was held to present an active mathematical topic to an audience at the beginning or intermediate graduate level (a list of participants should be provided).

   The resulting manuscript should not be just a collection of course notes, but should require advance planning and coordination among the main lecturers. The subject matter should dictate the structure of the book. This structure should be motivated and explained in a scientific introduction, and the notation, references, index and formulation of results should be, if possible, unified by the editors. Each contribution should have an abstract and an introduction referring to the other contributions. In other words, more preparatory work must go into a multi-authored volume than simply assembling a disparate collection of papers, communicated at the event.

3. Manuscripts should be submitted either online at www.editorialmanager.com/lnm to Springer's mathematics editorial in Heidelberg, or electronically to one of the series editors. Authors should be aware that incomplete or insufficiently close-to-final manuscripts almost always result in longer refereeing times and nevertheless unclear referees' recommendations, making further refereeing of a final draft necessary. The strict minimum amount of material that will be considered should include a detailed outline describing the planned contents of each chapter, a bibliography and several sample chapters. Parallel submission of a manuscript to another publisher while under consideration for LNM is not acceptable and can lead to rejection.

4. In general, **monographs** will be sent out to at least 2 external referees for evaluation.

   A final decision to publish can be made only on the basis of the complete manuscript, however a refereeing process leading to a preliminary decision can be based on a pre-final or incomplete manuscript.

   Volume Editors of **multi-author works** are expected to arrange for the refereeing, to the usual scientific standards, of the individual contributions. If the resulting reports can be forwarded to the LNM Editorial Board, this is very helpful. If no reports are forwarded or if other questions remain unclear in respect of homogeneity etc, the series editors may wish to consult external referees for an overall evaluation of the volume.

5. Manuscripts should in general be submitted in English. Final manuscripts should contain at least 100 pages of mathematical text and should always include

   – a table of contents;
   – an informative introduction, with adequate motivation and perhaps some historical remarks: it should be accessible to a reader not intimately familiar with the topic treated;
   – a subject index: as a rule this is genuinely helpful for the reader.
   – For evaluation purposes, manuscripts should be submitted as pdf files.

6. Careful preparation of the manuscripts will help keep production time short besides ensuring satisfactory appearance of the finished book in print and online. After acceptance of the manuscript authors will be asked to prepare the final LaTeX source files (see LaTeX templates online: https://www.springer.com/gb/authors-editors/book-authors-editors/manuscriptpreparation/5636) plus the corresponding pdf- or zipped ps-file. The LaTeX source files are essential for producing the full-text online version of the book, see http://link.springer.com/bookseries/304 for the existing online volumes of LNM). The technical production of a Lecture Notes volume takes approximately 12 weeks. Additional instructions, if necessary, are available on request from lnm@springer.com.

7. Authors receive a total of 30 free copies of their volume and free access to their book on SpringerLink, but no royalties. They are entitled to a discount of 33.3 % on the price of Springer books purchased for their personal use, if ordering directly from Springer.

8. Commitment to publish is made by a *Publishing Agreement*; contributing authors of multiauthor books are requested to sign a *Consent to Publish form*. Springer-Verlag registers the copyright for each volume. Authors are free to reuse material contained in their LNM volumes in later publications: a brief written (or e-mail) request for formal permission is sufficient.

Addresses:
Professor Jean-Michel Morel, CMLA, École Normale Supérieure de Cachan, France
E-mail: moreljeanmichel@gmail.com

Professor Bernard Teissier, Equipe Géométrie et Dynamique,
Institut de Mathématiques de Jussieu – Paris Rive Gauche, Paris, France
E-mail: bernard.teissier@imj-prg.fr

Springer: Ute McCrory, Mathematics, Heidelberg, Germany,
E-mail: lnm@springer.com

Printed in the United States
By Bookmasters